1/97

WOMEN AND CHILDREN IN HEALTH CARE

WOMEN AND CHILDREN IN HEALTH CARE

An Unequal Majority

MARY BRIODY MAHOWALD

New York   Oxford
OXFORD UNIVERSITY PRESS
1993

Oxford University Press

Oxford   New York   Toronto
Delhi   Bombay   Calcutta   Madras   Karachi
Kuala Lumpur   Singapore   Hong Kong   Tokyo
Nairobi   Dar es Salaam   Cape Town
Melbourne   Auckland   Madrid

and associated companies in
Berlin   Ibadan

Copyright © 1993 by Oxford University Press, Inc.

Published by Oxford University Press, Inc.,
200 Madison Avenue, New York, New York 10016

Oxford is a registered trademark of Oxford University Press

All rights reserved. No part of this publication may be reproduced,
stored in a retrieval system, or transmitted, in any form or by any means,
electronic, mechanical, photocopying, recording, or otherwise,
without the prior permission of Oxford University Press.

Library of Congress Cataloging-in-Publication Data
Mahowald, Mary Briody.
Women and children in health care / Mary Briody Mahowald.
p.   cm.   Includes bibliographical references and index.
ISBN 0-19-506346-5
1. Women's health services—Political aspects.   2. Child health
   services—Political aspects.   I. Title.
[DNLM: 1. Child Health Services.
2. Socioeconomic Factors.
3. Sociology, Medical.
4. Women's Health Services.
WA 305 M216w]   RA564.85.M35   1993
   362.1'98--dc20   M/DLC
   for Library of Congress   92-12912

987654321

Printed in the United States of America
on acid-free paper

In memory of my parents and my brother

Foreword

At midnight, I was awakened by a very sharp pain. The head nurse ... gave me an injection of scopolamine-morphine [sic]. ... I woke up the next morning about half-past seven ... the door opened, and the head nurse brought in my baby. ... I was so happy and so pleased, and I sat up in bed and had a wonderful breakfast.

Mrs. Cecil Stewart, quoted in Marguerite Tracy and Mary Boyd, *Painless Childbirth,* 1915

When first introduced into this country around the turn of the century, scopolamine-morphine was hailed as a gift to women, a drug that could liberate them from the Biblical injunction that they bring forth children "in sorrow." Not surprisingly, many women availed themselves of this gift, choosing to give birth in the scopolamine-induced, somnambulistic state known as "twilight sleep." Women eventually discovered, though, that this gift was a sort of serpent in the garden. Rather than eliminating the physical pain of childbirth, scopolamine-morphine merely masked it, while robbing a woman of her self-control and her memory. In twilight sleep, a woman surrendered control over an area of medical care in which she had always played the central role. Over time, women woke up to this fact—as well as to the other dangers of twilight sleep—and today American women are given a choice, many preferring to remain awake and in control during childbirth.

Women's journey from twilight sleep to contemporary "natural childbirth" parallels the journey they have charted in American society: a journey out of the home and into the work place. The medical community has not always kept pace with women's journey. In fact, physicians have sometimes lent scientific credence to the prejudices of their time. In 1882, for example, in a work entitled *Parturition without Pain: A Code of Directions for Escaping from the Primal Curse,* Dr. M. L. Holbrook expressed the opinion that "the Almighty, in creating the female sex, had taken the uterus and built up a woman around it." Other physicians of the time explained that, because the uterus was connected to the central nervous system, women must avoid overstimulation, such as the overstimulation that comes with advanced education. Like most nineteenth-century Americans, physicians considered childbearing and the rearing of children a

vii

woman's ultimate contribution to society, and "women's health" meant one thing: maternal health. Over the past hundred years, as women have redefined their contributions to society, physicians have been forced to redefine their views—and treatment—of women.

Just as women in American society over the past thirty years have journeyed out of the home and into the work place—or from twilight sleep to action—the study of women's health has moved from the shadows into the spotlight. Today, women are no longer seen as synonymous with their reproductive functions, and the term "women's health" has taken on a broader meaning. As a result, the American public has awakened to the important fact that, when it comes to women's health in its broader definition, the medical community is still very much in the dark.

The gaps in our understanding of the conditions and diseases that rob women of their health and their lives can be attributed to the fact that women have constituted a unequal majority when it comes to medical research and health care. Despite the fact that women's activities are limited by health problems 25 percent more days each year than are men's activities, that heart disease is more prevalent among women than among men, and that stroke accounts for a higher percentage of deaths among women than men at all stages of life, women have been inappropriately excluded from the study populations of many major clinical trials conducted during the past thirty years. As late as the 1980s, a major study on the preventive effects of aspirin in treating coronary disease involved 22,000 men, but not a single woman. Another study looked at the role of estrogen in preventing heart disease, but only in men.

Such clinical studies have been limited by the underlying assumption that men provide the normative standard for health, and they have served to reinforce the myth that heart disease is unique to men. In fact, heart disease has long been the leading cause of death for women in the United States, and half of all women, but only 31 percent of men, die within a year of suffering a heart attack. Even for conditions that affect only women, a knowledge base is lacking. For example, physicians have neither definitive guidelines for when hormone replacement therapy is indicated in the treatment of menopause nor a clear understanding of how such therapy might prevent osteoporosis. Thus, it would appear that biomedical research has been in a somnambulistic state, out of touch not only with the medical facts but also with the important social facts that women make up more than half the U.S. population and constitute 45 percent of the nation's work force, and that working women pay taxes, too.

In 1990, the National Institutes of Health (NIH) woke up to these facts and now is taking steps to fill in the gaps in the medical community's knowledge of women's health. Through the recently launched Women's Health Initiative, the NIH is supporting clinical trials that will study about 150,000 postmenopausal women nationwide to learn more about how to prevent heart disease and stroke, cancer, and osteoporosis, which are the major causes of death and frailty among older women of all races and socioeconomic groups in the United States. In addition to decreasing the prevalence of these diseases among women, the Initiative will develop practical guidelines on hormone replacement therapy, diet, and exercise that will be useful to a broad spectrum of women. By studying women from ethnic and socioeconomic groups that traditionally have been poorly served by the health care system, the Initiative will also yield information on ways to encourage women from such groups to lead healthier lives.

Through the recently established Office of Research on Women's Health, the NIH has developed an agenda for research into the causes, treatment, and prevention of disease and debility among women across their life span. In a recently published report, NIH outlines this agenda and calls on the biomedical community to pursue research into key areas of women's health. Among younger women, for example, the agenda identifies sexually transmitted disease (STDs) and contraception as high priorities for research.

Physicians and other health care professionals can provide a much needed social service in educating young women about the threat to their health, and to the health of their children, posed by STDs. Women must be encouraged to protect themselves from transmission of STDs, to recognize early symptoms of the more common STDs, and to seek early medical care if such symptoms occur. Just as other diseases, including cancer, tuberculosis, and Hansen's disease, lost their social stigma when public discussion began to center on metastasis and bacilli, so could educating the public about STDs have the same effect. For its part, the NIH is committed to supporting research aimed at improving the medical community's methods for detecting and treating STDs among women, with a view to safeguarding both women's and their children's health.

Sexually transmitted disease is one area in which changes in society have outpaced the medical community's capacity to respond, but in many areas of medicine and health care the opposite is the case. For example, medical breakthroughs in reproduction and fertility have often created needs for new laws and public policy. In vitro fertilization, surrogate motherhood, and the ability of physicians to closely monitor the developing fetus have presented challenges to ethics and public policy unimaginable twenty years ago.

For the first time in history, the essential link between women and children, to which Dr. Mahowald refers in the preface to this volume, may be redefined as contingent. If women were once defined by their reproductive capacity, developments in modern medicine may redefine motherhood in much the same way that medical advances of the recent past, such as contraception, have allowed women to redefine their roles in society. As always when there is change and redefinition, our society and the medical community will spend some time groping in the dark for enlightened, humane solutions, for the unequal majority deserves no less. Practices and policies considered enlightened today may, like twilight sleep, prove to be flawed, and better solutions will have to be created. This volume on women and children in health care is a significant effort to move out of the twilight into the dawn.

Bernadine Healy, M.D.
Director, National Institutes of Health

Preface

Women and children are commonly mentioned together—not only in the obvious context of birth and nurturing, but also in a social and political context in which they are viewed as a special group to be considered. We are familiar, for example, with the phrase "women and children first," expressing a priority for providing assistance in times of crisis or limited resources. History texts have treated "women and children" in special sections at the ends of chapters about wars and male-dominated politics. Discussions of social conditions and legislative requirements often allude to "women and children" as the group most affected by support programs sponsored by the government.

The phrase "men and children" is less commonly used, and less likely to connote a single group to be considered. This phrase would not surprise us in a social science text, where recent phenomena and attitudes are observed as somewhat novel. For example, "men and children" might be mentioned in relation to men opting to spend time at home as "househusbands," choosing to become fathers (even through artificial means), assisting in the birth of their children, and taking on the responsibility for child care. In all of these instances, however, the focus seems to be on the men rather than the children, and each is viewed as a separate group in relation to the other. In general, the link between women and children is perceived as essential, while that between men and children is perceived as gratuitous.

Why the difference? This book examines and critiques the discrepancy from the perspective of a concept of equality among human beings who inevitably and fortuitously possess unique characteristics as individuals. They are advantaged and disadvantaged in complex, overlapping ways; they are male and female, homosexual and heterosexual; they are very young, very old, and all of the ages between those extremes. The concept of equality that I propose in Chapter 1 starts with a recognition of these differences, and explores possibilities for respecting them while also maximizing respect for the choices of individuals.

Why concentrate on health care? There are two main reasons: (1) having worked in that setting for more than a decade, I am familiar with it; and (2) within that setting, the majority of health caregivers are women and the majority of patients are women and children. In Chapter 1, I develop a critical perspective that is applicable to diverse issues affecting women and children. This perspective involves a concept of equality as respect for relevant differences among individuals and groups. Chapter 2 deals with the gender stereotypes that the health care system generally reflects, and Chapter 3 considers obstetrics and gynecology as a target of feminist criticism in that regard. Other chapters fill out different dimensions of the relationships between women, children, and health care: issues relevant to fertility curtailment, fertility enhancement, severely disabled newborns, children's moral agency, the impact of gender socialization on adolescents, the feminization of poverty, and the elusive meaning of "family." I conclude with an examination of professional power as a means of promoting more egalitarian relationships.

In its present form, this book is the offspring of a long gestation, induced and sustained by various life experiences. Like many pregnancies, this one was unplanned but not unwanted. The experiences that nurtured it include teaching at different levels of education, from elementary and junior high through graduate and professional schools. They involve learning experiences ranging from life in the inner city to life in an affluent suburb, life as a single professional to life as a married woman and mother, and life in an academic environment to life in a busy hospital. My openness to the "pregnancy" was probably triggered by a natural philosophical orientation, that is, a tendency to reflect critically on my own experiences and ideas gleaned from others. It was fostered by my training in philosophy, and the opportunities that training afforded for careful analysis of concepts, their assumptions, and their implications. Only retrospectively, however, did I realize that one unifying theme had driven my overall critique. I have named that theme "an egalitarian perspective." Like other names, its meaning can only be adequately embodied in a full account of what it represents. The embodiment is the content of this book.

I had not realized I was "pregnant" until I noticed tell-tale signs, that is, essays and articles that I had written, each of which addressed some aspect of a general interest in women, children, and health care. Once I realized what had been developing in me and identified its orientation, I undertook to contribute to its development by extensive revisions and elaboration of my previous work, enriching the text with examples of cases I had encountered in my work in women's and children's hospitals. I added chapters necessary for a full sense of my views about women and children along the spectrum of their lives, and reworked all of the material to express more clearly my underlying perspective.

The issues examined here are particularly vital to those who study and live within the daunting environment of contemporary medical ethics, that is, students, teachers, and practitioners in the field. My intended audience also includes those interested in, or affected by, issues relevant to women and children. Obviously, the latter group comprises all of us.

Chicago   M. B. M.
April 1992

Acknowledgments

I am indebted to many people for help in preparing this book. First, I wish to thank those who facilitated my transition from the ivory tower to the market-place, when I moved from the world of academic philosophy to the world of the clinical setting in 1982. This group of people includes many philosophers from different parts of the country, with whom ongoing collaboration has been crucial to my remaining a philosopher while working in a hospital-medical school complex. It also includes my former colleagues at Case Western Reserve University, particularly those in the Departments of Pediatrics and Reproductive Biology. Among these, Miriam Rosenthal has been a special friend and support. Stephen Post came later, but continued to be a valued colleague after my move to Chicago. Joanna Rogers, a bioethics fellow at the Cleveland Clinic, offered constructive comments on my early chapters. Maureen Hack, a wonderful resource during my years in Cleveland, remained that when I needed to revise and update the material in Chapter 10. Claudia Coulton, another admired Cleveland colleague, helped me with the chapter on poverty.

Second, I want to thank my colleagues at the University of Chicago, who have contributed in various ways to my completing the book. The fellows and faculty of the Center for Clinical Medical Ethics have read substantial portions of the manuscript, providing me with useful comments and criticisms. These include Kenneth Iserson, Eugene Bereza, James Bowman, Ann Dudley Goldblatt, John Lantos, Marion Secundy, Stephen Toulmin, Bruce White, and September Williams. Another associate of the Center, Julie Gage Palmer, also provided helpful comments, as did Kevin Burke, Karen Geraghty, and Ruth Van Loon.

Arthur Herbst, who chairs the Department of Obstetrics and Gynecology at the University of Chicago, has generously facilitated my efforts, and other colleagues in the Department, Teresa Hadro, Kathleen Hanlon-Lundberg, and Patricia Mills, have taken time to read portions of the manuscript and offer constructive feedback. Hans van der Ven assisted with part of Chapter 6. Joan Hives could not have been a more generous, conscientious, and careful assistant in checking out references, helping with revisions, and preparing the index. Karin Bowman saved me from some glaring mistakes at the proofreading stage.

Faculty members from other departments at the University of Chicago have also been helpful. Robert Arensman from the Department of Surgery shared his considerable expertise about ECMO (extracorporeal membrane oxygenation). From the Department of Pediatrics, David Mendez-Soto pointed me to sources on surfactant therapy, Diana Woo advised me on resuscitation of preterm infants, and Maria Corpuz and Kwang-sun Lee were helpful with regard to the impact of adolescent pregnancy and drug use during pregnancy.

Third, a number of scholars from around the country spent time reviewing parts or all of the book: Marilyn Friedman, Joan Leguard, Anthony Mahowald, Mary Segers, and Rosemarie Tong. Joan Callahan's input was unstintingly thoughtful and thorough. If everyone had so constructively critical a colleague, academic writing would be much improved. Another constructive critic was the anonymous reviewer whose input facilitated my final revision. I am grateful to each of these people for the comments and criticisms that led, I believe, to a better book.

Also leading to a better book was the collective input of my editors at Oxford University Press: Jeffrey House, Edith Barry, and Melinda Wirkus. I wish to credit each of them for contributing appropriate expertise and advice.

Last, but most important, I want to thank my best friends, who support me in most of what I do by being who they are: Mike, Lisa, Maureen, and Tony.

Contents

List of Cases

WOMEN AND CHILDREN IN HEALTH CARE

1

An Egalitarian Overview

Equality is generally construed as a value or principle to be promoted in a democratic society. It is often associated with *justice, equity,* and *fairness,* all of which signify moral values or principles.[1] Recently, however, critics of equality have emerged from an unlikely source, namely, feminists.[2] Why does this surprise us? Because feminism throughout history has been associated with efforts to promote equality between women and men. As Christine Overall puts it, "feminism is committed to the further understanding and dissolution of sexual inequality."[3]

Understanding the difference between the concept of equality and its practice may be crucial to an understanding of feminist criticisms. The concept of equality is imbedded in an ethic of justice,[4] developed within a philosophical tradition that some view as essentially patriarchal or "masculist."[5] Various versions of this approach focus on individuals and their rights, upholding a theory of justice as necessarily blind or impartial with regard to differences that, for better or worse, characterize individuals in their lived situations. In contrast, the equality favored by some is a practical goal, passionately pursued through attention to the differences that may separate individuals from one another.[6] Contemporarily, the impartiality of traditional ethics is exemplified in a tendency to use gender-neutral language in addressing issues that differently affect women and men, such as those involving family and reproductive decisions. To feminists, justice demands elimination of social and economic disparities that arise from these differences.

Recent work on moral reasoning distinguishes between an approach based on justice and rights, and one based on care and attachment.[7] The justice model has been observed more often in men, and the care model more often in women; hence, the two approaches are sometimes referred to as "masculine" and "feminine."[8] Neither approach is applicable to all individuals of either gender. While the justice model tends to view individuals atomistically, the care model emphasizes their relatedness to one another through ties of blood, commitment, and affection. Clearly, an emphasis on care and attachment may be at odds with the practical goal of gender equality. But different concepts of justice offer different accounts of equality,[9] and not all of these are incompatible with "feminine" reasoning. If a concept of equality is not blind to individual differences and context, and if it extends beyond isolated selves to relationships among individuals, it is amenable to both egalitarian feminism and a care-based model of moral reasoning.[10]

Such an account of equality does not maintain that all individuals should be treated in the same way. While it departs from the classical distinction between equity and equality,[11] this usage is consistent with Michael Walzer's notion of "complex equality" (as opposed to "simple equality"),[12] and with Zillah Eisenstein's notion of "sexual egalitarianism."[13] It is also consistent with the popular usage of the term equality, particularly with regard to race and gender. Affirmative action, for example, is a practice whose goal is usually construed as promotion of equality or avoidance of inequality among different groups of people.

Walzer frames the issue of inequality versus differential treatment as a question of domination, or the "experience of subordination" by individuals or groups. In *Spheres of Justice,* he writes,

> The aim of political egalitarianism is a society free from domination. This is the lively hope named by the word *equality....* It is not a hope for the elimination of differences: we don't all have to be the same or have the same amounts of the same things. Men and women are one another's equals when no one possesses or controls the means of domination.[14]

Although Walzer's account of equality is congenial to the framework elaborated in this book, it lacks (I believe) adequate consideration of the role of relationships in moral decision making. The care model of moral reasoning pays more attention to relationships and attachments, but the need for care, and willingness to care, must be conjoined with an egalitarian critique in order to avoid domination or exploitation.[15]

In this chapter I outline the perspective that underlies my own efforts to combine an emphasis on equality with a "feminine" approach to issues in health care. I begin by describing the concepts of women, men, and children on which my view relies. Admittedly, these concepts do not (because they cannot) adequately convey the complexity of actual individuals; yet they are essential to a position on gender equality because they affirm a certain commonality on which that position is based. The link between women and children is integral to my account because the mother–child relationship figures significantly in feminist concerns and in the contextual framework of an ethic based on care and attachment.[16] After elaborating a concept of equality consistent with these views, I develop guidelines for applying an egalitarian perspective to situations of ethical conflict. Different positions on equality between women and men are critiqued for their compatibility with that approach, which in turn is seen as crucial to avoidance of pitfalls inherent in a care-based model of moral reasoning. I conclude with a discussion of *self-consciousness* as a concept and practical attitude required by an egalitarian perspective.

Underlying Concepts of Women, Men, and Children

Biologically, women are mature human beings typically capable during some portion of their lives of conceiving, bearing, and nursing children. In contrast to women, men are biologically mature human beings typically capable during part of their lives of fertilizing human ova. A child of either sex is a biologically immature member of the species.

Obviously, other male and female mammals have similar biological capabilities. As Aristotle and Thomas Aquinas observe, human beings are distinguishable from other mammals through their capacity for rationality and autonomy.[17] The source of our knowledge of these capabilities is experience rather than a priori judgments. Thus, human beings are generally observed to be capable during part of their lives of exercising reason and choice. However, just as biological capabilities may not be exercised by particular members of a species, so the capacity for reason and choice may never be exercised, and may even be absent, in some individuals. The specific meaning of human beings nonetheless depends on the presence of these developmental capacities in most members of the species.

Postmodernist critics might construe my attempt to delineate starting concepts as "foundationalist" or "essentialist."[18] According to these critics, it is impossible to identify *essential* meanings or *foundational* principles that adequately reflect reality because such generalizations fail to respect the uniqueness and dynamic character of individuals. Postmodernists themselves would attempt to bypass philosophical foundations in developing a model of social criticism. Consistent with the pragmatic tradition of philosophy that I have long supported,[19] however, I regard these concepts as instrumental to an egalitarian critique, and descriptive rather than definitive of actual human beings. Admittedly, whether and when the capacities described are present in individuals is often difficult and sometimes impossible to ascertain.

Clearly, biological maturity is not compelling evidence of rational maturation; some children are more rationally mature than some adults. Nor is biological impairment or immaturity compelling evidence of a lack of rational maturation:

some adults are more physically impaired or underdeveloped than their younger counterparts. The very meaning of *capacity* is controversial because it may or may not be identified with the potentiality of a healthy fetus or a newborn. But the certainty that such capacities are present in particular individuals is not crucial to determining whether they are human beings. A sufficient and necessary condition is that the individual belong to a class whose members generally possess those capabilities. Surely, one's womanhood (Freud to the contrary) is not negated by the fact that one has never become a mother, whether this is determined by choice, inability, or circumstance. A similar point may be made about manhood and humanhood.

As they occur in individuals, capacities for reproduction, reason, and choice are developmental rather than sudden, total achievements. In other words, they develop (that is, progress or regress) gradually from fertilization until death, whether these events occur uterinely or extrauterinely. Such capacities vary within the same individual as well as among different individuals, and each is relative to some ideal of its complete fulfillment or realization. Age is often a useful but inadequate indicator of developmental achievement.

The life experiences of individuals differ both quantitatively and qualitatively. By and large, however, the different experiences of women, men, and children are based on more than biological or life-span differences. As Nel Noddings observes, far more women than men have experience in areas such as teaching and caring for young children, caring directly for the elderly, nursing and teaching as professions, volunteer work, and raising one's own children.[20] Such experiences may at least in part explain the tendency of more women than men to base moral decisions on care and attachment to others rather than on considerations of justice and equality.[21] Although none of these activities is essentially tied to biological differences between the sexes, societal expectations often pressure women, but not men, to be involved in these areas.[22] If more men or fewer women experienced such activities on a regular basis, a care-based ethic might be found as prevalently in one sex as the other.

Although nonhumans as well as humans may be described as *individuals,* the term *person* ordinarily applies only to human beings. Nonetheless, I avoid the term person in characterizing men, women, and children because its meaning seems hopelessly mired in controversy, and the term is not crucial to fruitful discussion of the relevant concepts.[23] Important issues such as whether it is moral to end a pregnancy or to terminate life support can be effectively addressed without settling the issue of personhood. Accordingly, whether or not human individuals are persons, I argue that we have moral obligations to them. We also have obligations toward those who are neither persons nor human, such as our household pets. While rejecting the view that obligations to mature animals are greater than those toward immature humans, I believe it is wrong to cut off the food supply of nondangerous animals or to poison them needlessly. Similarly, even if a human fetus or comatose patient is not a person, deliberately terminating the life of either is not a morally indifferent matter. Neither is the deliberate continuation of pregnancy or continuation of life-support a morally indifferent matter.

Because human life is a developmental continuum, the level of development that an individual has reached or can reach, both physically and mentally, is significant to the assessment of moral decisions made by or about that person. Just as the fact that a patient is near death or irreversibly comatose is pertinent to decisions regarding treatment, so the fact that a fetus is nearly viable or seriously defective is pertinent to decisions regarding abortion.[24] Quality of life is a legitimate consideration for those who make decisions about extending life.

Another morally relevant factor about human individuals is that they are interacting entities. Although their interactions may be minimal or inadvertent, they occur at every level of development. Impaired or dying patients interact with their environment, their loved ones, and caregivers. So also do those whose competence has been compromised through pain or medication. Children are obviously dependent on others in varying degrees for survival and nurturance, but even a fetus interacts with a pregnant woman through biological and nutritional dependence on her. Moreover, many pregnant women (and others, sometimes) respond emotionally to fetuses, even before the final stage of pregnancy. They thus become attached to them.[25] Because human lives are inextricably related, these relationships are pertinent to moral decisions. Individuals are generally responsible for the relationships they develop or maintain, and for respecting such relationships in others. Even when relationships occur only by chance or necessity rather than by choice, the very fact that the relationships exist is morally relevant. For example, although children do not choose their parents, most have responsibilities to them. Professional ties also imply responsibilities, regardless of whether they are initiated by chance, a supervisor's assignment, or personal decision.

Conflicts inevitably arise because of the overlapping nature of interpersonal attachments and the scarcity of resources for responding to the needs or desires of others. Moreover, some relationships trigger behavior that is immoral or questionably moral. If equality is a good to be promoted, or even if it is simply an essential means to some good end, it represents a criterion by which to assess the competing values that emerge from an ethic of caring based on relationships to others.

The Concept of Equality

As already suggested, the term equality is subject to diverse interpretations. Among its political and economic meanings, we might subscribe to a literal interpretation that entails distribution of an identical share of the resources available to all individuals, regardless of their different needs and capabilities. Alternatively, we might maintain a laissez-faire view by which equality is defined as leaving all individuals equally alone or free to pursue their own interests according to their different talents. Or, we might assume a Marxist notion by which resources are to be equally distributed according to the distinctive traits of individuals, "from each according to ability, and to each according to need."[26] Although each of these views of equality has to do with treating people in the same way, their diversity shows that "the same way" may be interpreted quite differently in different political or economic contexts.[27]

An emphasis on differences leads to a rejection of equality as sameness. Martha Minow critiques this notion in her analysis of how the legal system in the United States has dealt with problems of inclusion and exclusion. "If to be equal one must be the same," she writes, "then to be different is to be unequal or even deviant. But any assignment of deviance must be made from the vantage point of some claimed normality."[28] Typically, the unstated reference point by which the norm is defined and differences are ranked is that of the dominant social group. This reference point "can remain unstated because those who do not fit have less power to select the norm than those who fit comfortably within the one that prevails."[29]

Regardless of who defines the reference point, different criteria may be used. From a purely biological perspective, every living, distinct human organism might be considered equal to every other regardless of its level of development, functional capacity, or health status. From a psychological perspective, only those who have achieved—or have the potential to achieve—consciousness, rationality, or autonomy might be considered as equals. The key questions, then, for determining a concept of equality that is applicable to diverse types of relationships and interactions are: On what basis are individuals declared comparable, and how is this evaluation reflected in a policy of distribution?

Here we are obviously dealing with the broader conception of justice. In effect, whenever we grapple with problems of fairness or equity we argue on the basis of specific conceptions of justice and equality, so that it seems impossible ever to speak of one without the other. This is particularly evident where distributive justice is the primary concern, as in John Rawls's treatment of justice, which calls for equal liberty and a minimization of the inequities that arise because of that liberty.[30] It is also evident in Ronald Dworkin's account of "distributive equality."[31] In such discussions, the terms equality and justice are often interchangeable, although the former may also be construed as a means to the latter.

Without attempting to explore further the conceptual and practical relations between equality and justice, I would like to suggest that either or both are means as well as ends. They are necessary means to the promotion of an ideal (egalitarian) society, that is, a community in which the potentials of all individuals and their interrelations are maximally supported and supportive. Such a community reflects essential aspects of an ethic of care as well as justice. Although the ideal is not fully attainable, it remains approachable. As approachable, it represents a moral paradigm in light of which alternative decisions may be assessed by individuals. The ethical framework in which the ideal is embodied is one of virtue or invitation rather than an ethic of obligation.[32] A virtuous person strives for maximal morality, which goes beyond adherence to legal or moral obligations.

In contrast with an ethic of obligation, the framework proposed here does not provide—or attempt to establish—a clear line of demarcation between right and wrong. Rather, it allows that there are various paths to, and degrees of, achievement of a common moral ideal. Most of the moral decisions that human beings make are of this type. In other words, we more often struggle to determine the better course of action among given alternatives than we attempt to discern what is morally necessary or obligatory. This is particularly true with regard to health care decisions, where circumstances sometimes force a choice among tragic alternatives, several of which may be morally right or justifiable. The notion of equality that is most helpful in addressing these issues is one that provides the possibility of applying a moral ideal to practical situations.

Consistent with the preceding account of human nature, I propose a concept of equality that respects biological, psychological, and experiential differences as these apply to human beings throughout their lives. Such a conception focuses primarily on individuals as such, insisting that attention to the differences among them is the only possible way of establishing genuinely egalitarian relationships. It is thus a conception that eschews generalizations based on age, sex, or gender,[33] or even on developmental stage, as adequately defining one's moral or social status. In this view, each human individual ought to be regarded not only as someone of a specific sex and age, but as unique within that context, with a unique set of attachments to others, needs, abilities, potential, and desires. Moreover, this notion of equality entails regard for all the differences, present and anticipated, in those affected by the decision. While treating different individuals differently *may* involve (unjust) inequality, it is also crucial to treating them with equal respect and dignity.

Dworkin distinguishes between equality of welfare and equality of resources, elaborating the difficulties involved in the former view.[34] Although he does not consider a biological view of equality, that notion is generally included in equality of welfare, where welfare is defined as biological fulfillment. The notion of psychological equality is another form of equality of welfare, one that Dworkin usually identifies with the fulfillment of one's preferences. Preferences are in part based on one's experience or knowledge of options. Both biological and psychological equality might be incorporated into a concept of caring for others, whether the caring arises from natural inclination or commitment. A useful and relevant vantage point from which to criticize these conceptions is Dworkin's proposed conception of equality of resources.

From that perspective, the unrestricted fulfillment of each one's preferences and desire to care for others is an impossible and unjust goal. Not only are the resources inadequate, but individual and group interests are bound to clash, introducing new restrictions for some. As for the biological model, which proposes that all living human individuals ought to be supported without regard to limitations in social resources, it is unrealistic and probably unjust. Surely, the quality as well as quantity of human lives deserves to be considered in making decisions regarding health care. So should the complexity of relationships that exist among individuals and social groups. However, despite these and other problems raised by an ideal of equal welfare, the ideal need not and ought not to be dismissed entirely. A possible and desirable alternative is a perspective that facilitates attention to differences, while critiquing unjust exploitation of those differences.

Obviously, traditional ethical approaches pay attention to differences and relationships. In general, however, their emphasis on universality and impartiality overshadow that attention. Inequality in family and gender relations has often been ignored on grounds of a distinction between public and private life. Despite his "difference principle," even Rawls evidences this flaw.[35] As Susan Moller Okin puts it, "In the most influential of all twentieth century theories of justice, that of John Rawls, family life is not only assumed, but is assumed to be just—and yet the prevalent gendered division of labor within the family is neglected, along with the associated distribution of power, responsibility, and privileges."[36] Okin thus identifies the gap in traditional theories that this book addresses. The egalitarian perspective developed in the next section is a strategy for reducing inequalities that arise because of gender and generational differences in health care.

An Egalitarian Perspective

A perspective, of course, is not a set of rules. Nor is it a fully developed theory. Ethics seems to have plenty of rules and theories already, and many of us involved in health care ethics at the clinical level find all of these wanting when it comes to particular cases. Accordingly, my aim is to provide a useful resource or critical strategy for addressing health care issues in which a key component is their differential impact on men, women, and children. Because this component has been relatively neglected in contemporary biomedical ethics, my focus is remedial. An egalitarian perspective is also applicable to other topics that require remediation, particularly those involving racism and classism, which are often related to injustice towards women and children.

This approach correlates, but is not congruent, with diverse principles, methods, and theories, including some that seem to conflict with each other. The contemporary canon of biomedical ethical principles (respect for autonomy, beneficence, and justice) and classical ethical theories (utilitarianism, deontology, natural law, virtue theory, contractarianism) are all relatable to equality, and to regard for differences among individuals and groups. Persistently, however, my approach also relies on critiques of traditional ethics found in pragmatism, feminism, Marxism, casuistry, and a care-based ethic. These critiques include a rejection of dogmatism, and an insistence on regard for relationships and the nuances of cases in making moral judgments.

Within that context, I propose the following set of guidelines for addressing issues from an egalitarian perspective:

1. **What already exists should not be destroyed.**
2. **Individual lives should neither be shortened nor terminated.**
3. **Individuals that can suffer should not be caused to suffer.**
4. **Individuals that can think and choose should not have their thoughts or choices ignored or impeded.**
5. **Individuals should not be misused or abused, that is, treated as other than who or what they are.**

As guidelines, these considerations operate within the levels of moral justification described by Tom Beauchamp and James Childress.[37] Their account reflects an increasing level of generalization as one proceeds from judgments and actions in particular situations, to rules that inform those actions, to principles from which the rules derive, to ethical theories. An individual facing an ethical quandary can look first to the rules, and need not go further if these are adequate to resolution of the question.

In solving practical ethical dilemmas, however, recourse to abstract levels of reasoning such as traditional ethical theories or metaethics is seldom if ever necessary. The guidelines proposed here may be viewed as middle-level generalizations. They are comparable to the maxims that Albert Jonsen and Stephen Toulmin identify as a central feature of classical casuistry.[38] Although medieval maxims were often stated more informally, as aphorisms, they did in fact function as guides for practical decisions.

All but one (4) of the proposed guidelines are applicable to others besides human beings, and even (1 and 5) to nonliving individuals. I believe that all five are self-evidently relevant to at least some situations. Some, such as 3, surely constitute prima facie obligations; others, such as 1, may be readily overridden in some circumstances. Destroying an ugly and useless pile of trash, for example, hardly deserves justification beyond recognition of its ugliness or uselessness. But if it were suspected that biochemical interactions within the pile had produced a new substance for treating disease, trashing it would be prima facie wrong. (The discovery of penicillin through the adventitious growth of bread mold on a discarded plate comes to mind.) Each guideline specifies a responsibility of caring for other individuals by avoiding certain harms to them; and any one guideline may be subordinated to the others if adequate reasons are present. Alternatively or additionally, the guidelines are analytically derivable from the principles already mentioned: numbers 1, 2, and 3, from the principle of beneficence; 4, from the principle of respect for autonomy, and 5, from both beneficence and respect for autonomy. (I deal more extensively with both principles in Chapter 3.)

To resolve the conflicts that inevitably arise for moral agents who wish to pay due regard to these guidelines, a sixth is needed:

6. Equality demands that all individuals have an equal share of the resources available, insofar as these are pertinent to their needs, desires, capabilities, and interests.

The obvious advantage of this guideline is the avoidance of merely literal equality, that is, the distribution of the same resources to each, regardless of their relevance to the individual, and of those to whom each individual is related. Instead, the equal shares entail the equal distribution of the resources relevant to those among whom (which) they are distributed.[39] Admittedly, the criteria of relevance are complicated and overlapping; for example, one might claim that a beautiful forest should not be destroyed, but one may also claim that a group of fire fighters should allow its destruction in order to save the homes of many inhabitants. It is therefore necessary to introduce a seventh guideline that acknowledges a priority among different kinds of individuals and their relationships to one another. The sequence of guidelines 1 through 4 already suggests that 1 is subordinate to 2, which is subordinate to 3, which is subordinate to 4, so that guideline 7 may be put as follows:

7. Human beings have a primary responsibility to distribute equal shares of pertinent resources to human beings, particularly those closest to them; a secondary responsibility to distribute equal shares of pertinent resources to other sentient[40] beings; a tertiary responsibility to distribute equal shares of pertinent resources to other living beings; and a quartic responsibility to distribute equal shares to nonliving beings.

This guideline might evoke the charge of speciesism. But speciesism, like other chauvinisms, occurs only where irrelevant reasons are used as grounds for discriminatory or unequal behavior toward individuals or groups.[41] Relevant reasons have been offered in support of the claim that human beings have a primary responsibility for distributing equal shares of pertinent resources to other human beings rather than to members of other species. Among these are the following:

    a. A natural, perhaps biologically rooted, inclination
    b. An a priori moral obligation towards one's kin
    c. Laws and customs

d. Religious views
e. Apparent superiority of human traits

Regarding (*a*) we need not go as far as Edward O. Wilson's thesis that the members of every species are genetically programmed to preserve one another.[42] If a strong form of that thesis is true, then there is no question of moral obligation with regard to such behavior. Short of sociobiological determinism, however, we can acknowledge what seems manifest, namely, that human beings, like members of most other species, are naturally inclined to value members of their own species more highly than members of other species. Actually, there seems to be a natural inclination based on proximity of blood, sexual, affective, and cultural ties, so that one is most inclined to preserve one's own life, then the lives of one's closest kin or friends, less close kin or friends, neighbors, other citizens of one's own country, and so on. Noddings elaborates the moral relevance of this natural inclination in her book on *Caring*.[43]

Unless we subscribe to a natural law ethic (which is discussed in Chapter 5), natural inclinations do not imply moral obligations. In fact, some natural inclinations seem to be a source of prejudice, including speciesism. Nonetheless, the natural inclination to support one's kin may be viewed as the empirical grounding for an a priori moral obligation that orders one's responsibilities toward others. On that account, (*b*), one does in fact have a graver responsibility to one's children, parents, and relatives than one does toward those to whom one is totally unrelated. Duties of fidelity are directly linked to kinship with one's family, race, species, and so on. Through a notion of psychological kinship, they are also linked to some individuals with whom one is not related by blood or even species. Ties of affection and obligations of commitment may thus extend to spouses, lovers, friends, and even to pets or other nonhuman animals.

As for (*c*), laws and customs support the seventh guideline. In the United States, as in other countries, legislation is not directed toward nonhumans, except insofar as its influence on them might affect human beings (for example, laws regarding the environment or corporations). Further, although those who live in modern cities may scarcely realize it, use of the environment to fill the needs of human beings has been sanctioned throughout history and prehistory. Many religions and religious teachings (*d*) have also reinforced the notion that the world and everything in it (except perhaps other human beings) were created for the benefit of humankind.[44] But justification for the use of nonhuman by human creatures does not constitute justification for their abuse or misuse. Regardless of whether raising or hunting and killing animals for necessary food is morally acceptable, needlessly killing or inflicting pain on them is not.

Perhaps the most convincing support for the claim that human beings have graver responsibilities to other human beings than to members of other species is the recognition that humans generally have traits that are more worth preserving than traits found in most members of other species (e). Some chimpanzees have been educated to a remarkable degree, and dolphins seem to have a rather impressive mode of communication, but humans have devised and practiced extremely advanced, complicated, and diverse modes of communication. This is not to say that an individual chimpanzee or dolphin may not be more intelligent than a particular (say, profoundly retarded) human adult. The important point is that the average chimp or dolphin (and probably the most precocious or educated chimp or dolphin) is surely less able than the average human being to reason and communicate.

Although the preceding reasons are relevant to the relationship between human beings and members of other species, they would not satisfy those who deny the moral relevance of species membership. They do suffice, however, to show the reasonableness of affirming its moral relevance. Even within the context of guideline 7, the priority of human beings' responsibility to other humans does not imply that commitments to nonhumans never take precedence over obligations to humans. It is possible to conceive of a situation in which the interests of a nonhuman animal might take precedence. Suppose, for example, a dog and a small child are both pinned against a building after an accident. The child can easily be rescued separately, but the dog can only be saved by minimally and temporarily inflicting pain on the child. An egalitarian perspective attempts to balance the burdens and benefits of alternative interventions in such a case, recognizing responsibility to the dog as well as the child. It would thus be morally justified to inflict pain on the child for the sake of the dog. In light of guideline 7, however, it would not be moral to rescue the dog if the child's interests would thereby be seriously jeopardized.

Even within the same species, different individuals and groups of individuals obviously have varying needs. All of these differences may be regarded and addressed in an egalitarian or inegalitarian fashion. Foremost among differences that have led to inequality are those between men and women.

Positions on Gender Equality

Equality between women and men may be viewed in the following ways:

1. Women and men are not equal: women are inferior to men.
2. Women and men are not equal: women are superior to men.
3. Women and men are equal as individuals.
4. Women and men are equal as members of the human community.

Regarding (1), the idea that women are inferior to men has, unfortunately, been around a long time. It is abundantly documented in records of the practices and customs of diverse cultures, and in the writings of (mostly male) philosophers throughout history.[45] Religious teachings have generally reinforced this view—for example, through their conceptions of God as a male authority figure. And women themselves, not surprisingly, have prevalently bought into the myth of male superiority and their own consequent secondary status. For example, Marabel Morgan's Total Woman movement of the 1970s has continued to draw a cult of devotees who enthuse over the traditional subservience of women to men.[46] Without reflecting awareness of the diversity of feminist theories, such a view is essentially, and often overtly, hostile or reactionary toward feminism.

The second position on gender equality represents an opposite view. It argues that the traditional inequality between men and women should be turned around, so that women rather than men exercise the controlling influence in society. Such an approach would constitute a form of matriarchal domination. The overturn of traditional patriarchy is called for because feminine qualities are seen as superior to masculine qualities—for example, through women's closer ties to nature and life, and their more pacific orientation. If women ruled the world, some feminists think, there would be no more wars.[47] In some instances, the notion that women are "better" than men amounts to a subtle put-down of women, excluding them from activities through which men prosper. Patriarchal religions have been particularly influential in perpetrating these practices.[48]

The third and fourth positions affirm equality between women and men. In position (3), however, equality is the starting point for social progress, a value to be upheld by "letting people be"; in position (4), equality is a goal or value to be actively promoted. The former approach is essentially individualistic, the latter socialistic or communalistic. A socialistic version of feminism challenges basic assumptions of our individualistic culture. Obviously the two approaches are at odds.[49]

In American society a liberal version of feminism is individualistic and reformist rather than radical or revolutionary; it endorses the underlying principles of the existent social order, while attempting to correct discriminatory applications of its principles.[50] For example, the Bill of Rights of the National Organization for Women (NOW) calls for the government to enforce "prohibitions against sex discrimination in employment . . . with the same vigor as it enforces the prohibitions against racial discrimination." It also maintains the right of women to be educated and to secure housing and family allowances "on equal terms with men."[51] Reformist feminists generally take issue with those who would destroy traditional institutions such as marriage and family, or the free market economy.[52] The reforms called for constitute an effort to provide women as individuals with as much access to positions of power, prestige, and income as men. Equal liberty is the paramount consideration.

Position (4) entails recognition that women and men may not be "created equal" and have not been treated as equals historically, but that gender equality is a crucial social goal. Alison Jaggar distinguishes Marxist and socialist variations on this theme.[53] Marx considered the quality of the man–woman relationship a measure of the degree to which society has been humanized.[54] Marxist feminism posits equality between the sexes as an essential component of the economic equality that defines the ideal of communism. Economic inequality is thus more basic than gender inequality. In contrast, socialist feminism maintains that gender inequality is the primordial oppression from which other forms of oppression proceed. It is basic to every expression of economic inequality because every area of life and work involves relationships between women and men.

My account of equality is consistent with a socialist or Marxist version of feminism because either version holds that equality is not a beginning point, but a communitarian ideal or goal. While liberal feminism insists that individual liberty should be equally distributed, socialist or Marxist feminism points to other goods as well that need to be equally distributed among individuals. Liberal feminism is less atomistic than traditional liberal theory because of its concern about the practical role of relationships in women's lives, but its attention to differences, relationships, and equality is still tied to the goal of maximizing individual liberty. In contrast, Marxist or socialist feminism attends to differences, relationships, and individual liberty in order to maximize social equality. Either of these versions captures the egalitarian component of feminism, which contemporary feminist scholarship has neglected somewhat in its emphasis on care and critique of traditional philosophy.

A "Feminine" Model of Moral Reasoning

At the outset I alluded to the apparent conflict between feminism and a care-based model of moral reasoning. Yet the preceding account of equality is not only compatible with feminism, but necessary for a morally adequate approach to moral reasoning. The "feminine" model elaborated by Carol Gilligan is but one moral voice.[55] Two are needed to produce a just health care ethic.

The descriptive studies of moral psychologists Gilligan and Lawrence Kohlberg highlight important differences between an ethic of justice and an ethic of care. Kohlberg's six stages of moral reasoning culminate in recognition of universal ethical principles based on justice.[56] Foremost among these principles are equality of human rights and respect for the dignity of human beings as individual persons. From a philosophical point of view, this approach identifies progress as movement from hedonistic through utilitarian into contractarian motivation for behavior. Although Kohlberg's original studies were based on interviews with male subjects about a hypothetical case, a liberal version of feminism is consistent with Kohlberg's final phase of moral development.

Gilligan's critique of Kohlberg is based on interviews with female subjects facing an actual ethical dilemma.[57] Her interpretation of the interview data reveals a different set of stages, beginning with an orientation to individual survival, progressing through a sense of moral goodness as self-sacrificial care, to an awareness of the importance of care for oneself as well as others. This approach does not fit neatly into mainstream ethical theories, but its consideration of context, attachment, relationship, and responsibility finds a congenial framework in nonmainstream theories such as phenomenology, pragmatism, and Marxism.

Some commentators on the Kohlberg–Gilligan debate treat their models as exclusive alternatives, and regard the care model as superior to the justice model. Gilligan herself, however, does not make either of these claims. In an article exploring the relationship between justice and care as models of moral reasoning, she uses the image of a drawing that may be seen as a duck or a rabbit. "Like the figure-ground shift in ambiguous figure perception," she writes,

> justice and care as moral perspectives are not opposites or mirror-images of one another, with justice uncaring and care unjust. Instead, these perspectives denote different ways of organizing the basic elements of moral judgment: self, others, and the relationship between them.[58]

Considerations of equality are crucial to relationships in a justice ethic, whereas attachment is the corresponding standard for an ethic of care. While justice has traditionally demanded impartiality with regard to objective differences, caring necessarily involves attention to context and interrelatedness. Men and women alike exhibit both modes of moral reasoning, but women are more likely than men to focus on care.

Gilligan delineates liabilities that accrue to individuals of either gender if they focus on care or justice exclusively. The potential error of a justice focus, she says, is "its latent egocentrism, the tendency to confuse one's perspective with an objective standpoint or truth, the temptation to define others in one's own terms by putting oneself in their place." The liability of a focus on care is that it tends "to forget that one has terms, creating a tendency to enter into another's perspective and to see oneself as 'selfless' by defining oneself in other's terms."[59] Historically, these liabilities have given rise to two common distortions of justice and care: the equation of human with male, and the equation of care with self-sacrifice. These liabilities are avoided and distortions are corrected in an ethic that incorporates both justice and care.

Self-Consciousness and the Experiential Gap

Throughout this book I attempt to apply an egalitarian perspective to issues that arise within the context of health care. I hope thereby to avoid reinforcement of the exploitative potential of an ethic based exclusively on care. To check the "latent egocentrism" to which I might fall prey through my accent on equality, I hope to maintain a sense of "self-consciousness" in my analysis of specific issues. By that I mean consistent awareness of the gap between my own possible and actual experience, and the experience of those who actually confront the issues about which I write.

Such self-consciousness is appropriate across a wide spectrum of differences among individuals and groups as a means of preventing imposition of narrowly defined values on others. Wendy Williams comments on that possibility with regard to legislation:

> A deliberate body made up exclusively of men—or whites, the rich, or Catholics— no matter how strong their desire to represent "all of the people," will, at least sometimes, inadequately discern, much less build into their laws, provisions that reflect the needs and interests of women—or nonwhites, the poor, or Protestants.[60]

The self-conscious acknowledgment of one's limitations in representing others' needs and interests does not imply endorsement of monolithic or stereotypic views of others. Rather, it argues for inclusion of a broad range of perspectives for policy making that affects us all. Williams makes the point with regard to women: "a diverse group of women would bring a somewhat different set of perceptions and insights to certain issues than would a similarly diverse group of men."[61]

Unfortunately, few male authors seem to write self-consciously about issues that mainly or only affect women.[62] Much of the time their views are valid, but only as developed from the standpoint of a nonparticipant in women's experience. This point applies even to men who may embrace a "feminist standpoint," that is, one that persistently critiques the domination of women by men.[63] Similarly, my views about men and about blacks are (I think) valid much of the time, but only as developed from the standpoint of a nonparticipant in men's or blacks' experiences. Because of the inevitable gap between their experiences and mine, I believe I have an intellectual as well as moral obligation to acknowledge the limitation of my perspective. Several years ago, an experience with an African American student impressed me with the need for such acknowledgment.

The student had come to talk about a paper he wanted to write on paternalism and censorship. On entering my office he asked whether I had seen *The Gods Must Be Crazy,* a movie scheduled to be shown on campus. In the film, the peaceful culture of the black Bush people of Africa is contrasted with the violence and chaos of an urban (black and white) ghetto. I had seen the film and liked it. However, while I thought the film provided an admirable view of black origins, my student saw the opposite: an image of blacks as naïve and ignorant. He argued that unless a showing of the film were accompanied by critical analysis and discussion that would overcome this negative interpretation, it should not be shown on campus.

I believe my arguments against censorship of the film were sound. Nonetheless, I did not and could not know what it was like to see the film from an African American's perspective. In other words, there are limits to empathy. My student's position had an experiential underpinning that I lacked. In speaking from that perspective, his credibility inevitably surpassed mine because my credibility could not be disassociated from the fact that I was an observer rather than a participant with regard to the experience of African Americans. Despite the relevance of my observations and arguments, self-consciousness remains an appropriate response to the realization of this experiential limitation.

From the standpoint of an ethicist working in a university hospital complex, the need for self-consciousness has been impressed on me in other ways. As a nonclinician I am not expected to implement ethical decisions that I facilitate and support. There is thus an important gap between the experiential base from which I and a clinician arrive at our decisions. Even if I were a clinician, however, I am uncertain that I could physically remove a patient from life-support. While I concur with the ethical arguments that justify the procedure in certain circumstances, it would be emotionally difficult to be the immediate or efficient cause of someone's death. I am self-conscious, therefore, about my relatively uninvolved status with regard to ethical dilemmas addressed by clinicians. The same point may be made with regard to the contrast between my experience and that of the patients and families directly involved in the situation. It also applies to the contrast between clinicians who are not responsible for particular patients, and those who are.

Despite these grounds for self-consciousness, experiences that I have had or may have are relevant to issues that I consider in this book. As a child, for example, I survived a life-threatening illness; as an adult I experienced a spontaneous abortion, an unplanned pregnancy, the births of two premature infants, and a diagnosis of (curable) cancer. As a teacher and mother of young children and teenagers I have accumulated further relevant experience regarding the issues discussed. Experiences that I could only have as a woman have enriched my perspective on issues concerning women, children, and health care. They do not supplant the need for self-consciousness because the experiential gap can only be reduced, not closed. They do, however, strengthen my views and arguments when I reflect and write about relevant topics.

A sense of self-consciousness can serve to forestall the easy categorizations that lead to stereotyping of individuals, their behaviors, ideas, and relationships. Stereotypes are generally unacceptable because they impede respect for the differences that characterize each human being as unique and dynamic. In the next chapter, I consider some of the stereotypes that arise in health care, examining their implications from an egalitarian perspective.

NOTES

1. *Justice, equity,* and *fairness* identify moral values or principles more definitively than *equality,* which is sometimes used in a morally neutral manner. All four terms and their cognates are used interchangeably, but when equality does not refer to a moral value it may be modified by one of the others, as in "fair (or 'equitable' or 'just') equality of opportunity." As developed in this chapter and used throughout the book, equality is a moral value or principle, and its cognates (equal, unequal, inequality, egalitarian, inegalitarian) relate to that meaning.

2. Sara Ruddick, for example, maintains that "the ideal of equality is a phantom," and those who engage in the "maternal thinking" that she advocates "would be slow to wish upon themselves or anyone they care for the fearful pursuit of equality." See her "Remarks on the Sexual Politics of Reason," in *Women and Moral Theory,* ed. Eva Feder Kittay and Diana T. Meyers (Totowa, New Jersey: Rowman and Littlefield, 1987), 252–53.

3. Christine Overall, *Ethics and Human Reproduction* (Boston: Allen and Unwin, 1987), 3. See also Alison M. Jaggar, *Feminist Politics and Human Nature* (Totowa, New Jersey: Rowman and Allanheld, 1983), 5.

4. The formal principle of justice is to treat like things alike. Note, however, that "like things" are not "same things." See Stanley I. Benn and R. S. Peters, "Justice and Equality," in *The Concept of Equality,* ed. William T. Blackstone (Edina, Minnesota: Burgess Publishing Co., 1969), 55: "Equals ought to be treated alike in the respect in which they are equal. . . . Injustice, said Aristotle, arises as much from treating unequals equally as from treating equals unequally." This principle underlies the egalitarian perspective that I apply to diverse issues. A nuanced concept of equality does not rest on Aristotle's distinction between equity and equality, but accords with popular usage of the term equality as a desirable social value.

5. For example, Caroline Whitbeck, "A Different Reality: Feminist Ontology," in *Beyond Domination,* ed. Carol C. Gould (Totowa, New Jersey: Rowman and Allanheld, 1983), 66–71.

6. Martha Minow, *Making All the Difference* (Ithaca, New York: Cornell University Press, 1990); and Elizabeth V. Spelman, *Inessential Woman: Problems of Exclusion in Feminist Thought* (Boston: Beacon Press, 1988).

7. Carol Gilligan's *In a Different Voice* is probably the best known work in this regard (Cambridge, Massachusetts: Harvard University Press, 1982), but her more recent article "Moral Orientation and Moral Development," in Kittay and Meyers, 19–33, provides a clearer delineation of both models. See also Nel Noddings, *Caring: A Feminine Approach to Ethics and Moral Education* (Berkeley: University of California Press, 1984). Noddings refers to the "masculine model" as that of "the father," and the "feminine model" as that of "the mother." Neither Noddings nor Gilligan use the term feminist to describe "feminine" or care-based reasoning.

8. Gilligan in Kittay and Meyers, 25. But a survey of the empirical studies that support this point concludes that the differences are explicable by differences in education and occupation rather than gender. See Lawrence Walker, "Sex Differences in the Development of Moral Reasoning," *Child Development* 55, no. 3 (1984): 677–91. Susan Sherwin also uses the term "feminine" to define an ethic based on women's experience. See her *No Longer Patient* (Philadelphia: Temple University Press, 1992), 42–49.

9. For example, Tom L. Beauchamp and James F. Childress, *Principles of Biomedical Ethics,* 3d ed. (New York: Oxford University Press, 1989), 261–70. See my "An Egalitarian Approach to Health Care," in *Revolution, Violence and Equality,* ed. Yeager Hudson and Creighton Peden (Lewiston, New York: Edwin Mellen Press, 1990), 265–82. Substantial portions of this article are incorporated into this chapter.

10. Some feminists explicitly identify "feminine" and feminist moralities; for example, see Sandra Harding, "The Curious Coincidence of Feminine and African Moralities," in Kittay and Meyers, 296–315, and Mary Finsod Katzenstein and David D. Lactin in Kittay and Meyers, 262. Sherwin distinguishes between the two but delineates important areas of overlap. See her *No Longer Patient,* 49–57.

11. As found, for example, in Aristotle's discussion of equity in relation to justice (*Nichomachean Ethics,* 1137b), and equality in relation to friendship (*Nichomachean Ethics,* 1158b, 1159a).

12. Michael Walzer, *Spheres of Justice* (New York: Basic Books, Inc., 1983), 3–30.

13. Zillah Eisenstein, *Feminism and Sexual Equality* (New York: Monthly Review Press, 1984), 206–8.

14. Walzer, xiii.

15. A number of authors have attempted such a reconciliation. See, for example, Michael Stocker, "Duty and Friendship," in Kittay and Meyers, 56–68, and Annette C. Baier, "The Need for More than Justice," in *Science, Morality and Feminist Theory,* ed. Marsha Hanen and Kai Nielsen (Calgary: University of Calgary Press, 1987), 41–56.

16. Shulamith Firestone, *The Dialectic of Sex* (New York: William Morrow and Company, Inc., 1970); Sara Ruddick, "Maternal Thinking" in *Mothering,* ed. Joyce Trebilcot (Totowa, New Jersey: Rowman and Allanheld, 1983), 213–30; Virginia Held, "Feminism and Moral Theory," in Kittay and Meyers, 112–28; Marilyn Friedman, "Care and Context in Moral Reasoning," in Kittay and Meyers, 190–204.

17. See Aristotle's *De Anima*, and Thomas Aquinas on intellect and will in *Summa Theologica* I, Questions 75–83. While some aspects of this account of human nature coincide with liberal political theory, my development of these concepts diverges from traditional liberalism in important respects, for example, by rejecting normative dualism and abstract individualism.

18. See Nancy Fraser and Linda J. Nicholson, "Social Criticism without Philosophy: An Encounter between Feminism and Postmodernism," in *Feminism/Postmodernism*, ed. Linda J. Nicholson, (New York: Routledge, 1990), 21–27.

19. See my *An Idealistic Pragmatism* (The Hague: Nijhoff, 1972).

20. Nel Noddings, "Feminist Fears in Ethics, " *Journal of Social Philosophy* 21, no. 2, 3 (Fall/Winter 1990): 24.

21. This tendency is documented in adults through studies of decision making by medical students, physicians, and lawyers. See Carol Gilligan and Jane Attenuci, "Two Moral Orientations," 76–86; Carol Gilligan and Susan Pollak, "The Vulnerable and Invulnerable Physician," 245–62; Dana Jack and Rand Jack, "Women Lawyers: Archetype and Alternatives," 263–88, all in *Mapping the Moral Domain,* ed., Carol Gilligan, Janie Victoria Ward, and Jill McLean Taylor, (Cambridge, Massachusetts: Harvard University Press, 1988).

22. Noddings, "Feminist Fears in Ethics," 26.

23. For different accounts of criteria for personhood, sometimes equated with humanhood, see Joseph Fletcher, *Humanhood: Essays in Biomedical Ethics* (Buffalo, New York: Prometheus Books, 1979); and John T. Noonan, Mary Anne Warren, and Michael Tooley, in *The Problem of Abortion,* 2d ed., ed. Joel Feinberg (Belmont, California: Wadsworth Publishing Co., 1984), 9–20, 102–34. For an account of persons as "*second* persons," see Annette Baier, *Postures of the Mind* (Minneapolis: University of Minnesota Press, 1985), 74–90.

24. Noting the impact of such knowledge on abortion decisions, two physicians who have operated an abortion clinic in New York wrote the following: "About 25% of our patients decide not to have the procedure done when they learn that a formed fetus will be aborted." Selig Newbardt and Harold Schulman, *Techniques of Abortion* (Boston: Little, Brown, 1977), 69.

25. See John C. Fletcher and Mark I. Evans, "Maternal Bonding in Early Fetal Ultrasound Examinations," *New England Journal of Medicine* 308, no. 7 (Feb. 17, 1983): 392–93.

26. See Karl Marx, "Critique of the Gotha Programa," in *The Marx-Engels Reader*, ed. Robert C. Tucker (New York: W. W. Norton and Co., Inc., 1972), 388.

27. Richard E. Flathman elaborates this point in "Equality and Generalization: A Formal Analysis," in *Nomos IX: Equality*, ed. James Roland Pennock and John W. Chapman (New York: Lieber-Atherton Publisher, 1967), 38.

28. Minow, 50.

29. Minow, 51.

30. John Rawls, *A Theory of Justice* (Cambridge, Massachusetts: Belknap Press of Harvard University Press, 1971), 60ff. My use of the term *inequality* is comparable to Rawl's use of the term *inequity*.

31. Ronald Dworkin, "What Is Equality?" *Philosophy and Public Affairs* 10, no. 3 and 4 (Summer/Fall 1981): 185–246, 283–345.

32. I am thinking here of an Aristotelian ethic of character or virtue, one such as Bernard Mayo describes as positive rather than negative morality. See his *Ethics and the Moral Life* (New York: St. Martin's Press, 1958), 200–18. Also see Alasdair MacIntyre, "Aristotle's Account of the Virtues" in his *After Virtue* (Notre Dame, Indiana: University of Notre Dame Press, 1984), 146–64.

33. Throughout this book I use the term *sex* to refer to (relatively) permanent biological characteristics and functions that distinguish men and women; the term *gender* refers to socially induced behaviors. *Sex-roles* are socially induced or gender-based behaviors.

34. Dworkin, 185–246, 283–345.

35. Rawls, 75–83.

36. Susan Moller Okin, *Justice, Gender, and the Family* (New York: Basic Books, Inc., 1989), 9.

37. Tom L. Beauchamp and James F. Childress, *Principles of Biomedical Ethics*, 2d ed. (New York: Oxford University Press, 1983), 5–8.

38. See Albert R. Jonsen and Stephen Toulmin, *The Abuse of Casuistry* (Berkeley: University of California Press, 1988), 252–53.

39. Peter Singer's "principle of equal consideration of interests" is another way of stating this guideline. However, Singer's account ignores relationships and stresses impartiality, and he would undoubtedly disagree with my next guideline. See Peter Singer, *Practical Ethics* (London: Cambridge University Press, 1979), 11, 22.

40. The meaning of *sentient* I have in mind here is "capable of feeling or perception," as particularly applicable to the ability to experience pain. See *Webster's New World Dictionary*, 2d college ed. (New York: Simon and Schuster, Inc., 1984), 1297; also see guideline 3.

41. Tooley and Warren are likely to level the charge of speciesism because they view membership in the human species as irrelevant to the morality of abortion decisions. Singer develops a more extended attack on speciesism in his *Practical Ethics*, 48–69.

42. See Edward O. Wilson, *Sociobiology* (Cambridge, Massachusetts: Harvard University Press, 1975) and *On Human Nature* (Cambridge: Harvard University Press, 1978).

43. Noddings, *Caring*, 79–81.

44. For example, Genesis 1.

45. See my *Philosophy of Woman* (Indianapolis: Hackett Publishing Company, 1983).

46. See Marabel Morgan, *The Total Woman* (Old Tappan, New Jersey: Fleming H. Revel Company, 1973). With her husband, Charlie, as business manager, Morgan has continued to lead this movement through workshops and publications. See her more recent *The Electric Woman* (Waco, Texas: Word Books, 1985).

47. Without going quite that far, Sara Ruddick argues for the relationship between maternal thinking and the peace movement. See Kittay and Meyers, 237–60. See my "What are the Connections between Concern for the Environment, Feminism and Peace?" *Journal for Peace and Justice Studies* 3 (1990): 103–8.

48. Mary Daly may be the strongest critic of patriarchal religion, but there are many others. See her *Gyn/Ecology, The Metaethics of Radical Feminism* (Boston: Beacon Press, 1978 and 1990), and *Pure Lust* (Boston: Beacon Press, 1984).

49. See my "Feminism: Individualistic or Communalistic?" *Proceedings of the American Catholic Philosophical Association* L (1976): 219–28.

50. My account of different versions of feminism is based mainly on Jaggar's, *Feminist Politics and Human Nature;* and Rosemarie Tong's, *Feminist Thought* (Boulder, Colorado: Westview Press, Inc., 1989).

51. National Organization for Women (NOW) Bill of Rights, in *Morality in Practice*, 2d ed. ed., James P. Sterba (Belmont, California: Wadsworth Publishing Co., 1988), 252.

52. As Betty Friedan put it, "it is a perversion of the new feminism for some to exhort those who would join this [feminist] revolution to cleanse themselves of sex and the need for love or to refuse to have children." From "Our Revolution is Unique," in *Voices of the New Feminism*, ed. M. L. Thompson (Boston: Beacon Press, 1970), 40. More recently, Okin's work provides an excellent account of the types of reforms that would be necessary for gender justice from a liberal perspective. See her *Justice, Gender, and the Family*, note 36.

53. Jaggar, 12.

54. Karl Marx, "Economic and Philosophic Manuscripts of 1844," in Tucker, 69.

55. See Gilligan, *In a Different Voice;* and Gilligan, in Kittay and Meyers, n. 2.

56. Lawrence Kohlberg, *The Psychology of Moral Development* (San Francisco: Harper and Row, Inc., 1984). Following Gilligan, however, Kohlberg acknowledges that "an orientation of care and response usefully enlarges the moral domain." See p. 340.

57. For example, in *In a Different Voice* Gilligan discusses and interprets women's explanations of their decision making when confronted with unwanted pregnancies. See 70–127.

58. Gilligan in Kittay and Meyers, 22.

59. Gilligan in Kittay and Meyers, 31.

60. Wendy W. Williams, "The Equality Crisis: Some Reflections on Culture, Courts and Feminism," *Women's Rights Law Reporter* 7, no. 3 (Spring 1982): 175, n. 2.

61. Williams, 175, n. 2.

62. The abortion issue is a particularly striking example in this regard. While there are exceptions, women writing on the issue have tended to view abortion as termination of the tie between a pregnant woman and fetus, as in Judith Jarvis Thomson's famous "A Defense of Abortion" in *The Problem of Abortion*, ed. Joel Feinberg (Belmont, California: Wadsworth Publishing Co., 1973), 173–87. Men addressing the issue have tended to bypass the fact that pregnancy and abortion occur within women's bodies, focusing exclusively on the moral status of the fetus.

63. Claudia Card reserves the term "feminist" for women, preferring to call men who support feminism "profeminist." See her "The Feistiness of Feminism," in *Feminist Ethics,* ed. Claudia Card (Lawrence, Kansas: University Press of Kansas, 1991), 5. Concerning the important difference between "women's standpoint" and "feminist standpoint," see Nancy Hartsock, *Money, Sex, and Power* (Boston: Northeastern University Press, 1983), 232. In general, my view of the need for self-consciousness accords with Hartsock's account. I believe, however, that the account needs to be applied to individuals as well as nondominant groups.

2

Sex-Roles and Stereotypes in Health Care

Most stereotypes contain an element of truth. The profession of medicine, for example, may be seen stereotypically as an inegalitarian relationship between doctors, who are mainly men, and patients, who are mainly women.[1] The term "inegalitarian" describes a relationship in which differences between parties are such that one is regarded as, or is superior to, the other in one or more respects. The differences provide a rationale for one party having power over the other. Such inequality may be temporary, as in the parent–child relationship in which the child is dependent on the adult for a limited period of time; or permanent, where disparities are not subject to change.[2] In either view, an inegalitarian relationship involves a ranking based on differences. The ranking can vary according to differences among the parties being ranked; for example, children generally rank lower than adults in terms of experience, strength, and knowledge, but higher than adults in terms of their expected life span and capacity for further development.

In temporarily inegalitarian relationships, the goal of either or both parties may be to reduce the disparity between them. The parent, for example, attempts to facilitate the child's progress toward independent adulthood, or the health practitioner strives to restore health to a point where medical treatment is no longer required. Unchangeable differences that are irrelevant to ranking sometimes become the basis of inegalitarian relationships in which one party maintains advantages or domination over the other. Racism and sexism are notorious examples in this regard.

Lest we infer from the fact that most patients are women that this constitutes grounds for averring their inferiority to men, we need to recognize significant reasons for women's involvement in the health care system. More patients are

24

women because women can do something that men cannot do, namely, bear and nurse children, and because women tend to live longer than men by an average of seven years.[3] Normal healthy women are likely to involve themselves in the health care system in connection with childbirth. There is no counterpart to this experience for healthy men. Even in infancy, females are more robust than males. Although the male birth rate is greater than the female birth rate, females are more likely to survive infancy than males.[4] Throughout their lives, women are generally more involved in the health care system on a preventive basis than are men, not only for themselves but in behalf of others such as children and elderly or disabled relatives. Their propensity to care for the health of others suggests a care-based model of moral reasoning.

Many reasons (valid and invalid) have been cited to explain why most physicians are male. For example, women supposedly have less aptitude for the skills of mind and hand required of the doctor, but have a naturally greater aptitude for skills of the "heart" such as those required of the nurse.[5] The role of the nurse is described as one of caring, while the doctor's role is one of curing.[6] Typically, the latter role is more esteemed, and rewarded accordingly in terms of income and prestige.[7] Obviously, this rationale reflects a stereotypic image of physician's and nurse's roles.

Many women who choose nursing over medicine as a profession do so in part because of a perception that nursing allows them to plan their lives more efficiently so as to meet the demands of home and family.[8] The same rationale often applies to those who choose allied health specialties such as medical technology, physical therapy, and dietetics. The majority of these health care workers are women.[9] A perceived conflict between the professional demands of medicine and the role of wife and mother that most women fulfill during some part of their lives has served as a deterrent to some women who might otherwise choose medicine.[10] Men experience less conflict because they are not essentially involved with childbirth or early nurturance, and there is a widespread social expectation that men will devote themselves primarily to work outside rather than inside the home.

A tension between feminism and medicine is evident in the contrast between feminist arguments for women's autonomy regarding their own bodies and the control of others' bodies that medical practice involves. While the physician exercises power in behalf of the patient, such control may simultaneously circumvent the expressed wishes of the patient, and constitute an invasion of her privacy. It may also circumvent the wishes of women who serve as proxy decision makers for their children, spouses, or elderly relatives.

Although women constitute only about one-sixth of all physicians in the United States, they make up more than a third of those enrolled in medical schools.[11] If the marked increase of women entering medicine continues, their influence could reduce the impact of sex-role stereotypes on the physician's relationship to patients and to other health professionals.[12] I explain later why this may not be the case. First, however, I wish to examine some of the essential features of gender or sex-role stereotypes and compare these with alternative models of the physician's role. On the basis of that comparison and different versions of feminism, I will propose a model of the physician–patient relationship and a version of feminism that are compatible, and preferable to the alternatives. A defense of this position is based on the notion of equality described in Chapter 1. Although my account focuses on the role of the physician, it is applicable to other health professionals to the extent that they participate in the hierarchical ordering of the professions in relation to each other and in relation to patients. In Chapter 15, I critique the power motif that informs this hierarchical arrangement.

Common Stereotypic Features

Since the early 1980s I have had regular classes with third- and fourth-year medical students during their clinical rotations in the obstetric-gynecological service of two women's hospitals. The proportion of women in these groups of students is higher than the national average in one medical school, and less than the national average in the other medical school. Typically, I begin a session by eliciting from the students their sense of what is unique about this specialty as compared with others. The features noted are obvious but significant: most patients are healthy, and an additional healthy patient, the newborn, is the usual outcome of obstetric practice. Another feature regularly mentioned is that all the patients are women, while most doctors (still) are men. Once this feature of the specialty is introduced, I ask the students to indicate qualities that are stereotypically attributed to women or men distinctively, allowing that the attribution is usually inaccurate for real individuals of either sex. The traits suggested are predictable: men are aggressive, women are passive; men are strong, women are weak; women are dependent, men are independent; women are emotional, men are rational. I then ask the students to outline the traits stereotypically attributed to physicians and patients. Not surprisingly, these correspond with those assigned to men and women.

Advertence to the stereotypic traits proposed by the medical students is preliminary to their recognition that these are commonly viewed not only as opposite to, but also as excluding the corresponding trait in the other partner of the relationship. Thus, to the extent that a woman is (perceived as) affective or emotional, she is (perceived as) lacking rationality; and to the extent that a man is (perceived as) aggressive, he is (perceived as) not passive or receptive. Yet all of these features may be used to describe the same personality, man or woman. A specific individual may be passive or vulnerable in certain respects, but strong and aggressive in others. For example, I cannot lift 200 pounds, but I have been told that I have a considerably high pain threshold—which makes me strong in one respect while weak in another. Women traditionally have been imputed to be more emotionally dependent than men, yet a number of studies have suggested the opposite—for example, the loss of a spouse is more devastating to a man than to a woman.[13] Clearly, women are fully capable of exercising intellectual rigor concurrently with their affectivity. Depending on their type and degree of illness, patients often possess a similar capability.

In her 1973 cross-cultural study of sex-roles, Jeanne Humphrey Block describes another difference that may be stereotypically attributed to men or women: men are "agentic," and women are "communal."[14] To be agentic is to promote one's own interests through self-protection, self-assertion, and self-expansion; to be communal is to promote the interests of others, even at times at the expense of one's own. These differences reflect the "masculine" and "feminine" models of moral reasoning described in Chapter 1. The inadequacy of such stereotypes is fairly obvious, both in their application to men and women, and in their extended application to physicians and their patients. Traditionally, women are perceived as interested in home and family more than in themselves or the wider community; men are perceived as acting in the interests of the public or the wider society. For either sex, however, the orientation may be interpreted as communal rather than agentic, or possibly as agentic on behalf of the community. Insofar as the communal interests of women are narrowly defined as embodied in the family, they may be defined as acting more agentically than men. Similarly, the traditional image of the (male) physician is communal in that it represents an altruistic commitment to the welfare of others. Nonetheless, with the rise of the "medical-industrial complex," the American public has become increasingly persuaded that medicine is more self-serving than other-serving.[15] Patients, too, may be either agentic or communal in their orientation. With good reason, many are exclusively concerned about their own welfare, but many apparently worry most about preservation of relationships and the impact of their condition on those they love.[16]

The recent controversy over models of moral reasoning is illuminating with regard to agentic or communal traits in individuals. Lawrence Kohlberg's approach involves three levels, each comprised of two stages, and proceeding from premoral to mature moral reasoning.[17] The mature level is illustrated by deontological argument, that is, an appeal to universal moral rules that govern the exercise of individual rights. Carol Gilligan's countermodel of responsibility is based on relationships and attachment to others rather than individual rights.[18] My own concern about these models is that either or both may be stereotypic. Prior to Gilligan's critique, Kohlberg's categories had become standardly invoked in some circles as applicable to all individuals.[19] Gilligan has already been interpreted somewhat stereotypically, and inaccurately, as maintaining the superiority of the "feminine" model.[20] As I have already suggested, both models yield valid insights about moral reasoning, but neither model is as adequate as one that incorporates insights from both.

Models of the Physician–Patient Relationship

On the basis of traits stereotypically assigned to women, men, patients, and doctors, certain models of relationships emerge. In assessing these models, however, an important difference should be noted; namely, that one relationship is temporary while the other is permanent. The power that temporary inequality involves for practitioners may be defended as a means of promoting permanent equality between patients and nonpatients, including the practitioners themselves. The goal of medical practice may in fact be defined as that of making itself unnecessary through improvement of the patient's status: illness, as a possible source of inequality, is to be eradicated.[21] A similar argument is inapplicable to the woman–man relationship. If being a woman and being a man are permanent conditions, an assumption of inequality between the two necessarily implies that one is permanently inferior to the other, at least with respect to behaviors or abilities based on gender. Whether inequality is permanent or temporary, it need not apply to all aspects of the relationship. Most patients, for example, have less knowledge of medicine than their doctors, but many have greater knowledge and skills in other areas.

The preceding point suggests another construal of the woman–man relationship, and this is also applicable to the physician–patient relationship: complementarity. A man may thus be viewed (or view himself) as needing a woman in order to prove his virility, a doctor as needing a patient in order to be a doctor, and vice versa in both types of relationships. But the woman or the patient is essentially a receptor of the other's strength or expertise or sexual drive. Accordingly, the supposed "complementarity" is in fact an inegalitarian relationship in which women or patients are passive and vulnerable to men or doctors. In the health care setting, this relationship is sometimes characterized as paternalism.

While other caregivers may act paternalistically, the term *paternalism* is most often used to describe the relationship between physicians and patients. Its meaning is commonly associated with its etymological root (from the Latin *pater* or "father"): a principle or system of behavior toward others resembling that of a father towards his child.[22] The specific role of the father is further defined as that of a protector, one who takes responsibility for another. In contrast to the child who depends on him, the father figure is independent. On such an account, paternalistic medical interventions are those that impute to the practitioner total responsibility for what is done or not done to patients. Where the patient's own wishes are overridden through these interventions, the principle usually invoked to justify violation of the patient's autonomy is that of beneficence, or at least, nonmaleficence. Traditionally, this principle is embodied in the Hippocratic corpus, which stipulates that physicians ought above all "to help, or at least, to do no harm."[23]

The serious limitation of beneficence as an ethical principle is that it may be invoked to justify horrendous intrusions on a patient's autonomy. It may thus reduce the roles played by physician and patient to the stereotypic extremes of aggressivity and passivity. Appeal to beneficence under the guise of commitment to the patient's best interests gives priority to the physician's rationality and expertise as opposed to the patient's emotional vulnerability and medical ignorance. Such a position strengthens the physician's role as an independent agent, while the patient is construed as essentially but dependently communal. In short, the paternalistic model of the physician–patient relationship expresses all of the inegalitarian aspects of sex-role stereotypes regardless of the gender of the physician or patient. The fact that most doctors are men and most patients women reinforces the paternalistic elements of this approach. Because of its emphasis on beneficence toward patients, the model of paternalism may be defended as an ethic of care. If we define care as encompassing respect for autonomy, however, this defense is hardly adequate.

In the past several decades the paternalistic model of the physician–patient relationship has been increasingly criticized and rejected in favor of a patient–autonomy model.[24] The key but controversial concept here is "informed consent." That concept conflicts with sex-role stereotypic qualities because it imputes to patients both rationality and agency. However, an emphasis on informed or proxy consent[25] may be so exclusive as to ignore the broader principle of autonomy applicable to others involved in health care decisions, for example, physicians and other professionals. Logically, if one interprets informed consent as morally indispensable to medical interventions,[26] an instrumental role for the physician is implied, which reverses the assigned sex-role or gender traits. The patient is then in the position of agent, and the physician is dependent on the patient in order to practice the art of medicine. On such an account, the patient is viewed as one who knows better than the doctor what should be done in her behalf.

If we wish to choose a label comparable to paternalism to describe the relationship between doctor and patient when patient autonomy is championed, we might call it *maternalism*. Admittedly, this term may be applied to women who practice paternalism in the sense defined above. It has also been used to describe a theory that points to biological differences between the sexes as grounds for affirming that women are equal or superior to men.[27] However, if we use the term in accordance with its etymological meaning (from the Latin *mater* or "mother"), it means the opposite of paternalism. One becomes a mother by nurturing a fertilized ovum through all of its natural stages of development, giving birth through a final push of the fetus from womb to world. Birthing essentially involves letting go of what has been a part of oneself. As soon as the infant begins to breathe on her own, the umbilical cord is cut, and the newborn achieves independence after nine months of total dependence on the pregnant woman.

The mothering task continues as the child develops, but to the extent that its paradigmatic meaning is fulfilled, it involves persistent attentiveness to the ability of the growing child to function less dependently, avoiding overprotectiveness that might impede that development. In time, the child becomes not only independent but autonomous. The essential task of nurturance or motherhood, therefore, is to foster another human being's developing autonomy. To be maternalistic implies respect for the other's (for example, the child's or patient's) autonomy, even at the expense of the nurturer's autonomy, and even where risk is involved for either party. The parent who refrains from intervening so as to allow the child to learn from his own mistakes is thus practicing maternalism. While the orientation is obviously different from paternalism, this model, too, may be construed as advocating an ethic of care rather than justice. The care exercised by the paternalist is a protective influence; the care exercised by the maternalist is the eliciting of greater freedom.

The very term *patient* is inappropriate in a maternalistic context because the word comes from the Latin *patior,* which means "to suffer, let happen, or be passive." Possibly this is why terms such as *client* or *consumer* (of health services) have been used more frequently by health care providers who wish to stress the importance of autonomy or agency on the part of the "patient." [28] For example, the 1976 code of the American Nurses Association substituted the term "client" for "patient" in referring to the nurse-other relationship.[29] In contrast, the 1989 code of the American Medical Association retains the traditional language of the "physician–patient" relationship, simultaneously acknowledging the importance of free and informed consent on the part of "patients."[30]

Despite the oxymoronic status of its literal meaning, the term *patient autonomy* is used broadly to signify that the receiver of health care is in fact a moral agent. Proposed substitutes for the term "patient" have drawn little broad support from doctors or from those they treat. "Client" is judged to impute too much power to the subject of health care, and "consumer" is criticized as modeled on a business mentality. The term I believe preferable to either of these is *subject,* which in some respects connotes autonomy (as contrasted with "object") and in other respects connotes passivity (as in "research on human subjects"). However, despite the use of "client" in the ANA code, the term "patient" is so imbedded in the consciousness and jargon of most health practitioners and recipients that it is unlikely to be dropped from routine usage. In what follows, therefore, I refrain from imposing my preferred term on the discussion.

Difficulties inherent in a maternalistic account of the health caregiver's role may be as obvious as those inherent in that of paternalism, and criticisms of both have been well-developed elsewhere.[31] Basically, the paternalistic model allows neglect or violation of patient autonomy. A maternalistic model (as I have defined it) permits neglect or violation of beneficence, that is, the traditional medical obligation of doing good and avoiding harm to the patient. However, an exclusive emphasis on informed or proxy consent also fails to uphold the principle of respect for autonomy, which is applicable to others besides patients. To treat health professionals only as instruments is surely to deny or at least ignore their right and responsibility to exercise their own moral autonomy in professional decisions. The patient-autonomy model also, then, allows inegalitarian applications.

In an effort to overcome the problems inherent in paternalistic or maternalistic (patient-autonomy) models of the doctor's role, David Thomasma has proposed a model based on the physician's conscience.[32] The elements to be considered in the physician's formulation of a conscientious decision or prudential judgment are complex and demanding. While emphasizing beneficence, Thomasma insists on paying due regard to "the existential condition" of the patient, including her autonomy. Each patient, he says, "must be handled individually," with a consensus developed through participation of other members of the health care team. As many as possible of the different values at stake are to be preserved, through adherence to the following moral rules:

1. Both doctor and patient must be free to make informed decisions.
2. Physicians are morally required to pay increased attention to patient vulnerability.
3. Physicians must use their power responsibly to care for the patient.
4. Physicians must have integrity.
5. Physicians must have a healthy regard for moral ambiguity.

Actually, Thomasma's use of the term *conscience* in describing this model is misleading because it suggests a subjectivism that his fuller elaboration does not support. Moreover, since he would allow for a patient-conscience model to complement this elaboration, an adequate account of the doctor–patient relationship is not possible until the matter is viewed from that perspective also. What happens if conscience judgments of different physicians, or of patients and their physicians, conflict? This question cannot be answered solely on the basis of the physician-conscience model.

My main complaint with Thomasma's model, then, is a point that he himself admits: the model reflects only the perspective of the conscientious physician. Since it does not reflect the perspective of the conscientious patient, it can hardly be said to describe the physician–patient relationship as a set of interactions or behaviors transpiring *between* two individuals. The relationship itself is crucial not only to an ethic of care, but to therapeutic effectiveness (at least in some cases). Even though physician and patient alike may be conscientious and may concur in decisions regarding health care, they inevitably have different perspectives. Their relationship is not observable from either standpoint exclusively, but it is describable from a perspective that attends to differences in roles, responsibilities, and values.

Insofar as Thomasma focuses on the physician's conscience as the basis of judgment, he invokes a framework by which the physician may move beyond a stereotypic construal to view the physician–patient relationship more critically. Although *conscience* suggests a purely subjective and individualistic basis for decision making, an enlightened or informed conscience is probably indispensable to avoidance of the sexist or reverse sexist tendencies of paternalist or maternalist models. In the clinical situation, after all, what counts as paternalistic or maternalistic is bound to vary. What is called for is a view of the physician–patient relationship that represents both needs and responsibilities, defined as conscientiously and objectively as possible, and applicable simultaneously to unique individuals on both sides of the relationship. Since such a model would be free of the stereotypes inherent in either paternalist or maternalist models, an appropriate label for it would be *parentalism.*

Admittedly, a disadvantage of the parentalist model is its susceptibility to another stereotype that health practitioners often find attractive: that of the parent of either sex who is wiser and more powerful than the child (patient). In effect, this interpretation is equivalent to paternalism. In what follows, I develop the strengths that the parentalist model can sustain if this interpretation is avoided.

A Parentalist Model

The term parentalism has been proposed in some quarters as a nonsexist substitute for paternalism.[33] As already suggested, it then raises the same ethical questions as paternalism regarding respect for autonomy. Others have argued that use of the term *parenting* for behaviors usually attributed to mothers or fathers fails to overcome the sexist implications of paternal and maternal roles. According to Susan Rae Peterson,[34] the term "parenting" dangerously masks "the motives and goals of those opposed to feminist values," namely, to ignore the real contributions of women as women, irreplaceable by men. In other words, the gender neutrality of the term avoids recognition of differences that must be addressed in order to preserve or promote equality.

While I share Peterson's concerns, my proposal of the term parentalism is partly motivated by a desire to avoid sexist assumptions and implications of either maternalism or paternalism, and partly by a desire to critique the narrow notion of "parenting" and "family" relationships that prevail in today's society.[35] The narrow notion is that parenting is a maternal *or* paternal relationship to children who are chronologically quite young. Such parental relationships are based on biological or legal ties. Yet real parenting, and so parentalism, is practiced by many who are neither biological nor adoptive parents, and in some instances is not practiced by those who are. It is a behavior and an attitude that entails both nurturance and protection across the entire span of life, promoting the autonomy of the other even beyond the point where the nurturer's interventions are needed or desirable. It is egalitarian in its acknowledgement of differences between parents and children, while maintaining a commitment to overcome temporary inequalities. Obligations to ensure survival, to comfort or prevent pain, and to respect the developing autonomy of others reflect the guidelines proposed in Chapter 1; different obligations have priority at different stages of a child's development.

Parenting thus construed is both life begetting and life sustaining, with an understanding of life as unfolding and developing towards fulfillment of each individual's unique potential. On this account, parentalism is a model for all of our human relationships, no matter what our age or biological or legal ties to one another as we move toward an ideal of equality. Just as my mother parented me when I was small by doing for me what I could not do for myself (feeding, clothing, etc.), she continued to parent me by encouraging my independence through most of my life. In her advanced age I parented her by doing for her what she could not do for herself (feeding, clothing, etc.) and also by respecting and encouraging her independence as much as possible. An egalitarian perspective suggests that we are all similarly bound to protect others from harms and to do for them what they cannot do for themselves, while respecting and encouraging their autonomy. Obviously, we have different degrees of obligations to different individuals, depending on our relationships and commitments to them, their need, and our capacity to respond. Moreover, conflicts inevitably arise between our obligations to different individuals. Nonetheless, the parentalist model is useful as a moral ideal or paradigm for addressing the changing needs and capacities of unique individuals with unique relationships to one another.

A clear advantage of this model for the physician–patient relationship is its applicability to a variety of medical situations. Consider, for example, the types of disease that Albert Jonsen, Mark Siegler, and William Winslade have categorized through the acronyms ACURE, CARE, and COPE.[36] ACURE refers to diseases that are acute, critical, unexpected, responsive to treatment, and easily diagnosed; for example, the otherwise healthy individual who contracts bacterial meningitis. CARE refers to diseases that are critical, active, recalcitrant, and eventually fatal; for example, a patient with cancer in its terminal stage. And COPE refers to diseases that are chronic, treated on an outpatient basis with palliative and efficacious therapies; for example, a patient with diabetes mellitus or rheumatoid arthritis.

In the ACURE situation, there may scarcely be time, possibility, or need to check on the patient's wishes before acting beneficently in his behalf, just as one would attempt to rescue a friend from sudden danger. Such actions are sometimes considered justifiably paternalistic. I agree that they are justifiable but not that they are paternalistic so long as the intervention is assumed to accord with the patient's or friend's own wishes. In the CARE situation, there is a persuasive case for respecting the patient's wishes over those of others, including clinicians. To the extent that this position overrides what the physician considers best for the patient, the intervention may be construed as maternalistic. The American legal system generally supports this approach through its insistence that no competent patient shall be coerced to undergo treatment (except for the sake of dependent minors). In the COPE situation, the challenge to both physician and patient, like that which parents and children typically face, is to act both beneficently and autonomously toward one another, or in cases of conflict, to violate either principle as little as possible.

A parentalist model of the clinician's role is applicable to all three kinds of situations because it involves ascertaining the degree of protective (paternal) influence that is necessary, while also attempting to maximize expressions of autonomy on the part of all those concerned. Indeed this model is applicable beyond the physician–patient relationship to relationships that occur between physicians and nurses, patients and nurses, and beyond the health care setting to the full spectrum of human interactions.

Whether applied in the clinical setting or more broadly, the parentalist model avoids the stereotyping that either maternalist or paternalist models reinforce. It does so by rejecting the idea that the physician is solely the instrument of the patient, or that she is totally superior to the patient. For most of their life spans, after all, parents *and* their children are adults. Throughout their lives, just as parents learn from their children, so physicians who subscribe to this model learn from their patients, whose rationality is not necessarily distorted but may even be sharpened by the time for reflection which sickness and hospitalization sometimes afford. Moreover, such physicians feel no need to cut off their emotional life when they put on their white coats and wear their stethoscopes around their necks; they may also construe their role not as independent agents but as collaborators with other health professionals as well as with their patients. In other words, parentalist physicians as well as their patients may manifest passivity and aggressivity, emotion and reason, community and agency — all of the attributes stereotypically assigned only to one sex or the other.

Feminist Criticisms

A parentalist model of the physician–patient relationship necessarily entails rejection of sex-role stereotypes. It is thus compatible with feminism, but not all varieties of feminism. Unfortunately, many people interpret feminism stereotypically, as an ideology that denies sex or gender differences, or that proclaims the superiority of women over men. Neither of these interpretations is compatible with the egalitarian approach that I recommend. Accordingly, I conclude this chapter by relating different versions of feminism to alternative models of the physician–patient relationship. Preliminarily, it should be noted again that feminists, whatever their political commitments, emphasize the validity and significance of women's experience. They share both the empirical view that women generally occupy a subordinate position to most men in contemporary society and the normative view that this is not the way society should operate.[37]

The empirical point is well documented in the comparative data on economic distribution, and by laws and statutes illustrating sex discrimination throughout history and continuing through our own day.[38] The normative point is more controversial. In general, however, women as well as men who enter medicine tend to support either view less often or less strongly than others support them.[39] If we were to speculate on reasons for the reticence, and disagreement in some cases, about basic feminist positions, several possible explanations are apparent. One is that the very accomplishment of those women who enter medicine is enhanced by the fact that so few of them do so. Another is that their success has been achieved by and may yet be dependent on male-defined criteria of accomplishment. A third reason is that an attitude of solidarity with other women is difficult to develop or maintain for individuals who have been socialized to view women as generally inferior to men, and in that context to view themselves as exceptional. A fourth reason is that individual differences are greater than those between the sexes.[40] Since male and female medical students are more similar to one another than female medical students are to other women, it is less likely that the female medical students see themselves as subordinate or inferior to their male counterparts.[41]

As I have already suggested, physician–resistance to feminist arguments might be reduced through recognition that feminism has a variety of forms. A *liberal feminism* entails a reformist critique of sexual discrimination as impeding the rights of individual women to achieve their full potential. The potential of women, it is claimed, cannot justifiably be confined by legislation based on sex-role stereotypes. Applying that critique to the physician–patient relationship, the liberal feminist would reject medical paternalism as reflecting the legally reinforced social paternalism that has kept women "in their place" (at home raising children, dependent on men), and so stifled their potential throughout most of history. Liberal feminism obviously stands opposed to situations where physicians are allowed to violate the autonomy of their patients (whether women or men), even though this is defended on grounds that they thus promote the patient's good. But if this version of feminism is consistent with its liberal foundation, a patient autonomy model is not acceptable either, because exclusive emphasis on patient autonomy might conflict with the liberal's insistence on respect for the autonomy of physicians, a growing number of whom are women.

Different versions of *radical feminism* have emerged since the 1960s, but all share the belief that the oppression of women is at the root of all other systems of oppression. To counter this oppression, radical feminism calls for rejection of all patriarchal institutions or social structures, including the "institution" of heterosexuality.[42] Development of women's culture and feminist alternatives to male dominated forms of literature, music, spirituality, health services, and sexuality are encouraged as means of maintaining feminine values.[43] Applying this orientation to the medical profession, radical feminism challenges its male dominance, particularly with regard to reproductive health. Although the critique extends to all aspects of medicine as embodying a kind of "patriarchal religion,"[44] it specifically advocates efforts to place the care of women's health in the hands of women themselves. Accordingly, the model invoked is a maternalism that seeks to empower women as women. To the extent that women replace men in a hierarchical system, medicine may then be matriarchal. In general, however, radical feminists (and other feminists as well) affirm cooperation rather than competition as modalities for their interactions.

A *Marxist interpretation of feminism,* whether in its classical or *socialist version,* clearly challenges the inequality signified by paternalist or maternalist models of the physician–patient relationship. Both versions are consistent with parentalism. Karl Marx's motto at the end of the *Critique of the Gotha Programma* exemplifies the parentalist attempt to balance response to individual needs with social contributions proportionate to the differing talents of individuals: "From each according to ability, to each according to need."[45] This orientation is plainly opposed to the individualist economic orientation often attributed to the American medical establishment. To the extent that physicians subscribe to that orientation, whether they happen to be women or men, endorsement of either socialist or traditionalist Marxist feminism is improbable.

It is possible, however, for physicians to join a critique of economic or other kinds of individualism as occasioning disparities and inequality among groups of people requiring health care.[46] A corresponding version of feminism is then compatible with their view of the physician's role. Either Marxist or socialist feminism exemplifies this compatibility, and both may be called egalitarian. Either view would take account of the differing needs, relationships, preferences, and abilities of individual women and men, relying on the principles of beneficence and respect for autonomy as well as equality. The balancing of these principles occurs in the context of an ethic of (health) care and responsibility.

The importance of considerations of justice to the parentalist model of the physician–patient relationship is apparent when we recognize that most parents have more than one child, and most physicians have many patients. Equal care for each child or patient does not imply the same treatment. Since time, expertise, and material resources are surely limited (and psychological or immaterial resources are limited as well), parents and physicians alike have to monitor their distribution of resources so as to treat their different children or patients with equal care, that is, care that respects their different needs and interests. As never before, through government curtailment of reimbursement for services and through escalation of costs of new and old services, physicians are forced to confront the ethical dilemmas arising from problems of limited resources. Some interpretation of distributive justice is inevitably applied to these situations. The challenge is to choose among alternative interpretations the one most defensible from a moral point of view, and feasible for statutes and clinical practice.

An egalitarian version of feminism is compatible with a parentalist model of the physician-patient relationship, because it entails treating individuals on either side of the relationship as unique and dynamic composites of needs, abilities and preferences. Equality, imbedded in the principle of justice, respects these differences. Alternative models are inegalitarian if they neglect relevant differences that apply to individuals on either side of the relationship. The patient autonomy model of maternalism ignores the autonomy of the physician; the paternalistic model bypasses the autonomy of the patient.

An egalitarian perspective provides a basis for overcoming the frustrating influence of sex-role stereotypes and stereotypic images of the physician, the patient, and the relationship between them. While liberal feminism appears compatible with the patient autonomy model, it is so only if we limit the autonomy of the physician, and thus violate the view of equality on which a liberal philosophy is based, namely, equal liberty for men as well as women. Radical feminism is compatible with the patient autonomy model also, but its promotion of women's values may also be inegalitarian. Stereotypes occur in versions of feminism that give undue weight to gender differences.

Different Roles in Different Contexts

Most feminists extend their critique of gender stereotypes to stereotypic conceptions of other roles and relationships. Crucial to this critique is the sense of context that is supported by care-based and case-based models of decision making.

Stereotypic conceptions of roles are inadequate to resolution of conflicts that arise because of the multiple roles that individuals sustain. Women, for example, are always daughters, usually citizens, and often mothers, spouses, sisters, friends, employees, and employers. Each of their roles involves responsibilities that may be at odds with other role-based responsibilities. Rigid adherence to stereotypes exacerbates the conflicts. Attention to context facilitates their resolution.

As an example of such conflict and its context-based resolution, consider the following case:

A neonatologist is called to the delivery room to determine whether to resuscitate a very premature infant whose parents are uncertain about whether their child should be treated aggressively. The physician knows that the odds for survival through successful resuscitation are 50/50. She also knows that the chance of serious disability for survivors is 20 percent, and that, even without a serious permanent impairment, the infant is likely to spend at least three months in the intensive care unit, incurring costs of at least $300,000. The doctor believes that the expenditure necessary to save this baby would have a greater therapeutic impact if it were used to fund nutrition programs for pregnant women in the inner city. Her belief is based on recognition that poor nutrition during pregnancy is a major cause of preterm birth.

So what does the good doctor do? Multiple relationships suggest obligations to the infant, the family, future patients, and the wider community. She is a doctor with a special commitment to promote the best interests of her infant patient. She is also a doctor whose expertise and position in the community confer a special responsibility for fostering equitable and effective distribution of scarce medical resources. Not only is she a doctor, however; she is also a citizen who is concerned about equal distribution of scarce social resources on which medical costs may impinge. Responsibilities that follow from her relationship to the wider society and with future patients make it impossible to view her commitment to the patient stereotypically.

An egalitarian perspective calls for a seemingly schizophrenic ethic with regard to the physician's relationships to the patient and to society. The physician who wants to be faithful to both relationships pursues ends that may finally be incompatible: beneficence towards the infant and justice towards the wider community. While recognizing the latter obligation, the physician will (probably) intubate the baby in order to maximize his chance of overcoming the hurdles his life already involves. This is an appropriate resolution to the apparent dilemma. Within the context of the delivery room, the physician has control over one aspect but not over the other. She has power to make a lasting positive difference for one child and his family, but she has no way of insuring that costs saved by not resuscitating the infant might be used to promote greater therapeutic benefits for more people. To discharge her responsibility to the larger society, the doctor needs to use her influence to effect changes that will foster a more equitable distribution of support for basic human needs. Lobbying, voting, and other political activities may in time reduce support for acute problems in order to spread the benefits of health more widely. Thus the physician can keep faith with both the infant patient and with society, recognizing that the relationship through which one has the greater impact is more binding, but that both engender practical responsibilities. A stereotypic understanding of the caregiver's role is inadequate to that realization.

Stereotypic thinking places individuals in multiple jeopardy when they fit into more than one stereotypic category. For example, an individual patient may be very old, female, and black, prompting (prejudiced) judgments that entail ageism, sexism, and racism, as well as the medical chauvinism that professional caregivers may be inclined to exercise. In the next chapter I deal with a health care specialty that is uniquely positioned to reinforce stereotypic thinking with regard to women patients, namely, obstetrics and gynecology.

NOTES

1. The total number of physicians in the United States in 1988 was 585,600; of these 15.8% were women, up from 11% in 1975. See U.S. Bureau of the Census, *Statistical Abstract of the United States: 1991,* 111th ed., (Washington, D.C.: U.S. Government Printing Office, 1991), 103. Regarding women as patients, see a study of physician contacts by the U.S. Department of Health and Human Services, *Vital and Health Statistics,* Series 10, No. 176, DHHS, Publication No. (PHS) 90-1504 (Hyattsville, Maryland, October 1990), 115. Although the rates did not differ significantly for males and females in the youngest and oldest age groups, for the age span from about 18 years of age to 64 years of age, women had a higher rate of physician contacts than men did. A portion of this difference is related to visits concerning childhood and pregnancy, particularly in the 18- to 44-year age bracket. Pregnancy-related visits should also be considered relevant to physician contacts for healthy women less than 18 years old.

2. Jean Baker Miller, *Toward a New Psychology of Women* (Boston: Beacon Press, 1976), 4–6. I have also used this distinction in "Sex–Role Stereotypes in Medicine," in *Hypatia* 2, no. 2 (Summer 1987): 21–38, substantial portions of which are incorporated here.

3. For example, for the year 1988 life expectancy at birth for women in the United States was 78.3 years, for men 71.5 years. See U.S. Bureau of the Census, *Statistical Abstract of the United States: 1991,* 111th ed. (Washington, D.C.: U.S. Government Printing Office, 1991), 73.

4. The death rate for male infants in the United States in 1988 was 1,114 per 100,000 population; for female infants the rate was 898 per 100,000. These data apply to infants up to one year of age. See U.S. Bureau of the Census, *Statistical Abstract of the United States: 1991,* 111th ed., 75.

5. Mila Aroskar, "The Fractured Image: The Public Stereotype of Nursing and the Nurse," in *Nursing Images and Ideals,* ed. Stuart F. Spicker and Sally Gadow (New York: Springer Publishing, 1980), 20.

6. Barbara Bates, "Doctor and Nurse: Changing Roles and Relations," *New England Journal of Medicine* 283 (1970): 129: and Joanne Ashley, *Hospitals, Paternalism and the Role of the Nurse* (New York: Teachers' College Press, 1976).

7. Vicente Navarro, "Women in Health Care," *New England Journal of Medicine* 292, (1975): 398–402; and J. Weaver, "Sexism and Racism in the American Health Care Industry: A Comparative Analysis," *International Journal of Health Services* 8 (1978): 677–700.

8. I base this claim on an unpublished set of interviews of nurses enrolled in the School of Medicine at Case Western Reserve University, Cleveland, Ohio, in 1983–1984. The interviews were conducted by Stuart J. Youngner from a set of questions developed by Julie Hambleton, him, and me.

9. U.S. Bureau of the Census, *Statistical Abstract of the United States: 1987,* 107th ed., (Washington, D.C.: U.S. Government Printing Office, 1987), 91.

10. Marilyn Heins, "Medicine and Motherhood," *Journal of the American Medical Association* 249, no. 2 (Jan. 14, 1983): 209–10; and Marilyn Heins, Sue Smock, Jennifer Jacobs, and Margaret Stein, "Productivity of Women Physicians," *Journal of the American Medical Association* 236, no. 17 (Oct. 25, 1976): 1961–64.

11. See U.S. Bureau of the Census, Statistical Abstract of the United States: 1991, 111th ed., 103. Also see Carola Eisenberg, "Medicine Is No Longer a Man's Profession," *New England Journal of Medicine* 321, no. 22 (Nov. 30, 1989): 1542–44.

12. See note 1. See also George Silver, "The Feminisation of Medical Practice," *The Lancet* 335 (May 12, 1990): 1149–50.

13. Selby Jacobs and Adrian Ostfield, "An Epidemiological Review of the Mortality of Bereavement," *Psychosomatic Medicine* 39 (1977): 344–57.

14. Jeanne Humphrey Block, "Conceptions of Sex Role: Some Cross-cultural and Longitudinal Perspectives," *American Psychologist* 28 (1973): 64–78.

15. Paul Starr, *The Social Transformation of American Medicine* (New York: Basic Books, Inc. 1982), 428.

16. For example, Carol Gilligan's studies illustrate these concerns on the part of pregnant women and college students. See her *In a Different Voice* (Cambridge, Massachusetts: Harvard University Press, 1982).

17. Lawrence Kohlberg, "Stages of Moral Development as a Basis for Moral Education," in *Moral Education,* ed. Clive M. Beck, Brian S. Crittenden, and Edmund V. Sullivan (Great Britain: University of Toronto Press, 1971).

18. Carol Gilligan, "Moral Orientation and Moral Development," in *Women and Moral Theory,* ed. Eva Feder Kittay and Diana T. Meyers (Totowa, New Jersey: Rowman and Littlefield, 1987), 19–33.

19. James Rest, Douglas Cooper, Richard Coder, JoAnna Masanz, and Douglas Anderson, "Judging the Important Issues in Moral Dilemmas of Development," *Developmental Psychology* 10 (1974): 491–501.

20. Martha Saxton, "Are Women More Moral than Men? An Interview with Psychologist Carol Gilligan," *Ms* (December 1981): 63–66.

21. Mary B. Mahowald, "The Physician," in *The Power of the Professional Person,* ed. Robert Lawry and Robert Clarke (New York: University Press of America, 1988), 119–31.

22. *Webster's New World Dictionary* (New York: Simon and Schuster, Inc., 1984), 1041.

23. Stanley Reiser, Arthur J. Dyck, and William J. Curran, eds., "Selections from the Hippocratic Corpus," in *Ethics in Medicine: Historical Perspectives and Contemporary Concerns* (Cambridge, Massachusetts: MIT Press, 1977), 7. That our modern notion of informed consent never occurred to Hippocrates is evident in this piece of advice to physicians: "Perform all this calmly and adroitly, concealing most things from the patient...." (p. 8).

24. For example, see H. Tristram Engelhardt, *The Foundations of Bioethics* (New York: Oxford University Press, 1986), 279–84; Robert Veatch, *A Theory of Medical Ethics* (New York: Basic Books, Inc., 1981), 195–97; and American Medical Association Council on Ethical and Judicial Affairs, *Current Opinions* (Chicago: American Medical Association, 1989), 32.

25. Note that informed consent and proxy consent are not identical concepts, either legally or morally. Although the terms are sometimes used interchangeably, *proxy* consent is imputed to a sur-

rogate, that is, someone who decides in behalf of another, while *informed* consent refers to the consent one gives for oneself.

26. Paul Ramsey, *The Patient as Person* (New Haven, Connecticut: Yale University Press, 1970), 1–58.

27. Gerda Lerner, *The Creation of Patriarchy* (New York: Oxford University Press, 1986), 26.

28 I have recently heard a term that even more forcefully articulates respect for the patient's understanding and choices, regarding his or her own condition. An internist at Oak Forest Hospital, Chicago, calls her patients "customers."

29. American Nurses' Association, *Code for Nurses with Interpretive Statements* (Kansas City, Missouri, 1976).

30. American Medical Association, "Principles of Medical Ethics," *Current Opinions of the Judicial Council of the American Medical Association* (Chicago, 1989).

31. James Childress, *Who Should Decide?* (New York: Oxford University Press, 1983); John Kleinig, *Paternalism* (Totowa, New Jersey: Rowman and Allanheld, 1984; Mary B. Mahowald, "Against Paternalism: A Developmental View," *Philosophy Research Archives* 6 (1980): 340–57; and David Thomasma, "Beyond Medical Paternalism and Patient Autonomy: A Model of Physician–Patient Relationship," *Annals of Internal Medicine* 98 (Feb. 1983): 243–48.

32. Thomasma, 243–48.

33. Martin Benjamin and Joy Curtis, *Ethics in Nursing* (New York: Oxford University Press, 1981), 48–58.

34. Susan Rae Peterson, "Against 'Parenting'," in *Mothering,* ed. Joyce Trebilcot (Totowa, New Jersey: Rowman and Allanheld, 1984), 68.

35. I deal more extensively with the problem of defining "family" in Chapter 14.

36. Albert R. Jonsen, Mark Siegler, and William J. Winslade, *Clinical Ethics,* 2d. ed. (New York: Macmillan Publishing Co., 1986).

37. See Alison M. Jaggar, *Feminist Politics and Human Nature* (Totowa, New Jersey: Rowman and Allanheld, 1983), 5; and Rosemarie Tong, *Feminist Thought* (Boulder, Colorado: Westview Press, Inc., 1989), 1.

38. Ivan Illich, *Gender* (New York: Pantheon Books, 1982), 24.

39. Eli Ginzberg and Edward Brann, "How the Medical Student Views His Profession and Its Future," *Inquiry* 17 (Fall 1980): 195–203; and Jane Leserman, "The Professional Values and Expectations of Medical Students," *Journal of Medical Education,* 53 (1978): 330–36.

40. Simone de Beauvoir, *The Second Sex* (New York: Bantam Books, 1952).

41. Carola Eisenberg, "Women as Physicians," *Journal of Medical Education* 58 (July 1983): 534–41. See also Marilyn Heins, Sue Smock, Lois Martindale, Margaret Stein, and Jennifer Jacobs, "A Profile of the Woman Physician," *Journal of the American Medical Women's Association (JAMWA)* 32, no. 11 (Nov. 1977): 421–27.

42. The word *institution* is important here because it suggests the power that may be exerted on individuals through social definition of roles. Heterosexuality is an institution because it is "normal" or "expected" behavior. Homosexuality and lesbian sexuality are not institutions because they depart from the "norm" of heterosexuality. See Jaggar, 271–72.

43. Jaggar, 84.

44. Joan G. Raymond, "Medicine as Patriarchal Religion," *Journal of Medicine and Philosophy* 7 (1982): 197–216.

45. Robert Tucker, ed., *The Marx-Engels Reader* (New York: W. W. Norton and Co., Inc., 1972), 388.

46. The concept of individualism I have in mind is one that Josiah Royce described as a "tendency to hold that the ideal of life is the separate happy man" (Josiah Royce, *The Religious Aspect of Philosophy* [Boston: Houghton Mifflin, 1885]1, 201). It should be distinguished from individuality, which refers to the uniqueness of individuals. Paradoxically, individuality can only be developed by transcending the ideal of the "separate happy self" through interaction with others, that is, by rejecting individualism by promoting the relational self. For a more extended feminist critique of individualism, see Naomi Scheman, "Individualism and the Objects of Psychology," in *Discovering Reality,* ed. Sandra Harding and Merrill Hintikka (Dordrecht: D. Reidel, 1983), 225–44.

3

Obstetrics-Gynecology: A Unique Specialty

As public support for the goal of gender equality has grown, so have criticisms and mistrust of medical power and technology. From a feminist egalitarian perspective, the criticisms have been especially directed toward the practice of obstetrics and gynecology as epitomizing the paternalistic model of the physician's role toward women.[1] Manifestations of such criticisms and response to them include the growth of the women's health movement, along with books such as *Our Bodies, Ourselves;*[2] Suzanne Arms' *Immaculate Deception;*[3] Adrienne Rich's *Of Woman Born;*[4] Diane Scully's *Men Who Control Women's Health;*[5] and Sue Fisher's *In the Patient's Best Interest.*[6] Physicians who have written provocative accounts of clinical practice in obstetrics-gynecology include Michelle Harrison (*A Woman in Residence*)[7] and John M. Smith (*Women and Doctors.*)[8] As with any critical movement, there may be excesses, inaccuracies, inconsistencies, and vagueness in its expressions, but the movement also provides health caregivers with an opportunity for growth in self-understanding and professional sensitivity through analysis of what is valid and invalid in the criticisms.

In this chapter I attempt to facilitate that analysis through discussion of factors and issues peculiar to the practice of obstetrics and gynecology. I discuss the main ethical principles through which issues have been addressed, and suggest questions that may guide the application of those principles in specific cases. My intent is to provide an interpretive egalitarian account that incorporates both care and justice perspectives of moral reasoning.

42

Role of the Obstetrician-Gynecologist

As already discussed, the physician–patient relationship is traditionally paternalistic, mimicking the father–child relationship in which one party is clearly powerful and autonomous, the other vulnerable and dependent.[9] Because "father knows best" what is in the child's (patient's) best interest, he has not only the right but at times the responsibility to override the wishes of the child. Moreover, the father's decisions are not simply made in behalf of an incompetent child (since some children are surely competent), or in behalf of an unconscious or extremely ill child (since most children are neither), but mainly in behalf of normal, healthy children.

In some situations, the practice of obtaining informed consent has undoubtedly shifted the paternalistic tendency to an emphasis on patient autonomy. Indeed, some have argued that the shift has led to a betrayal of basic medical values, particularly that of beneficence towards the patient.[10] Paternalism continues to be practiced and defended, but more critically than in the past.[11] Even when the literal requirement of informed consent is fulfilled, the autonomy of patients is at times undermined by efforts to direct their decisions toward the goals defined by practitioners.

The father-knows-and-does-what-is-best image of the physician is especially applicable to the obstetrician-gynecologist's role toward the patient. Most obstetric patients are healthy, competent, and conscious, yet their relationship to the physician is one of vulnerability. Because our socialization process tends to encourage men to be dominant and women to accept that dominance, the fact that the physician is usually male while the patient is always female serves to reinforce inegalitarian aspects of the relationship.[12]

Further compromising the situation of the woman-as-patient is the fact that matters to be discussed routinely with an obstetrician-gynecologist are often quite intimate (life-style, sexual practices, and intimate relationships), and no similar personal revelations are required or expected on the part of the physician. The conditions treated are typically more integral to the patient's self-concept than are conditions treated by other specialties. Not infrequently, they raise important moral questions for both patient and clinician. Moreover, when the patient is pregnant, her situation is significantly complicated by the fact that her physician may view his primary responsibility as extended to two individuals whose interests and needs are sometimes incompatible.

Precisely because the obstetrician may be perceived to have two patients, the woman and the fetus, from the beginning (if not before) to the end of the pregnancy, ethical decisions in this specialty are inseparable from legal and social questions. For example, the *Roe v. Wade* decision of the U.S. Supreme Court (1973) challenges the traditional two-patient concept of obstetrics by affirming the right of a woman to terminate a pregnancy before 24 weeks of gestation.[13] Legally, whether the nonviable fetus counts as a patient depends on whether the pregnant woman wants it to be counted as such.[14] Thus, the role concept of the professional may be at odds with the legal system. The individual states have authority to override the autonomous decision of a pregnant woman whose fetus is viable. Accordingly, the technology that has advanced viability into the second trimester may occasion changes in laws that permit later access to abortion.

Decisions regarding fertility and infertility constitute a major portion of the ethical decisions made by patients and clinicians in an obstetrics-gynecology practice. Such decisions have more impact on women's lives than on men's, even if we allow that fathers are equally responsible for their children. It is inevitably and exclusively women's bodies that are affected by pregnancy, childbirth, and lactation; it is not inevitable but it is empirically true that most women bear most of the responsibility for raising children.[15] The feminization of poverty, which is considered in Chapter 13, is but one result of the greater burden women bear in this regard.[16]

Since the participation of men in the process of childbirth is a comparatively recent phenomenon, it is not surprising that some women resist the medicalization of the event on grounds that it involves masculine control of an essentially feminine capability. Indeed, some have read the motivation of men choosing the specialty of obstetrics-gynecology as a way of reducing the lack of power they experience in confronting the origin of life.[17] The history of obstetrics reveals a drive on the part of the medical establishment to displace the previously dominant role of female midwives in assisting in childbirth.[18] (I develop this point further in Chapter 7.) Currently, the medicalization of childbirth and resistance to the revival of midwifery support the suspicion that men wish to maintain a high level of control over reproduction.[19]

In general, feminists decry a double standard regarding the social responsibilities and expectations of women and men. Those women who work outside the home return to domestic responsibilities in which some men assist, but few share equally.[20] That women's work, whether in or outside the home, is prevalently viewed as less important than the work of men is reflected in salary discrepancies between the sexes, and in the low pay and regard given for child care and housework.[21] This attitude may be expressed in the physician–patient relationship through disregard for women's time: the assumption that women who accompany patients (for example, children or parents) or who are patients themselves must wait because the physician's time is more valuable. Men who are patients or companions of patients must wait also, but this occurs less frequently, because fewer men are either patients or primary caregivers of family members.[22]

Another frequent assumption to consider is that women are more concerned about their appearance than are men.[23] Where the assumption proves true, whether the trait is natural or nurtured, it may negatively affect a woman's self-concept during the course of a pregnancy, exacerbating the psychological toll of surgical interventions such as breast removal for metastatic cancer. A double standard is also evident when physicians use first names for patients, who address their physicians by title and last name. Admittedly, this practice is often reinforced from both sides of the relationship.

In support of the double standard, it may be argued that the physician–patient relationship is necessarily inegalitarian, and that the practice of obstetrics-gynecology merely illustrates that relationship without any ethical implications, sexist or otherwise. In other words, the fact that most obstetrician-gynecologists are male and their patients female does not imply sexism; and inequality, if based on fact, does not constitute a moral problem. The soundness of this argument hinges on the concept of equality that it assumes.[24] As already suggested, if equality is construed as sameness (and inequality, correspondingly, as difference), then obviously the physician–patient relationship, the obstetrician-gynecologist's relationship to a patient, or, for that matter, any person's relationship to any other person, is inegalitarian because all of us are different. Surely the differences in themselves are morally neutral; it is the ranking assigned to them that moves us into an ethical context, where moral conflicts may arise.

On such an account, inequality in the physician–patient relationship only presents a moral problem if differences between the two lead to one side dominating the other, or either party is not valued by the other as a person. Where sexism occurs, it means that sex or gender is used irrelevantly as a criterion for valuing one individual or group over another. From the perspective outlined in the first chapter, an ideal of equality provides a useful framework for assessing whether empirical disparities on either side of the relationship are justified. While this perspective may not yield definitive answers, it constitutes a critical strategy for examining ethical dimensions of issues raised by the practice of obstetrics and gynecology.

Persistent Ethical Issues

Ethical issues in obstetrics and gynecology[25] run the gamut of those raised in other medical specialties, for example, questions regarding disclosure, confidentiality, consent to experimental therapy, criteria for determination of death, and allocation of scarce medical resources. But issues peculiar to obstetrics and gynecology have mainly to do with reproduction, either curbing or enhancing it. Recurrent themes include the right and responsibility to assist others to reproduce or not to reproduce, and rights and responsibilities concerning sexual expression and gender identity. Complicating some of these themes is the age range and competence of patients. The fact that the physiological capability for reproduction is present in children as well as in retarded or mentally ill adults expands the responsibility of care for them to their potential offspring. Motherhood constitutes legal emancipation for some minors, but does not establish that the new mother is mature enough to assume total responsibility for her newborn.[26]

Ethical questions concerning curtailment of pregnancy, especially the issue of abortion, have been dealt with extensively in philosophical literature. With regard to the practice of sterilization and contraception, the concerns invoked are often based on the principle of informed consent, the right to express one's sexuality without risk of pregnancy, and responsibility to the larger society, including other family members and future generations.[27] Natural law arguments are especially pertinent to obstetrics because they view reproduction as a natural process that ought not to be obstructed.[28] On the question of abortion, the moral status of the fetus is usually perceived as crucial to assessment, and the right of the pregnant woman to control what transpires within her own body is also stressed.[29] This issue challenges the practitioner to confront the apparent contradiction between his commitment to both fetus and pregnant woman. (All of these issues are considered in more detail in subsequent chapters.)

An obstetrician-gynecologist may wish to separate her own moral stance from that of the patient, allowing the latter to choose a course at variance with her own. This approach implies some tolerance for what the clinician perceives as wrong, or else ambivalence regarding what she believes is right. Unless we posit a dichotomy between professional and moral responsibility (by which the physician sheds one skin for another in commuting between hospital and home), there surely is need to reconcile the two. In other words, it is necessary to determine how to act both morally and professionally with patients with whom one has ethical disagreements. Typically, such disagreements reflect moderate rather than extreme positions on the part of clinicians — for example, one who considers early abortion for an unwanted pregnancy morally justified, but considers it wrong to choose or perform a second trimester abortion solely because the fetus is not of the desired sex. There is admirable consistency on the part of the physician who not only refuses to perform abortions but organizes lobbying efforts to eliminate liberal abortion statutes; there is similarly admirable consistency on the part of the clinician who operates a free abortion clinic and lobbies against restrictive abortion laws. But consistency is not a sufficient basis for honoring the practices of either individual. Beyond consistency, each of us needs to confront and evaluate the reasons that support our practice.

Legally, there are limits to what patients and clinicians may do, but these do not resolve the moral dilemmas of individuals. Even for those who consider the abortion issue settled, other ethical questions are inevitable. For example, how should a physician respond to the queries of the mother of a sexually active teenager, also his patient, concerning the daughter's apparent practice of contraception? Or how should a clinician deal with a patient's request to sterilize her moderately retarded child? In such situations, there are no predefined, fixed, or exclusive ways of acting morally; yet this does not imply that whatever one does is morally acceptable. An ethic of care requires consideration of the unique relationships that apply to each situation; an egalitarian perspective introduces a critical dimension for resolving conflicts. Distributing the benefits of health care equally among those affected requires consideration of basic ethical principles or values.

Recent critics of contemporary biomedical ethics have rejected methodologies based on traditional ethical theories and principles. Colleen Clements and Roger Sider, for example, have accused philosophers of using the "autonomy principle" as "a way of separating ethical thinking from the empirical world and placing it in the rationalist realm of metaethics."[30] Such an approach, they maintain, constitutes an assault on the traditional medical values of adaptation, best interests, and function of the patient. Clements and Sider attribute the latter values to a "naturalistic ethics" consistent with the Hippocratic tradition; this view competes with the "formalist ethics" of Kant and contemporary biomedical ethicists.[31] The model they propose to replace the formalist emphasis on patient autonomy is one that proceeds inductively from an examination of specific cases, rather than deductively from a priori principles. The choice of the individual patient is defined as one among other biological functions to be promoted through clinical expertise.

More recent critics of "philosophical ethics" are Albert Jonsen and Stephen Toulmin, whose study of casuistry entails not so much a rejection of traditional theories and principles as an exposition of a more practical method for addressing the ethical quandaries that arise in particular situations.[32] Apparently, the authors learned the method from living it. While working on the President's Commission for the Study of Ethical Problems in Medicine and Biomedical and Behavioral Research, they discovered that the commissioners "often disagreed seriously over the formulation of principle, while they quickly reached agreement on particular cases."[33] This experience spurred the two to examine the history and practice of casuistry more carefully, and eventually elaborate a contemporary version applicable to issues in medical ethics. The main features of the method are reliance on paradigm cases, reference to broad consensus and maxims, and acceptance of "probable certitude." The relevance of principles to particular cases may be seen in analogies with the paradigm cases and consideration of applicable maxims.

Such critics of reductionist theories seem to attack a straw person, namely, one who is purely deductive in approaching clinical or ethical dilemmas. In reality, neither deductive nor inductive reasoning is practiced exclusively in the clinical setting.[34] The more characteristic situation is one of straddling both case-based and theoretical approaches, utilizing whatever strategies are available to facilitate resolution. Such a method is pragmatic in the sense that it is practically oriented, rejecting a dichotomy between theory and action. To the pragmatist, theory and action are inseparable.[35] This view is consistent with the emphasis on context and the critique of absolutism embodied in a care-based model of moral reasoning and feminist critiques of traditional ethics. Classical pragmatists also point to the reality and relevance of relationships;[36] this too is evident in feminist and care-based reasoning, and in various accounts of the health caregiver's responsibilities. (I return to pragmatism in Chapter 15.)

While principles are not adequate to resolution of particular questions, the application of reason to the nuances of the situation, especially when conjoined to the rational, collaborative input of others, brings us closer to the correct solution. Basic ethical principles and guidelines may thus be viewed as diagnostic strategies for addressing clinical dilemmas. Just as therapeutic prescriptions are never certain to be successful, although that is their intent, so ethical decisions are never certain to provide correct answers to the questions asked, although that is their intent.

Applicable Principles

What are the ethical principles or "systems" to be checked out in determining ethical "prescriptions"? In recent years prominent biomedical ethicists have proposed various combinations of principles that they consider crucial. Robert Veatch,[37] for example, considers six different principles: beneficence, contract keeping, autonomy, honesty, avoiding killing, and justice. Contract keeping involves the obligation of confidentiality in the doctor–patient relationship. From differing ethical perspectives (utilitarian and deontological), Tom Beauchamp and James Childress[38] primarily discuss four principles: autonomy, beneficence, nonmaleficence, and justice; in addition they consider veracity and confidentiality as important ethical components of the caregiver–patient bond. Laurence McCullough and Tom Beauchamp,[39] and H. Tristram Engelhardt[40] focus on beneficence and respect for autonomy as the foremost ethical principles, dealing with justice as a necessary principle for situations requiring allocation of health care resources. Although Albert Jonsen, Mark Siegler, and William Winslade are resistant to the language of "principles," they propose four "topics" that exemplify those principles as a list of essential considerations for case analysis: medical indications, patient preferences, quality of life, and contextual factors.[41] The "contextual factors," labeled "socioeconomic factors" in previous editions of their book, reflect the principle of justice; medical indications and quality of life illustrate the principle of beneficence, and patient preferences are obviously included in the principle of respect for autonomy.

The meanings embodied in a broad list of principles can be encapsulated in three: respect for autonomy, beneficence, and justice. Honesty, veracity, and confidentiality may be viewed as essential to respect for autonomy; beneficence may be seen as encompassing nonmaleficence and the avoidance of killing. In teaching future doctors I often compare these principles to the basic bodily systems that must be checked out if one is to conduct an adequate assessment of the patient's health before determining a specific plan of treatment. Unlike clinical principles, however, the ethical principles refer not only to the relationship of practitioner to patient, but also to the patient's relationship to practitioners, practitioners' relationships with one another, with family members, and so on. As moral agents, all of those involved in particular cases are responsible for respecting others' autonomy, acting beneficently, and acting justly.

As already suggested, the first principle, beneficence, is essentially embodied in the Hippocratic imperative: "to help, or at least to do no harm."[42] Accordingly, this principle consists of both positive and negative duties towards the patient. The latter are sometimes described under the rubric of nonmaleficence, which represents a more stringent obligation than simple beneficence. Practically, however, it may not only be difficult but unhelpful to observe this distinction because the very "doing of harm" sometimes constitutes the "doing of good" to particular patients. Insofar as medical and surgical interventions are therapeutic, they are all of this type. It remains the task of individuals to determine the proportion of harms and benefits that is expected so that the minimal criterion of an equal balance between the two is observed. Beyond that point of obligation, virtue invites the practitioner to do more good, that is, to do more than what is minimally required legally or morally in behalf of the patient as well as others.[43] Whether applying the principle of beneficence on the basis of virtue or obligation, however, the question it entails is the following: "Is the proposed intervention (or nonintervention) justified on the basis of the good result expected primarily for the patient(s), and secondarily for others affected?" The answer to this question provides an important consequentialist or utilitarian component in the decision-making process.

Because benefits and burdens keep shifting, constant calculation is required in order to apply the principle of beneficence faithfully. Burdens to be weighed include pain, medication, surgery, dietary restrictions, loss of privacy, incurrence of costs, loss of income, and separation from one's family. Expected benefits that may justify imposition of the burdens include maintenance, promotion, or restoration of health. To the extent that others are affected, an egalitarian perspective calls for assessment of burdens and benefits to them as well as the patient.[44] In obstetrics-gynecology the calculation is complicated by concerns about the fetus and the pregnant woman's reproductive capacity. The overall (egalitarian) goal, however, is to render the patient or patients free of any burden of health care interventions, that is, to make professional assistance no longer necessary.

The next principle, respect for autonomy, is essentially deontological, that is, based on nonconsequentialist rules or laws that arise from the very meaning of human nature or from some a priori source of moral knowledge. While philosophical deontological principles are discerned by reason and religious deontological principles are based on faith, in either case the principles are considered universally applicable and binding. In our own day, the principle of autonomy embodied in the concept of informed consent has widely replaced the traditional paternalistic model of the physician's role. Important as it is, this concept applies the principle of respect for autonomy or self-determination to only one of those influenced by clinical ethical decisions, namely, the competent patient. In obstetric and other reproductive decisions, however, there is often another autonomous individual affected, namely, the potential father. In nonreproductive medical decisions also, other autonomous persons besides the patient (such as close friends or other relatives) are influenced by the decision. Moreover, clinicians themselves surely do not surrender their own autonomy at the door of the treatment or operating room. They are responsible for what they do and do not do regardless of what the patient chooses. To base decisions exclusively on the patient's "informed consent" is to ignore that ongoing responsibility and subscribe to a purely instrumentalist interpretation of the physician's role. To base decisions on the broader principle of respect for autonomy is to ask the question: "What course is preferred by each of the individuals influenced by this decision?" Where there are conflicting answers, an egalitarian perspective argues that these should be weighed according to the degree in which the individual will be affected. Among patients themselves, who are generally much more significantly affected than others, that significance is not always of equal weight.

Since autonomous decisions can only be made on the basis of correct understanding of alternatives and their implications, respect for autonomy involves an obligation of disclosure or "truth-telling." There is a sense, of course, in which truth-telling is an impossible goal because of the inevitable limitations of knowledge and ability to communicate. What remains possible is truthfulness, honesty, or veracity, that is, an attempt to communicate to others accurately and adequately. The obligation of disclosure does not imply that every individual has a right to specific information in all circumstances. A clearly justified limitation to truth-telling or the obligation of disclosure is recognition of the right to privacy of the individual about whom information is known. In other words, there is also an obligation to observe confidentiality, which is especially applicable to the kinds of information that an obstetrician-gynecologist possesses. The relevant question for the clinician, therefore, is two-faceted: "What information do I owe my patient, those close to her, and to other health professionals, and what information am I obligated not to disclose?" Another practical question immediately follows: "How might I most effectively communicate or refrain from communicating where circumstances call for either approach?" Comparable questions apply to others who possess pertinent information, including nonclinicians, family members and the patients themselves.

The third ethical "system" to be checked out in determining responses to ethical dilemmas is the principle of justice. This may be construed as deontological or utilitarian, and as applicable either to individual decisions or social policy. In either case, it is the notion of distributive (rather than retributive) justice about which we are concerned. Inevitably, this entails an interpretation of equality as applicable or inapplicable to individuals or groups. Macroallocation decisions regarding distribution of health care resources are mainly a matter of social policy, but microallocation decisions are made every day through determinations regarding time, space, equipment, expertise, and medication.

If justice is considered in obstetric practice, some determination needs to be made regarding the status of the fetus, so that conflicts that pit the alleged rights of the fetus against those of the pregnant woman may be addressed in that light. If fetus and pregnant woman count equally as patients or persons, justice requires that neither has the greater claim against the other, and a random way of determining care for one as opposed to the other may be acceptable.[45] Even then, however, duties toward others and effects on them are morally relevant when the rights of persons are in conflict. On the other hand, justice does not require equal consideration of fetus and pregnant woman if either of the following positions is maintained: rights to life and liberty are proportionate to the developmental status of the individual, whether that individual counts as person or not; or the personhood by which an individual is judged worthy of equitable treatment with other persons commences at birth or some later point of life.

Since 1973, the United States Supreme Court has upheld a view not quite like either of the above positions.[46] According to that ruling, abortions are permissible on the basis of a woman's choice prior to viability. Subsequently, the pregnant woman's choice may be subordinated to the state's interest in "potential life." The concept of justice invoked here is one that involves equal treatment for all whose lives may legally be protected by the state. When viability is achieved, the states may proscribe termination of pregnancy unless its continuation threatens the life or health of the pregnant woman. In other words, life and liberty are not viewed as equal values of those whose interests are relevant before the law; justice implies that life is the more basic right or value, whose denial may not be permitted in the name of another's liberty.

Newer Ethical Issues

Subsequent chapters deal with continuing as well as newer ethical issues in obstetrics and gynecology. In Chapter 8, for example, I consider court ordered cesarean sections of pregnant women, which represent even greater restrictions on women's autonomy than do statutes prohibiting termination of pregnancy.[47] Again, an interpretation of justice as entailing equal treatment of all persons, extended to viable fetuses as "potential lives," may be invoked to justify the intervention. Positions that affirm the rights of the pregnant woman over those of the fetus may also be maintained under the aegis of justice, and these seem more defensible in a society that prevalently upholds the rights of individuals to control their own lives and bodies. In deciding among conflicting interests, therefore, the pertinent question for clinicians and patients alike is the following: "Which approach is most likely to result in an equal distribution of harms and benefits among those affected?"

Recent developments in fertility research have evoked recognition of ethical questions surrounding marital and sexual relationships, parenthood, and the right to biologically related progeny. Some of these issues have been with us for a long time, without raising much public stir — perhaps because the relevant practices (such as artificial insemination by husband or donor) are usually implemented covertly. With newer means of facilitating reproduction, successful interventions can hardly be covert — for example, when a woman who has long been infertile is manifestly pregnant, or when a nonpregnant woman brings home the newborn of her husband and a surrogate. In time, as techniques are refined and routinized, confidentiality may be better maintained in cases of ovum transfer and in vitro fertilization, as in artificial insemination. However, in all of these situations, a rather new metaethical question is raised, namely, "Whose child is this?"

In an individualistic society such as ours, the concept of parenthood is often quite traditional and biological: a heterosexual married couple who conceive a child through sexual intercourse, each contributing 23 chromosomes, one contributing her uterus, along with the physical and emotional costs of pregnancy and childbirth, both partners contributing thereafter to the child's nurturance. In some respects, the child is then looked upon as the property of the parents: he belongs to them, is named by them, is kept and cared for by them unless "given up" for adoption, or until he symbolically gives himself to another in marriage. Admittedly, parental rights over children have legal as well as moral limits, but they remain extremely significant determinants of children's lives and identities. Moreover, through the nuclear families that our modern transient life-style encourages, the family unit has come to represent an end itself, a kind of self-justifying system. Within that framework it is difficult to appreciate a more complicated and extended concept of parenthood.

To grapple adequately with the ethical questions that advances in reproductive technologies present, we need to examine our traditional concept of parenthood in light of the fact that in many cases it takes more than two individuals to develop a child. This fact points to rights and responsibilities on the part of the developers, and to conflicts in the exercise of those rights and responsibilities. An egalitarian perspective suggests that such conflicts might be resolved through examination of differing contributions of parent figures, including the duration, depth, and demands of their relationship to the child. A "priority among parents" might be determined on the basis of their contribution to the child, whether physical or psychological, or both. The nurturant input of those who care for a child beyond birth, regardless of their biological relationship, marital status, or sex, surely gives them a higher priority than those who never actually cared for the child. (I discuss this topic further in Chapter 7.)

The principle of beneficence is also relevant to the new reproductive technologies, encompassing benefits owed to nonautonomous as well as autonomous individuals. Although newborns and very young children are not autonomous, they are legally regarded as persons rather than property, and most moralists support that view. There is thus no right to have a child, if *having* means "ownership" or "possession." From a moral point of view the right to parenthood is limited by the extent to which the parental relationship will be beneficent or at least not maleficent towards the child. The same may be said for the right or responsibility to assist another in having a child.

Justice is related to the principle of beneficence because it involves an equal distribution of benefits as well as harms. Applying justice to reproduction requires examination of the costs of developing and implementing the technologies, and their accessibility to the general population. As long as certain ways of facilitating reproduction are available only to a limited number of those in need (those who can pay for them), it may hardly be claimed that biological reproduction is a positive, universal right.[48] Limited application of technology may be justified on the basis of its experimental status, but the egalitarian perspective implied by the principle of justice demands that discrepancies in accessibility be reduced as much as possible. That the affluent receive the benefits of the research may be defended on utilitarian grounds: current discrimination is essential in order to make the technology eventually available to others. But this argument is hardly adequate because the ideal of justice might be approximated even now. Since the technology is of such great benefit to a wealthy few, part of what those few pay might be directed to the purchase of its availability to some of the poor. Although this view may be incompatible with the self-interested motivation of a free-enterprise system, it is not inconsistent with the democratic demands of an egalitarian ethic. Similar arguments have been made regarding the availability of abortion and contraceptive procedures.

Assessing the Alternatives

At the beginning of this chapter, I suggested that analysis of current feminist criticism of the medical profession in general, and of obstetrics and gynecology in particular, might yield helpful insights for clinical practice. The same may be said for the women's movement. In fact, I believe the question "Whose children are they?" is crucial to understanding and assessment of different versions of feminism, as well as antifeminism.[49] Advances in reproductive technologies have not only increased the capacity of individuals to control their reproductive lives, they have greatly complicated the expression of that capacity. Individualistic versions of feminism and medicine (those focusing on individual women or patients) accent the former, while "communalistic" versions (those concerned with the larger community) accent the latter.[50] The question "Whose child?" is answered individualistically as "Mine." From a communalistic perspective, the answer is "Ours." The first response implies that the answerer has full rights and total responsibility for the child; the second implies shared rights and responsibilities. Increasingly, the first answer is unrealizable, while the second is empirically unavoidable but morally problematic.

As medicine has moved toward a more socialized understanding of itself, and health care professionals have observed their responsibility to patients as extending to the broader society, American feminism has offered a critique of individualism, even while emphasizing the reproductive rights of the individual. Conflicting ideologies, whether in medicine or feminism, are not likely to be resolved any sooner than the perennial metaphysical problem of "the one and the many," but some of the tension between them may be reduced and rendered constructive through recognition of the validity and inadequacy of mutually exclusive positions.[51] To the question "Whose child is this?" the answers "Mine" and "Ours" are *both* valid and inadequate if one subscribes to an ethic that is care-based as well as feminist, one that stresses relationships as well as rights. "Mine" is least inadequately applied to the biological mother who has nurtured her child beyond pregnancy to some stage of self-sufficiency; but even here, the man through whom the child was conceived and who shares in the child's nurturance may also call the child "Mine." Thus the couple together, less inadequately, say "Our." To the extent that others share the joy and burden of nurturing children, they too, even less inadequately, join in the "Our."

Thus feminism is compatible with the physician–patient relationship so long as the mutuality of rights and responsibilities among individuals is respected in both contexts. Such an account reflects the concept of equality described in Chapter 1 as well as the parentalism described in Chapter 2. The emphasis on nurturance that characterizes health care practice provides further grounds for compatibility between this approach and a model of reasoning based on attachments and relationships. The emphasis on context or care-based reasoning eschews rigid adherence to unchanging principles, even those that have become well-established in contemporary biomedical ethics. Like clinical concepts and generalized treatment plans, these provide a useful framework for understanding the complexity of a changing situation, but they do not yield definitive answers to questions raised by unique cases. The cases themselves are instructive regarding interpretation of the concepts and modes of treatment. As already suggested, this interplay between theoretical and practical considerations illustrates the notion of pragmatism that William James wrote about in 1907. Using an image proposed by the Italian pragmatist Giovanni Papini, James compares the method of pragmatism to a corridor in a hotel:

> Innumerable chambers open out of it. In one you may find a man writing an atheistic volume; in the next someone on his knees praying for faith and strength; in a third a chemist investigating a body's properties; in a fourth a system of idealistic metaphysics is being excogitated; in a fifth the impossibility of metaphysics is being shown. But they all own the corridor, and all must pass through it if they want a practicable way of getting into or out of their respective rooms.[52]

Alternative theories, concepts, principles, and strategies are "owned" simultaneously but used discriminately in practical problem solving. An adage often used in the clinical situation is the following: "When in doubt, look at the patient." Obstetrics-gynecology is a unique specialty in many respects, but each case within the specialty is unique as well, and different contexts raise issues that occur in other situations. Although general considerations are never adequate to the peculiarities of cases, it would be irresponsible not to draw on the meaning and relevance of basic principles and guidelines in addressing specific questions. The next chapter does this with regard to what is probably the most controversial of all issues in obstetrics and gynecology: abortion.

NOTES

1. Helen B. Holmes, "The Birth of a Women-Centered Analysis," in *Birth Control and Controlling Birth: Women Centered Perspectives,* ed. Helen Holmes, Betty B. Hoskins, and Michael Gross (Clifton, New Jersey: Humana Press, 1980), 5–6.

2. Boston Women's Health Course Collective (Boston: New England Free Press, Boston, 1971); see also *The New Our Bodies, Ourselves,* Boston Women's Health Book Collective (New York: Simon and Schuster, 1984); and Paula Brown Duress, Diana Laskin Siegel, and the Midlife and Older Women Book Project, *Ourselves, Growing Older* (New York: Simon and Schuster, 1987). These books are intended to assist women in providing for their own health without recourse to the medical profession.

3. Suzanne Arms, *Immaculate Deception* (New York: Bantam Books, 1975).

4. Adrienne Rich, *Of Woman Born* (New York: Bantam Books, 1977).

5. Diane Scully, *Men Who Control Women's Health* (Boston: Houghton Mifflin, 1980).

6. Sue Fisher, *In the Patient's Best Interest: Women and the Politics of Medical Decisions* (New Brunswick, New Jersey: Rutgers University Press, 1986). Philosophical accounts that have surfaced recently include Christine Overall's *Ethics and Human Reproduction: A Feminist Analysis* (Boston: Allen and Unwin, 1987), and two special issues of *Hypatia,* one edited by Helen B. Holmes, "Feminist Ethics and Medicine," 4, no. 2 (Summer 1989) and one edited by Laura M. Purdy, "Ethics and Reproduction," 4, no. 3 (Fall 1989). Several works have focused critically on new reproductive techniques, for example, Patricia Spallone, *Beyond Conception* (Granly, Massachusetts: Bergen and Gervey Publishers, Inc., 1989); Michelle Stanworth, ed., *Reproductive Technologies* (Minneapolis: University of Minnesota Press, 1987); and Mary Anne Warren, *Gendercide* (Totowa, New Jersey: Rowman and Allanheld, 1985).

7. Michelle Harrison, *A Woman in Residence* (New York: Random House, 1982).

8. John M. Smith, *Women and Doctors* (New York: Atlantic Monthly Press, 1992).

9. James F. Childress, "What is Paternalism?" In *Medical Responsibility: Paternalism, Informed Consent and Euthanasia,* ed. Wade L. Robison and Michael S. Pritchard (Clifton, New Jersey: Humana Press, 1979), 15–27.

10. For example, Colleen D. Clements and Roger C. Sider, "Medical Ethics Assault on Medical Values," *Journal of the American Medical Association* 250, no. 15 (Oct. 21, 1983): 2011–15.

11. For example, James F. Childress, "Paternalism in Health Care," and Bernard Gert and Charles Culver, "The Justification of Paternalism," both in Robinson and Pritchard, 21–24; 1–14. Strong paternalism, in which the wishes of competent patients are clearly overridden, is more difficult to defend than weak paternalism, which generally involves incompetent or questionably competent patients. In my view, "weak paternalism" is not really paternalism because interventions undertaken in such circumstances are based on beneficence without the necessity or intent to override autonomy. See my "Against Paternalism: A Developmental View," *Philosophy Research Archives* 6 (1980), 340–57.

12. American Medical Association, Division of Survey and Data Resources, *Physician Characteristics and Distribution in the U.S.,* 1987 ed., 8, 42.

13. *Roe v. Wade* majority opinion, United States Supreme Court, Jan. 22, 1973, 410 U.S. 113, 93 S. Ct. 705. Further legal and ethical challenges to the traditional two-patient concept of obstetrics are discussed by many authors, but especially in feminist writings on reproductive issues. See, for example, Janet Gallagher, "Eggs, Embryos and Foetuses: Anxiety and the Law," in Stanworth, ed., *Reproductive Technologies,* 139–50. Other essays in this volume are relevant, as are some in other collections such as Rita Arditti, Renate Klein, and Shelley Minden, eds., *Test-Tube Women* (Boston: Pandora Press, 1989); and Elaine Baruch, Amadeo D'Adamo, and Joni Seager, eds., *Embryos, Ethics, and Women's Rights* (New York: The Haworth Press, Inc., 1988).

14. A distinction may be made between the physician's obligation to the fetus, and to the child-that-a-fetus-may-become. But even if a woman intends to continue her pregnancy, her right to abortion deters the physician from treating the fetus as independently a patient.

15. This is true not only because the majority of single parent families are headed by women, but also because women bear most of the child-rearing responsibility in families where both parents are present. See Alice Rossi, "Children and Work in the Lives of Women," a paper delivered at the University of Arizona, Feb. 7, 1976, cited in Rossi, p. 36; and Naomi Breslau, "Care of Disabled Children and Women's Time Use," *Medical Care* 21 (1983): 620–29.

16. Of all female-headed families, 33% live below the poverty level. In contrast, only 8.2% of families headed by two parents or by a single male parent live below the poverty level. See Sara Rix, ed., *Women 1990–91* (New York: W. W. Norton and Co., Inc., 1990), 396.

17. For example, Rich, 51–56.

18. Barbara Ehrenreich and Deirdre English, *Witches, Midwives and Nurses: A History of Women Healers* (Old Westbury, New York: Feminist Press, 1973), 12–15.

19. Arms, 192–206.

20. See Hilda Scott, *Does Socialism Liberate Women?* (Boston: Beacon Press, 1974), 199–200.

21. Ivan Illich, *Gender* (New York: Pantheon Books, 1982), 24, 274; and R. S. Ratner, *Equal Employment Policy for Women: Strategies for Implementation in the USA, Canada and Western Europe* (Philadelphia: Temple University Press, 1978), 20–23.

22. That physicians view their own time as more valuable than patients' and caregivers' time is the general conclusion to be drawn here. Whether scheduling practices would change if more men were patients and primary caregivers of family members is an open question.

23. The assumption is supported by the fact that women spend so much more time and money on their appearance. It is arguable, however, that the expenditure of time and money is more motivated by social pressures than by genuine concern. See Sandra Bartky, *Femininity and Domination: Studies in the Phenomenology of Oppression* (New York: Routledge, 1990), 65–82.

24. See Chapter 1, and William Blackstone, ed., *The Concept of Equality* (Minneapolis: Burgess Publishing Company, 1979); Amy Gutman, *Liberal Equality* (New York: Cambridge University Press, 1980); and Michael Walzer, *Spheres of Justice* (New York: Basic Books, Inc., 1983).

25. Much of the material developed in this chapter is drawn from my earlier article, "Obgynethical Issues — Present and Future," *Advances in Psychosomatic Medicine* 12 (1985): 166–81. It also appears, with minor changes, as "Ethics and the Practitioner," in John Patrick O'Grady, ed., *Obstetrics: Psychological and Psychiatric Syndromes* (New York: Elsevier Science Publishing Co., 1992), 31–43.

26. Sanford L. Leikin, "Minors' Assent or Dissent to Medical Treatment," *Journal of Pediatrics* 102, no. 2 (Feb. 1983): 169–76.

27. Rosalind Petchesky, "Reproductive Freedom," *Signs* 5 (1980): 661–85.

28. Charles Curran, *Contemporary Problems in Moral Theology* (Notre Dame, Indiana: Fides, 1970), 97–158, 159–88.

29. L. W. Sumner, *Abortion and Moral Theory* (Princeton, New Jersey: Princeton University Press, 1981), 26–33, 40–123.

30. Clements and Sider, 2011.

31. Clements and Sider, 2011–13.

32. Albert R. Jonsen and Stephen Toulmin, *The Abuse of Casuistry* (Berkeley: University of California Press, 1988).

33. Albert R. Jonsen, "Casuistry and Clinical Ethics," *Theoretical Medicine* 7 (Feb. 1986): 65–73.

34. Despite their emphasis on cases, Jonsen and Toulmin appeal to general maxims embodied in the applicable paradigms, and Clements and Sider apply the general principle of beneficence to cases they consider.

35. H. Standish Thayer, *Meaning and Action: A Critical History of Pragmatism* (Indianapolis: Hackett Publishing Company, Inc., 1981), 424–30.

36. For example, William James, "The Thing and Its Relations," in *The Writings of William James,* ed. John J. McDermott (New York: Modern Library, 1967), 214–26.

37. Robert Veatch, *A Theory of Medical Ethics* (New York: Basic Books, Inc., 1981), 141–287.

38. Tom L. Beauchamp and James F. Childress, *Principles of Biomedical Ethics,* 3d ed. (New York: Oxford University Press, 1989), 67–306.

39. Laurence McCullough and Tom L. Beauchamp, *Medical Ethics* (Englewood Cliffs, New Jersey: Prentice Hall, Inc., 1984), 13–51.

40. H. Tristram Engelhardt, *The Foundations of Bioethics* (New York: Oxford University Press, 1986), 66–103.

41. Compare Albert R. Jonsen, Mark Siegler, and William J. Winslade, *Clinical Ethics* (New York: Macmillan Publishing Company, 1986), 2d ed., 5–8, with the third edition of this book. In the third edition, the term "topics" replaces "categories," and "contextual" replaces "socioeconomic." See Jonsen, Siegler, and Winsdale, *Clinical Ethics* (New York: McGraw-Hill, Inc., 1992), 3d ed., 4–10.

42. Stanley Reiser, Arthur J. Dyck, and William J. Curran, eds., "Selections from the Hippocratic Corpus," *Ethics in Medicine: Historical Perspectives and Contemporary Concerns* (Cambridge, Massachusetts: MIT Press, 1977), 7.

43. See Daniel Callahan, "Minimalist Ethics," *Hastings Center Report* 11, no. 5 (Oct. 1981): 19–25.

44. A patient-centered ethic can be egalitarian in that it calls for treating all patients with equal concern and respect. Ordinarily, an egalitarian perspective also calls for greater attention to balanc-

ing burdens and benefits to patients because their benefits and burdens are greater than those of others affected by health care decisions.

45. James F. Childress, "Who Shall Live When Not All Shall Live?" *Intervention and Reflection,* ed. Ronald Munson (Belmont, California: Wadsworth Publishing Company, 1983), 501–04.

46. *Roe v. Wade,* see note 13.

47. George J. Annas, "Forced Cesareans: The Most Unkindest Cut," *Hastings Center Report* 12, no. 3 (June 1982): 16–17, 45.

48. It may well be a universal negative right, but I deal with this more in Chapter 6.

49. Alison M. Jagger and Paula Struhl, eds., *Feminist Frameworks* (New York: McGraw-Hill, 1978), 206–59.

50. Mary B. Mahowald, "Feminism: Individualistic or Communalistic?" *Proceedings of the American Catholic Philosophical Association* L, (1976): 219–29.

51. William James, "The One and the Many," in *Pragmatism* (Cleveland: Meridian, 1955), 89–108.

52. William James, "What Pragmatism Means," in *Pragmatism,* 47.

4

Abortion: Complexity and Conflict

Abortion is an issue with which many people feel uncomfortable, regardless of the position they take on its morality. The majority of obstetrician-gynecologists in private practice avoid the issue by referring most of their abortion patients to clinics or teaching hospitals.[1] Some acknowledge psychological conflict stemming from the apparent inconsistency of assisting some women to maintain their pregnancies while terminating the pregnancies of others. Theoretically, the conflict can be resolved by viewing either situation as one in which the physician respects the autonomy of the pregnant woman. But a long-standing perception of the physician as one who saves and supports life seems opposed to this. Traditionally, the obstetrician is seen as one who promotes the health of the pregnant woman *and* the fetus.

The self-consciousness mentioned in Chapter 1 is especially appropriate to those inclined to judge the abortion decisions of others. It is one thing to agree that a woman in a particular set of circumstances may morally terminate her pregnancy; it is another to be the practitioner who performs the termination. It is one thing to support a decision to continue a pregnancy; it is another to be the woman who experiences the consequences of the decision. Differences in perspective, role, and experience imply different moral responsibilities, and respect for the differences is a moral responsibility in its own right.

In this chapter, I describe a range of cases that illustrate the complexity of abortion decisions, and critique the labels that have become prevalently associated with different positions on abortion. I also consider views about the pregnant woman, fetus, and the relationship between them, as well as different concepts of abortion and their ethical implications. My analysis draws on the egalitarian perspective elaborated earlier to support a position on the morality of abortion that falls between positions of total condemnation of all abortions and acceptance of abortion for any reason whatsoever.

A Range of Cases

Persons with opposing positions in the abortion debate often have (I think) simplistic images of the circumstances and reasons that prevail in specific cases.[2] For example, militant "pro-lifers" envision abortion during late gestation for reasons of convenience by a healthy, economically secure adult. "Pro-choice" activists, in contrast, picture abortion during early gestation for, say, a poor junior high school student whose health is threatened by continuation of her pregnancy. In fact, the majority of abortions occur early in gestation and most are not undertaken for health-related reasons. The following cases are intended to provide a better sense of the range of reasons and circumstances associated with abortion decisions. Each is based on situations I have encountered in a women's hospital.

CASE I: PREGNANCY SEVERELY THREATENS A WOMAN'S HEALTH.

Ann Brown is a 32-year-old woman with multiple sclerosis who became pregnant despite use of a diaphragm. Her husband is opposed to abortion on religious grounds, but he is also extremely concerned that continuation of the pregnancy would further compromise his wife's health and her ability to care for their children (ages 4 and 6 years). Following her last pregnancy Ms. Brown was no longer able to walk unassisted, and her condition has deteriorated since then. She is three weeks pregnant when the couple meets with the obstetrician-gynecologist to determine how best to deal with the situation.

CASE II: PREGNANCY DUE TO RAPE.

Carol Day is a 14-year-old with a history of suspected child abuse. She became pregnant after being raped by her 19-year-old cousin, an incident that she did not report because she was embarrassed and ashamed. About four and a half months into the pregnancy, Ms. Day comes to the emergency room, tells the nurse she thinks she is pregnant and that she wants an abortion. She also asks that her family not be informed about her pregnancy and her request for an abortion.

CASE III: PREGNANCY IS (SEEN AS) A MEANS TO INDEPENDENCE.

Eve Flynn is a sexually active 16-year-old, who apparently harbors an extremely hostile attitude toward her parents and other authority figures. She deliberately became pregnant as a means of expressing and achieving greater independence. On learning of the pregnancy, her parents tried to convince her to have an abortion.

CASE IV: PREGNANCY AS A HINDRANCE TO PROFESSIONAL ADVANCEMENT.

Gail Hoy is a 28-year-old healthy, successful woman who discovered she was pregnant despite consistent practice of contraception. Her main reasons for considering abortion are that motherhood would interfere with her professional advancement, she never planned on having children, and she doubts that she would be a "good mother."

CASE V: FETUS MAY BE DISABLED.

Jill Katz and her husband, who would like to be parents, recently learned of Jill's pregnancy. They also recently learned that Jill's mother is a carrier for Down syndrome resulting from a chromosomal translocation. The mother's condition was detected in family studies done by the Genetics Center after her other daughter gave birth to an infant with Down syndrome. Ms. Katz is now scheduled for testing to determine if she is a carrier. If so, there is a 15 percent chance that her child will also have Down syndrome.[3] The couple feel insecure about their ability to raise a disabled child, and ambivalent on the issue of abortion.

These cases illustrate circumstances in which a pregnancy may be either terminated or continued for self-interested reasons; they also illustrate the opposite for either situation. The different circumstances reflect diverse and morally relevant ways in which someone can become pregnant—unintentionally, violently, defiantly, or lovingly; and they demonstrate that abortion is often an option that involves considerable moral ambiguity for those who have a legal right to choose it. While none of the cases mention techniques of abortion, the first two indicate the duration of gestation; both factors may be morally relevant to the decision. Taken together, the five cases dramatize the fact that abortion decisions are more complex than some commentators on the issue acknowledge. An egalitarian approach to the issue must take account of that complexity.

A Critique of Labels

Wittgenstein describes philosophy as a "battle against the bewitchment of our intelligence by means of language."[4] One way to reduce the bewitchment of abortion rhetoric is to consider the basic affirmations embedded in the available arguments. Some are truly pro-choice, others pro-life, and others still pro-abortion, pro-woman, or pro-fetus. On careful analysis, each of these positions has implications that might startle those who use the labels uncritically.[5]

A genuinely *pro-abortion* position views abortion itself as either morally neutral or positively recommended because pregnancy is undesirable. The fetus may be regarded as an invasive growth, like a wart or a tumor, better expelled than preserved, at least in some cases. Indeed, even without assuming that status for the fetus, one might argue that abortion is morally obligatory in certain situations, for example, where world hunger and overpopulation threaten to deny the necessities of life to those already born, or when the fetus is known to have a devastating incurable disease.[6] But a pro-abortion position could not logically entail the converse, namely, that in situations of depleted populations or known fetal health, abortions ought to be proscribed. Generally speaking, a pro-abortion position is only indirectly an affirmation. It affirms the good of abortion as a means to some good end, but not as an end in itself. All of the cases previously described permit a pro-abortion position in this instrumental sense.

A genuinely *pro-choice* position would impute to the pregnant woman the right to decide whether to terminate a pregnancy regardless of her reasons. To be consistent, however, a pro-choice position would also take account of the choices of others affected by abortion decisions, for example, the prospective father and health professionals who might assist in the procedure, or the children of the woman described in Case I. It might further be argued that individuals whose taxes or insurance premiums contribute to the support of seriously disabled newborns deserve to have their choices weighed in decisions regarding abortion. Needless to say, those popularly labeled as pro-choice advocates are not usually concerned about the autonomy of others in addition to the pregnant woman.

A *pro-woman* position may focus on the interests of pregnant women or women in general. If the former is the focus, the position means that the interests of the pregnant woman are the exclusive determinant of whether her pregnancy should be terminated or continued. One need not then be pro-abortion to be pro-woman, because one may actually view abortion as morally wrong while endorsing the right of the pregnant woman to choose it. This position is more clearly related to a pro-choice view because it affirms the right of the pregnant woman to choose her own interest over that of the fetus. It is possible, however, that the woman's choice and her (other) interests are at odds. Moreover, in cases where the autonomy of the pregnant person is questionable or lacking (for example, when those who are pregnant are children, as in Case II, or mentally retarded, insane, or comatose), affirmation of interests is obviously not equivalent to affirmation of choice.

If a pro-woman position focuses on all women, then interests other than those of pregnant women must be considered. "Women" here may mean future women, including some fetuses, as well as female children or adults who are adversely or favorably affected by a decision regarding abortion. To be fully affirming of all women's interests even implies concern for women of future generations, who are probably less affected, but inevitably affected nonetheless, by decisions presently made.

A *pro-fetus* position would stand at the opposite end of the spectrum from a pro-abortion view. It basically affirms that the interests of the fetus are the primary determinant of the morality of abortion decisions. Typically, this view involves a belief that the right of the fetus to continue developing in utero supersedes the rights and interests of all others affected by the decision—regardless of developmental immaturity or possible or actual defect in the fetus, or threat to the life or health of the pregnant woman. If fetal life is to be preserved even at the cost of the pregnant woman's life, one might characterize this position as anti-woman. Opposition to abortion in Case I suggests this sentiment. Some might argue, however, that a pro-fetus position entails the possibility of allowing or requiring abortion for severe fetal defect, defending this action in the name of fetal euthanasia.[7]

In contrast to the popular view that is typically labeled "pro-life," a genuinely *pro-life* position would reject all of the preceding views as unjustifiably limited in their affirmation of human life. The pro-abortion position negates the value of fetal life outright; the pro-choice position makes fetal life subordinate to liberty; and both the pro-woman position and the pro-fetus position ignore the relevance of other lives affected by abortion decisions. Because life is not lived by individuals in isolation from one another, but as an ongoing, complex system of interpersonal relationships, a genuinely pro-life position does not affirm the life of the fetus alone or the pregnant woman (or even all women) alone; it also affirms the life of the community in which they both participate. In other words, it affirms not only individual rights but human relationships. On that account, termination of pregnancy in situations such as Case I may be more life affirming than continuation of pregnancy.

In what follows I attempt to capture valid elements of different affirmations regarding abortion, utilizing the concept of equality considered earlier. Specific views of the pregnant woman, fetus, and the relationship between them are necessarily involved in that attempt.

The Pregnant Woman, the Fetus, and Their Relationship

Despite numerous attempts to clarify and justify the moral status of the fetus, this issue has remained a pivotal point of disagreement in the abortion debate. In contrast, the status of the pregnant woman as a person is generally uncontested. Most of the leading arguments have focused on the fetus, justifying abortion on grounds that the fetus is not a person, or rejecting it on grounds that this is not the case.[8] Most of the early authors of these views were men.[9] Even before the *Roe v. Wade* ruling, however, Judith Jarvis Thomson published her classic argument on the moral permissibility of abortion even if the fetus is a person. The novelty and significance of her view was that it did not focus on the moral status of the fetus but on the relationship between the pregnant woman and her fetus.[10]

Although both the pregnant woman and the fetus are human, the differences between them are morally significant. Like other women, the pregnant woman is a biologically mature human being capable of bearing a child; unlike some other women, she is actually exercising that capability. Ordinarily, the pregnant woman is also a moral agent because she chooses behaviors recognized as subject to moral evaluation. The fetus, in contrast, is an immature human being having the potential (rather than actual) capacity for reason, choice, reproduction, and moral agency.

The main difference between a viable human fetus and a newborn child is that the latter exists outside of the mother's uterus.[11] Both fetuses and infants are entirely dependent on others for survival. However, the "other" on whom a nonviable fetus depends is singular and definite, that is, the pregnant woman, whereas the crucial needs of children may be addressed by an indefinite number of others. For example, an infant born at seven months' gestation in the United States is usually more dependent for survival on nurses and doctors than on either parent. If a viable fetus is removed or delivered from the uterus to be nurtured by others, it is no longer a fetus but a newborn infant.

Just as children of either sex represent a range of ages and abilities, so do fetuses. Some fetuses have not yet developed brain function or sensitivity to pain; some have. Some fetuses are previable, some will never be viable, and some are already viable. The prima facie obligation not to deliberately terminate lives applies to all living human fetuses. The prima facie obligation not to inflict pain on sentient beings applies to most fetuses at some point during gestation.

Pregnant human beings also represent a range of ages and abilities. Some are legally children, and some are emotionally immature despite their biological maturity. Some pregnant adults are not capable of moral agency and some never will be. Some are economically independent, healthy, and married; others are not. Some are already mothers. A prima facie obligation not to harm others, either by shortening their lives or causing them pain, and the prima facie obligation to respect the autonomous wishes of others apply to most pregnant women in most of these varied circumstances.

Objectively, human pregnancy is the generic relationship between a human fetus (or embryo) and a pregnant woman, but different pregnancies involve different relationships (even for the same woman), with different burdens or benefits for each.[12] Although differences among individual fetuses and among individual pregnant women influence the relationship, the relationship itself ordinarily affirms the health of the pregnant woman. It may even lead her to take better care of her health, and elicit a sense of wonder at the thought and feeling of new life growing within her. In some cases, however, pregnancy constitutes a positive endangerment to women—physically, emotionally, or socially. For the nonviable fetus, in contrast, pregnancy is indispensable to survival. Its relationship to the woman is parasitic, causing obvious and significant changes in her body and often in her life-style.

Pregnancy may be unplanned but not unwanted, planned and very much wanted, planned but not now wanted, or unplanned and unwanted. Ambivalence about their pregnancies is common in newly pregnant women, but the longer a pregnancy continues, the more difficult it is to choose or undergo abortion. Her fetus may be viewed by a pregnant woman as a means of saving a marriage or maintaining a relationship with a man; as a way of demonstrating her adulthood, separating herself from her parents, or obtaining financial independence. As the outcome of rape or incest, it may be a constant reminder of the ultimate intrusion. In a situation of commercial surrogacy, the pregnant woman may relate to her fetus as producer to product; the fetus is the potential child she has contracted to deliver for payment of her labor. For some pregnant women the fetus is regarded as an ultimately prized possession, a gift offered to another, or a perpetuation of herself in the world.

The preceding account does not exhaust the unique features of pregnancy for women, fetuses, and their relationships. From an egalitarian perspective, not only these differences are morally relevant, so are all the differences, present and anticipated, in those affected by the decision.

Operative Definitions of Abortion

In the cases recounted earlier, different motives seem to be operative among those considering abortion. Ann Brown (Case I), for example, might choose abortion to avoid further physical deterioration, so that she might continue to fulfill current responsibilities towards her husband and children. While Eve Flynn (Case III) chooses pregnancy in order to achieve independence, Gail Hoy (Case IV) views it as an obstacle to professional advancement.[13] If Jill Katz (Case V), who purportedly wants to be a mother, now chooses abortion because the fetus is diagnosed with Down syndrome, she will apparently be motivated by a desire to terminate this fetus rather than a desire to terminate pregnancy. She would prefer to be pregnant if the fetus were not disabled.

The differences in motives for abortion point to different operative definitions of the procedure itself. In clinical texts, the definition is usually framed in terms of the fetus' relationship to the pregnant woman. Consider the following examples: "termination of pregnancy by any means before the fetus is sufficiently developed to survive,"[14] "premature expulsion from the uterus of the products of conception,"[15] "giving birth to an embryo or fetus prior to the stage of viability."[16] All three of these definitions apply to spontaneous as well as induced abortion. If the primary responsibility of the medical practitioner is to affirm rather than deny life, it is not surprising that clinical definitions do not mention fetal death as a goal of elective abortion. "Termination of pregnancy before the fetus is viable" rather than "premature termination of fetal life" is also a more apt definition for physicians who view their primary patient as the pregnant woman rather than the fetus.[17] As Case V suggests, a probable exception to this occurs when abortion is performed for fetal indications, where the clear intent is to terminate a particular fetus.

In contrast, the definition of abortion that is generally assumed in popular debate is one that focuses on the fetus rather than the pregnancy. Abortion is then construed as termination of a fetus, either before or after it has achieved viability. If in fact the fetus is nonviable or previable, the method of abortion is irrelevant to its survival ex utero. If the fetus is possibly viable, a method may be chosen to ensure death prior to expulsion from the uterus; or it may be chosen to maximize the possibility of survival. Regardless of the procedure, if it results in a living survivor, the event is defined as a preterm birth, and the survivor as a preterm infant, whether viable or not. The term *killing* may be used as an intelligible substitute for *termination* when applied to fetal life, but not when termination applies to pregnancy. When we end or terminate a relationship such as pregnancy, we do not ordinarily use the term "kill" to describe what we have done. Moreover, some abortions are spontaneous and unavoidable (rather than elective or induced), and the term killing is inappropriate in these cases because no active intervention has occurred.

In some clinical situations, the term *therapeutic abortion* is used to signify any elective or induced abortion.[18] Unfortunately, this use masks the distinction between abortions undertaken for medical reasons and those undertaken for nonmedical reasons. The broader meaning may be intended to suggest that all induced abortions are performed for medically compelling reasons, reducing prejudices on the part of those opposed to abortions for nonmedical reasons. In the clinical setting, however, therapeutic abortions are often assumed to be performed for reasons unrelated to maternal health. The failure to distinguish between nontherapeutic and therapeutic grounds for terminating a pregnancy is misleading, and denies clinicians information that may be relevant to treatment.

Although abortion is defined in most clinical situations as termination of pregnancy rather than termination of the fetus, practitioners are sometimes forced to choose between the two types of termination. From a legal point of view, a woman's right to terminate a pregnancy does not imply the right to ensure the death of a fetus if the two are separable events. For example, in 1975 a young obstetrician-gynecologist named Kenneth Edelin was prosecuted and convicted of manslaughter for preventing survival of a living fetus after elective abortion.[19] After saline infusion failed to induce the abortion, Edelin performed a hysterotomy, a procedure comparable to cesarean section delivery except that the fetus is assumed to be nonviable. According to court testimony, Edelin separated the placenta from the uterus, and waited three or four minutes before removing both the placenta and the fetus from the uterine cavity. The separation allegedly induced fetal anoxia, causing fetal death before the fetus was actually taken from the woman's body. On appeal, the conviction was overturned on grounds that the fetus was already dead when removed from the uterus. Presumably, had the fetus been alive at that time, and had Edelin failed to attempt treatment, the conviction would have been sustained.

To avoid such legal repercussions, physicians may choose a technique that ensures fetal demise in utero. From a moral point of view this requires justification beyond the right of a woman to terminate a pregnancy. The question of responsibility for fetuses is then unavoidably separable from the right to terminate a pregnancy.

Applying an Egalitarian Perspective

In the guidelines elaborated in Chapter 1, responsibilities to others are outlined on the basis of their existence, life, and capacity to suffer, think, or choose. Although human fetuses cannot think or choose (as far as we know), they can probably experience pain during late gestation.[20] Moreover, so long as they continue to develop, they are obviously living rather than dead.

Although living human fetuses may not be persons, their humanness implies the probability that they will in time develop the ability to think and choose. I use the word *probability* here instead of potentiality to connote a stronger meaning than the latter.[21] A potentiality may remain permanently unfulfilled unless certain steps are initiated in order to actualize it—for example, my potentiality for learning Russian, which I do not intend to actualize even though ability and opportunity for doing so are already present. A probability means that an event will occur unless its occurrence is prevented. The application of this probability–potentiality distinction to the abortion issue is clear enough. Under normal circumstances, unless the process is interrupted, a human fetus will develop to term, be born, and eventually reach undisputable personhood. In contrast, human ova and spermatazoa may have the potentiality but not the probability of similar development. (I deal more with the distinction between gametes and embryos in

Chapter 7.) Although alive and human, germ cells do not develop into persons without the initiative of fertilization.

Another distinction between the human germ cell (egg or sperm) and a human zygote, embryo, or fetus is also relevant to the abortion issue: the former does not constitute *a* human life, whereas the latter does. (A fertilized ovum constitutes at least one human life; it may of course constitute more than one, as twinning remains possible until about two weeks after fertilization.) Of itself, a human germ cell is no more a human being than is any other human cell, such as skin or blood cells; in fact, we might say it is less human because it represents only half the chromosomal complement of normal human somatic cells. If human cloning from nongerm cells ever becomes feasible, then we might argue that every human cell represents a possible human life, but it would not constitute a probable human person until initiation of development in that direction through the technology of cloning, which is thus comparable to the initiative of fertilization.

Although it seems clear that a human life is of greater value than mere parts of a human life (for example, blood or germ cells), this does not imply that every human life, at every stage of its development, is of greater value than the life of any or even every member of other species, at any point in their development. We might consider, for example, the way human beings treat other sentient animals as well as human fetuses in light of guidelines 7 and 3. Guideline 7 acknowledges a primary responsibility to distribute equal shares of pertinent resources to other human beings, and secondary responsibility for doing so with regard to other sentient beings. Guideline 3 implies a responsibility not to cause suffering to sentient beings. Our capability or resources for reducing or eliminating pain is not as pertinent to a zygote or an embryo as it is to late-gestation fetuses and other sentient animals. Accordingly, an early abortion intended to reduce or eliminate the pain or suffering (physical, mental, or both) of a pregnant woman may be justifiable, whereas a late abortion — that is, one that occurs after the fetal nervous system has developed to a degree sufficient for the experience of pain (surely during the third trimester and possibly during the second trimester)[22]— may be unjustifiable if the fetus' pain would be greater than that of the pregnant woman.[23] In the latter case, guideline 3 argues for choosing a means of avoiding pain to both. The consideration of pain would include psychological as well as physical pain, and would apply to future as well as present experience.

Just as a permissive view of capital punishment does not imply the moral acceptability of any means of inflicting that punishment (for example, torture), so the fact that abortion is probably justified in certain cases does not imply that a specific abortive procedure is morally justified. Accordingly, the available techniques need to be assessed in light of the third guideline, which applies to those affected by the procedure.[24] Hysterotomy, by which the fetus is removed intact from the uterus, is probably the least destructive for the fetus, but the least safe for the woman. In contrast, dilatation and evacuation (D&E) involves mutilation and removal of the parts of the fetus by means of instruments inserted into the uterus through the vagina of the anesthetized woman. In addition to the obvious violence and possible pain inflicted on the fetus, some clinicians report that this is an an extremely uncomfortable experience for them; nonetheless, it is probably the safest among the available techniques for second trimester abortions.[25] Infusions of saline, urea, oxytocin, and prostaglandin are less extreme in their effects; these agents (used separately or in combination) induce preterm labor and the delivery of a nonviable fetus. Because of their toxicity, saline and urea are likely to damage the fetus in utero, causing pain if the fetus is sentient; oxytocin and prostaglandin induce delivery without similarly affecting the fetus.

The preceding techniques are mainly used for late (second trimester) abortions, when the possibility of a sentient or viable fetus is a morally appropriate consideration.[26] However, even at a very early stage of pregnancy, a new abortion technique raises further ethical questions about responsibilities to human zygotes or embryos. Use of the drug RU 486, now legally available in France, England, and China, allows pregnant women to safely abort themselves within several days of conception.[27] Although medical supervision is required, the abortion decision can be made and implemented in private.[28] The pregnant woman's autonomy is then totally determinative of whether an induced abortion will occur. This factor, along with the fact that the development of the zygote or embryo is so limited, is both morally relevant to an abortion decision. Guideline 3 suggests that responsibility to the nonsentient embryo is weaker than that towards a sentient (more developed) fetus, and guideline 4 suggests that responsibility to respect the autonomy of the pregnant woman is stronger than it would be if her decision required the autonomous cooperation of others.

The moral relevance of different techniques of abortion will probably be more fully appreciated as the understanding of fetal development increases, and as reproductive technology advances to the point at which artificial wombs may be a reality. Even now, however, consideration of the moral justifications for alternative procedures is not only consistent with guideline 3, but also with the *Roe v. Wade* decision of 1973. In upholding the right of women to choose abortion, the Supreme Court assumed a definition of abortion as termination of pregnancy rather than termination of a fetus. In light of that distinction, if it were prevalently possible to end a pregnancy safely without seriously injuring the fetus, procedures such as D&E might be legally prohibited in the United States. Moral arguments for the right to terminate pregnancy are not equivalent to moral arguments for the right to terminate fetuses.[29]

While supporting the claim that we have responsibilities toward human fetuses as possible or potential persons and as sentient beings at some stage of their development, an egalitarian perspective does not justify a pro-fetus or anti-abortion position. Guideline 4 as well as 3 applies to the pregnant woman. To interpret these guidelines in situations of conflict, we need also to recall our concept of human beings as developing, psychosomatic entities. Typically, uterine development, as well as development during infancy and childhood, is almost exclusively progressive, consisting mainly of increments in one's physical and mental powers. After a certain point (the "prime of life?"), however, human development involves a larger component of regression or deterioration of one's powers. The life lost through an abortion is therefore likely to be a more progressive stretch of life than that lost in practicing euthanasia toward an elderly comatose patient. Nonetheless, there is more than individual development to be concerned about because important interactions occur constantly among all developing individuals. In other words, human beings, like other species, are essentially social animals. One could further argue that the early, painless abortion of the healthy fetus of a woman whose pregnancy places her life or health in jeopardy is justified because the pregnancy also threatens the developing lives of those who depend on her. Obviously this rationale is applicable to the case of Ann Brown (Case I).

Although both the pregnant woman and the fetus are developing, psychosomatic individuals, the responsibility of giving due regard to others' thoughts and choices (guideline 4) is applicable to the pregnant woman but not to the fetus. This, of course, is the point at which the debate over abortion often seems most heated, since woman's choice is pitted against the fetus' life as irreconcilable values. One possible way of resolving the dilemma is through the egalitarian route of guideline 6, which maintains that equal shares of pertinent resources ought to be distributed to the fetus and pregnant woman. Health and life are pertinent resources to both fetus and pregnant woman, but autonomy is pertinent only to the latter. Hence, the pregnant woman's autonomy ought to be respected at least to the extent that this respect is compatible with her and the fetus' equal share of health and life. The autonomy of the woman in circumstances where both she and the fetus are at risk thus tips the scale in favor of the woman. If those circumstances are not present, the woman's autonomy alone does not justify the termination of fetal life. In other words, abortion justified solely on the basis of the woman's choice at any stage of pregnancy cannot be defended on egalitarian grounds as long as we allow the reach of equality to include nonautonomous human individuals. According to guideline 3, for example, fetal sentiency or the possibility of fetal sentiency is a morally relevant consideration as well. Viability is also relevant. (I deal with this point more in Chapter 8.)

Other lives besides those of the pregnant woman and the fetus are obviously influenced by abortion decisions, as we all are affected by the lives and deaths of others. The potential father, other family members, and close friends are often significantly affected. Although it may be difficult to assess the impact, in some instances we can predict with a high degree of certitude that no one, including the fetus, will benefit by declining an abortion. Consider, for example, fetuses diagnosed through chromosomal or biochemical assays as having Tay-Sachs disease, trisomy 13, or trisomy 18.[30] In all of these cases, where the prognosis includes not only failure to survive infancy, but also a painful and painfully slow dying process, guideline 3 is applicable.

Beyond families and friends, clinicians are not only affected by, but at least partially responsible for, abortions or births in which they assist. Others who are totally uninvolved in specific cases are also influenced by such decisions, for example, by contributing voluntarily or involuntarily to the support of children who are not aborted, or through the enjoyment of the subsequent social contributions of those children. Nonetheless, the impact of abortion decisions on clinicians or other autonomous individuals is so much less than the impact on the pregnant woman herself that their contrary input could hardly justifiably negate her choice.[31] If we lived in an ideal world, where the responsibilities of pregnancy, abortion, childbirth, and child rearing were shared by all, the situation would be otherwise, because then the autonomy of everyone would be an equally pertinent resource.[32]

The fact that we live not in an ideal world, but in one in which the pregnant woman is overwhelmingly more affected by these events than anyone else (except the fetus), is obviously a departure from egalitarianism. And the inequality is scarcely reduced by legalizing abortion. In fact, the law's insistence on the (practically) exclusive right of the pregnant woman to decide the fate of her fetus places a great and ultimately solitary burden on many women, some of whom are still children. As an unavoidable option, that right is not one that can be exercised or not; rather, it represents an unavoidable and absolute responsibility. To impute to the pregnant woman such exclusive responsibility is not sufficiently warranted by her biological role because there are always social factors that greatly affect the decision and may severely limit the autonomy of the individual making it. If society were welcoming toward unwed mothers and disabled infants as well as supportive of the option of abortion at least in certain cases; if fathers and others really shared in child raising; if extrauterine means of reproduction were available; if overpopulation were not a matter of world concern— then we would have at least some of the conditions necessary for a genuinely egalitarian approach to the abortion issue.

Again, since we do not live in an egalitarian society, we have to deal with the inequality that exists while working toward an egalitarian, communal goal. Guideline 6 suggests a criterion by which individuals may assess their own choices; applying this to abortion, a woman might decide whether or how to terminate her pregnancy by considering to what extent equal shares of pertinent resources will thereby be distributed to those affected. Or a couple might decide together whether, when, and how to have a child and to share in the responsibility for its rearing—in light of the same guideline. Obviously, the more individuals and couples base their decisions on this guideline, the more egalitarian society will become, reflecting a recognition by others besides pregnant women and parents of the responsibility for uterine as well as extrauterine life.

Feminism and Abortion

Although feminists generally support a woman's right to choose or refuse abortion, that view is readily associated with a liberal version of feminism. Because a liberal version of feminism focuses on the rights of individuals, it affirms the right of individual women to terminate pregnancy at any stage for any reason. Presumably, this right extends to decisions to terminate fetuses because fetuses can hardly be viewed as having competitive liberty rights.

A radical feminist commitment to advocacy for women and overthrow of patriarchy also supports women's right to choose abortion at any stage of pregnancy for whatever reason. This support probably extends to the right to terminate fetuses as well as pregnancies. However, if advocacy for women extends to female fetuses, a further argument may be made in behalf of the latter, namely, that the interests of female fetuses should take precedence over those of male fetuses.

A socialist version of feminism attempts to pay due regard to the different needs, rights, and relationships of individuals. This view implies that the fetus is morally relevant to abortion decisions, but not exclusively determinative of their morality. At the same time, socialist feminism insists that the mere assertion by law or social attitude that individual women are "free" to choose abortion does not mean that they are actually free to do so. In fact, until and unless men as well as women share equally in the responsibilities for children, there shall be no equality between the sexes and no genuine liberation for women. Moreover, until and unless the fulfillment of those responsibilities is generally esteemed by others, and rewarded accordingly, the decisions of couples regarding the option of parenthood will be less than fully free. In general, under present circumstances, decisions to have children often place a greater burden on individuals than decisions not to have them—at least as far as their material situation is concerned. A socialist feminist approach would transform that situation into one where reproductive decisions were truly free, so that they might also be moral, both subjectively and objectively.

Admittedly, such a transformation would constitute a drastic shift in our way of viewing and valuing the "private" and the "public" spheres of life.[33] Minimally, it would require the dissolution of the prestige and power gap between the two. Those who are better at nurturing than others might continue to be women, but they would then not need to leave the workplace of their home to achieve equality with men. Some might combine roles of nurturing and material productivity; others might engage primarily or exclusively in the latter. And men might choose to be primary nurturers without suffering the disesteem of others. In such a turned-around world, feminism would no longer be necessary because sexual equality would be a reality. Abortions might still be moral in certain cases, but they would be an expression of, rather than a means of promoting, an egalitarian society. Moreover, because both men and women would utilize methods of contraception or sterilization more responsibly and consistently, and because the care of children would be more widely shared, there would actually be fewer abortions.

I end this chapter by citing the suggestion of a child who learned that her favorite baby-sitter, an unmarried college coed, had just had a baby. When told that the new mother had quit school, needed financial assistance, and would now be less available to play with her, the little girl spurted out with enthusiasm, "I know what we should do. We should take the baby, because we have room, and could pay for what he needs, and could take care of him. And his mom could go back to school and come here anytime." In terms of an egalitarian framework, the child was surely on target. Unfortunately, none of those concerned were "free" or virtuous enough to implement her suggestion. So much for the gap between our own ideals and practice.

NOTES

1. Jonathan B. Imber, *Abortion and the Private Practice of Medicine* (New Haven, Connecticut: Yale University Press, 1986).

2. Kristin Luker reports differences between "pro-life" and "pro-choice" activists in her "Abortion and the Meaning of Life," in *Abortion: Understanding the Differences,* ed. Sidney Callahan and Daniel Callahan (New York: Plenum Press, 1984), 26–45.

3. R. J. M. Gardner and Grant R. Sutherland, *Chromosome Abnormalities and Genetic Counseling* (New York: Oxford University Press, 1989), 60.

4. Ludwig Wittgenstein, *Philosophical Investigations,* trans. Gertrude Elizabeth Margaret Anscombe (New York: Macmillan Publishing Co., 1968), 109.

5. My critique of the labels was first developed in "Abortion and Equality," in ed. Callahan and Callahan, 177–96.

6. Laura M. Purdy, for example, argues along these lines in "Genetic Disease: Can Having Children Be Immoral?" reprinted in *Biomedical Ethics,* ed. Thomas Mappes and Jane Zembaty (New York: McGraw-Hill Book Co., 1981), 468–75.

7. For example, Susan Nicholson, *Abortion and the Roman Catholic Church* (Knoxville, Tennessee: Religious Ethics, 1978), 81–82.

8. For example, Joel Feinberg, ed. *The Problem of Abortion* (Belmont, California: Wadsworth Publishing Company, 1973); and John T. Noonan, ed., *The Morality of Abortion* (Cambridge, Massachusetts: Harvard University Press, 1970).

9. For example, all of the authors in Feinberg except Thomson. A later edition of Feinberg (1984) added articles by Mary Anne Warren, Jane English, and Sissela Bok.

10. Judith Jarvis Thomson, "A Defense of Abortion," *Philosophy and Public Affairs* 1, no. 1 (1971): 47–66.

11. While acknowledging the moral relevance of fetal sentiency, Mary Anne Warren provides an excellent account of morally relevant differences between fetuses and newborns. See her "The Moral Significance of Birth," *Hypatia* 4, no. 3 (Fall 1989): 58–63.

12. My use of the term *relationship* (between pregnant woman and fetus) does not imply that the fetus is an entity distinct from the pregnant woman, anymore than a relationship between me and my body implies that my body is an entity distinct from me.

13. With regard to Gail Hoy's rationale, it should be noted that motherhood is not equivalent to pregnancy and adoption is another means of avoiding the constraints of motherhood. The constraints of pregnancy are a separable consideration.

14. *Williams Obstetrics,* 18th ed., ed. F. Gary Cunningham, Paul C. MacDonald, and Norman F. Gant (Norwalk, Connecticut: Appleton and Lange, 1989), 489.

15. *Dorland's Illustrated Medical Dictionary,* 27th ed. (Philadelphia: W. B. Saunders Company, 1988), 5.

16. *Stedman's Medical Dictionary,* 23rd ed. (Baltimore: Williams and Wilkins, 1976), 3.

17. Reinforcement for the view that clinicians perceive abortion as premature termination of pregnancy rather than premature termination of fetal life was provided by a class of nurses pursuing a graduate degree in obstetrics and pediatrics. When I asked them to write their operative definitions of abortion, 68% described it in terms of termination of pregnancy or expulsion of the contents of a pregnant uterus, 32% described it as terminating fetal life.

18. Note that I refer here to clinical situations rather than clinical texts. My statement is based on ten years of participation in weekly meetings where mention of abortions is a recurrent feature of case histories. In clinical texts, "therapeutic" abortion is defined as abortion undertaken for the sake of the pregnant woman's life or health, but other situations are also viewed as justifying "therapeutic" abortion: when pregnancy has resulted from rape or incest, and when continuation of the pregnancy is likely to result in birth of a child with severe disability. "Elective" abortion is defined as "the interruption of pregnancy before viability at the request of the woman but not for reasons of impaired maternal health or fetal disease." See *Williams Obstetrics,* 18th ed., 501.

19. See William A. Nolen, *The Baby in the Bottle* (New York: Coward, McCann and Geoghegan, Inc., 1978).

20. K. J. S. Anand and P. R. Hickey, "Pain and Its Effects in the Human Neonate and Fetus," *New England Journal of Medicine* 317, no. 21 (Nov. 19, 1987): 1321–46.

21. I have also used this distinction in my article, "Abortion and Equality," in Callahan and Callahan, 189, but note that I have applied it to fetuses and not necessarily to early embryos. As many as 15% of all recognized pregnancies are spontaneously aborted. The great majority of these occur during the embryonic rather than fetal period of development. See Keith L. Moore, *The Developing Human,* 3d ed. (Philadelphia: W. B. Saunders Co., 1982), 1, 49, and James W. Knight and Joan C. Callahan, *Preventing Birth* (Salt Lake City: University of Utah Press 1989), 183. The distinction between probability and potentiality does not apply to embryos in unrecognized pregnancies because apparently up to 75% of all human conceptions are aborted spontaneously at this very early stage.

22. See Nobuo Okardo and Tokuzo Kojima, "Ontogeny of the CNS: Neurogenesis Fibre Connection Synaptogenesis and Myelination in the Spinal Cord," in *Continuity of Neural Functions from Prenatal to Postnatal Life,* ed. Heinz F. R. Prechtl, *Clinics in Developmental Medicine,* 94. (Oxford: Blackwell Scientific Publications, Ltd., 1984), 31–45; see also Anand and Hickey, note 20.

23. Peter Singer would argue that painless killing of humans and other animals is acceptable in certain circumstances, but he does not address the possibility that abortion may involve pain for some fetuses. See his *Practical Ethics* (London: Cambridge University Press, 1979), especially Chapters 6 and 7. He would probably dispute my claim that individual lives as such have a value (see guideline 2).

24. See "Techniques for Abortion," in *Williams Obstetrics,* 18th ed., 502–06; and Knight and Callahan, 183–99.

25. Centers for Disease Control, *Abortion Surveillance Annual Summary 1981,* (Washington, D.C.: U.S. Department of Health and Human Services, November 1985); 10, 41.

26. In particular cases, viability is a retrospective judgment even for third trimester fetuses. The possibility of viability obviously increases as gestation progresses, and eventually becomes highly probable. See Anand and Hickey, note 20; and Okardo and Kujima, note 22.

27. Ronald Kotulak, "French Abortion Pill Locked Out of U.S. by Political Fighting," *Chicago Tribune* (Dec. 29, 1991), sect. 4, 4. As yet (April 1992), the drug is not approved for use in the United States. However, privately funded trials of its use are expected in the United States, and more than twenty other countries are considering approval of RU 486. See Kotulak, 4. For an account of political aspects of the exclusion of RU 486 from the American market, see Lawrence Lader, *RU 486* (Reading, Massachusetts: Addison-Wesley Publishing Company, Inc., 1991).

28. Medical supervision is required to ensure that contraindications are not present and that the abortion has been completed without threatening side effects. Although the drug initially required administration by injection, it can now be taken orally in combination with a second pill containing sulprostone. See Lawrence K. Altman, "A Simpler Way to Employ RU 486 is Reported," *New York Times* (April 9, 1991), B6.

29. This distinction is more fully elaborated in my "Concepts of Abortion and Their Relevance to the Abortion Debate," *Southern Journal of Philosophy* 20, no. 2 (Summer 1982): 195–207.

30. See Richard M. Goodman and Arno G. Motulsky, eds., *Genetic Diseases among Ashkinazi Jews* (New York: River Press, 1969), 217–31; and Jean de Grouchy, *Clinical Atlas of Human Chromosomes* (New York: Wiley Medical Publications, 1977), 127–32, 160–64.

31. This does not imply that physicians are obligated to perform abortions, but rather that they are obligated not to interfere with pregnant women's choices in that regard.

32. An excellent elaboration of this argument is in Alison M. Jaggar's "Abortion and a Woman's Right to Decide," *Women and Philosophy,* ed. Carol Gould and Marx Wartofsky (New York: Putnam and Sons, 1976), 347–60.

33. See Jean Bethke Elshtain's ideas in *Public Man, Private Woman* (Princteon, New Jersey: Princeton University Press, 1981).

5

Fertility Curtailment: Selected Issues

The term *fertility* derives from the Latin *ferre* meaning "to bear" or "carry." *Sterility* is its opposite — the inability to bear. Despite the fact that women alone are capable of childbearing, both terms are used for men as well as women in the context of reproduction. Elective means of fertility curtailment include contraception, sterilization, and abortion. In this chapter, I focus on selected aspects of each of these types of interventions. Although contraception is relatively uncontroversial for most people, I briefly consider the natural law argument against it. With regard to sterilization, I examine the special case of the retarded as a possible example of discrimination against disabled persons. Preconceptive and postconceptive genetic counseling are considered in the context of sex selection, and fetal reduction in multiple gestation is also discussed. A number of these issues illustrate the fact that decisions about fertility curtailment and decisions about fertility enhancement are sometimes intertwined.

Contraception, Sterilization, and Natural Law

Despite its limitations, natural law thinking represents a well-developed tradition that still appeals to some clinicians and patients. The appeal is evident, for example, in their references to the distinction between "ordinary" and "extraordinary" treatment as a criterion for determining whether treatment may be withheld or withdrawn.[1] The principle of double effect, also elaborated within that tradition, has been found useful in addressing cases that involve inevitably tragic outcomes.[2] Basically, natural law reasoning is based on the Aristotelian notion that what is natural is good or moral. Applying this rationale to human reproduction, sexual intercourse is recognized as a natural drive of human beings, and conception is viewed as a natural result of intercourse. While natural circumstances frequently prevent or impede conception, technology provides means through which the natural process can be deliberately curtailed. Abortion, contraception, and sterilization constitute deliberate interferences with the natural propensity of human beings to reproduce their own kind.

Since my egalitarian perspective rests on a basic understanding of what it means to be a human being (see Chapter 1), a natural law analysis of reproduction might be thought consistent with that approach. In *Reproductive Ethics,* Michael Bayles formulates the natural law argument as follows:

1. Good ought to be pursued and evil avoided.
2. Good ends are those toward which people are naturally inclined.
3. People are naturally inclined to sexual intercourse.
4. A natural purpose of sexual intercourse is reproduction.
5. Therefore, reproduction ought to be pursued in sexual acts.
6. To act intentionally against the good of reproduction is evil and ought to be avoided.[3]

Like Bayles, I have problems with this argument on a number of grounds. The second premise is problematic because there are some apparently natural human behaviors that are patently immoral. For its advocates, however, natural law provides a rationale for rejecting any form of "artificial" interference with the reproductive process. Although the Roman Catholic Church has clearly stated its opposition to technologies that promote or obstruct fertility on this basis,[4] its position relies on a questionable concept of nature as essentially opposed to technology or "artificial" interference. The concept is questionable because it may just as reasonably be claimed that the technologies facilitate rather than impede natural processes. For example, contraception may be considered natural in the sense that it is more likely for ova not to be fertilized in each act of intercourse than for fertilization to occur. It is also more likely, and in that sense more natural, for embryos that have been implanted to be born as viable infants than that they be spontaneously aborted.[5] On such an interpretation, fertility curtailment prior to implantation may be regarded as more natural than fertility curtailment after implantation.

Ironically, the view that "natural reproduction" is impeded by "artificial technologies" is questionable on religious grounds as well. From that point of view, God has designed nature according to certain laws, and morality consists of conformity with the natural functions that those laws express. But the ability of human beings to transform the world, including themselves and their capacity for reproduction, is surely part (perhaps the most important part) of the divine design. To frustrate that human ability by proscribing the development and use of contraception or the practice of sterilization is to interfere with the natural capacities of individuals. Indeed, to presume that humans can obstruct God's overall plan for nature may be construed as blasphemous because it attributes to humans a supernatural power.

Despite its opposition to "artificial" impediments to reproduction, the Catholic Church supports at least two methods of contraception, both based on abstinence from intercourse during periods when ovulation is likely to occur.[6] If sexual union is a natural inclination of a man and woman, abstinence can hardly be described as a "natural" approach to fertility curtailment. Other means of contraception may be seen as less obstructive of the natural relation between them and as simultaneously morally responsible toward others. The others affected include possible embryos or fetuses. The guidelines elaborated in Chapter 1 do not mention possible or potential beings such as gametes, embryos, or fetuses, but guideline 1 alludes to moral responsibilities regarding the existent conditions that affect future offspring.[7] Our responsibilities toward possible beings are based only on the anticipated actualization of potential, and cannot be as compelling as responsibilities toward comparable existent beings. To the extent that actualization is improbable, the responsibility diminishes. Although gametes, embryos, and fetuses are already existent, they represent quite different probabilities regarding the likelihood of achieving the characteristics for which other overriding guidelines (3 and 4) are applicable (sentiency, ability to think and choose).

By opposing artificial means of curtailing or facilitating reproduction, natural law reasoning also opposes "artificial" means of promoting equality regarding reproductive decisions. The opposition to equality as a social goal is consistent with the views of Thomas Aquinas, from whom natural law theology has developed. In discussing whether men and women would have been equal before the fall of Adam and Eve, Aquinas maintained that inequality between the sexes would then have prevailed. "The things which are of God," he wrote, "are well-ordered. But order consists chiefly in inequality.... Therefore in the first state, which would have been most proper and orderly, inequality would exist."[8] Aquinas assumes a concept of equality as sameness, arguing not only that women are different from men but that the differences require their subservience to men. Like Aristotle, he regards individual women as "defective and misbegotten," claiming that

> the good of order would have been wanting in the human family if some were not governed by others wiser than themselves. So by such a kind of subjection woman is naturally subject to man because in man the discernment of reason predominates.[9]

The "natural order" through which inequality among persons is morally acceptable to Aquinas extends not only to differences between women and men, but also to differences in age, physical strength, beauty, and ability that prevail among all human beings. Refusal to accept natural advantages or disadvantages that result from such differences is wrong because it implies refusal to conform to God's plan for the world. "Artificial" means of contraception interfere with the natural inequality that God has established between men and women.

Because the concept of equality supported in this book distinguishes between differences and the ranking imputed to differences, the argument of Aquinas for social inequality is unacceptable within that framework. Equality remains a social ideal that individuals and groups are morally obligated to pursue. Moreover, the pursuit of equality through reproductive interventions is neither unnatural nor morally inappropriate. Indeed, practicing contraception may be morally obligatory if it is the only means through which harmful inequalities can be avoided.

Responsibility for Contraception

From an egalitarian perspective, responsibility for contraception should be shared by sexual partners in a manner that respects the values and preferences of each, and the disproportionate burdens and benefits that pregnancy involves for each. It may be argued, for example, that men have a stronger obligation to practice (accept the burden of) contraception because the burden of pregnancy falls on women rather than men. If pregnancy is seen as a benefit mainly to women, the opposite claim could legitimately be made.

Consider, for example, the following two cases:

CASE I: CONTRACEPTION FOR WOMAN'S HEALTH.

A 25-year-old woman suffering from heart disease has been told by her physician that pregnancy would exacerbate her health problem. Her husband is aware of the risk that pregnancy would impose on his wife. Neither he nor she desires to have children, and both consider abortion morally akin to murder.

CASE II: CONTRACEPTION FOR AVOIDANCE OF PARENTHOOD.

An 18-year-old youth knows that he would be legally responsible for support of a child produced through intercourse with his girlfriend. The 16-year-old girl is interested in becoming pregnant because parenthood would make her an emancipated minor, allowing her to leave her parents' home and supervision. While the two teenagers care for each other, there are no plans to marry.

All pregnancies entail a certain risk for women, but the risk for the woman in Case I is obviously greater than the risk of pregnancy for the 16-year-old in Case

II. Moreover, the teenager apparently considers independence from her family a benefit that overrides the risk of pregnancy. She may or may not be considering the benefit of having a child as well, but this factor may be viewed as an additional burden. Presumably, most women who choose to be or to remain pregnant regard the benefits of doing so as outweighing the burdens. Benefits and burdens to others as well as themselves are usually considered.

Ordinarily, men and women are equally responsible for initiating a pregnancy, and therefore should be equally responsible for the practice of contraception. But the men in the two cases above are in opposite positions regarding the impact of possible pregnancies of their partners on themselves. Their different positions place a greater or less burden on them regarding contraceptive practice. If we are morally bound to respect the interests and preferences of others in decisions that primarily affect them, and if the obligation not to harm supersedes the positive obligation of beneficence, then the husband in Case I and the sixteen-year-old in Case II have the greater responsibility to avoid the burden that pregnancy would involve for their partners.

Sterilization is another option in Case I. Here, as with contraception, the cost, risk, and invasiveness of the procedure are relevant egalitarian considerations in determining which partner should be sterilized. By these criteria, vasectomy or male sterilization is the fairer alternative. Reversibility is also relevant, however, and if the woman in Case I is permanently unable to safely undertake a pregnancy because of her cardiac condition, tubal ligation or female sterilization may be more appropriate. To be fair to future possibilities for each, if the couple were to divorce or separate the woman would probably continue to avoid the health threat of pregnancy, while the man might wish to father children with another partner.

Both contraception and sterilization constitute interferences with the "natural" process of reproduction, if that process is defined in association with uninterrupted sexual intercourse. But even among those who generally find sterilization morally acceptable, further questions are raised when the procedure is considered for the retarded.

Sterilization of the Retarded

Mental retardation covers a great range of possibilities among individuals who meet the standard criteria of the American Association of Mental Deficiency: "significantly subaverage general intellectual functioning existing concurrently with deficits in adaptive behavior manifested during the development period."[10] The range extends from those who may be classified as mildly retarded (who are educable and may reach at least a 10- to 11-year mental age), to moderately retarded (who are trainable and may mentally reach the level of an 8- to 9-year-old), to severely retarded (who may reach a 6-year-old level), to profoundly retarded (whose mental age range is somewhere below 3 to 4 years old).

Obviously, mental retardation may compromise an individual's ability to function as a parent and capacity to make autonomous decisions. But even where individuals have been declared legally incompetent and placed in the custody of guardians because of mental retardation, this action does not imply that they are legally incompetent to make *any* decision affecting them. Permission to perform specific procedures such as sterilization or abortion ordinarily must be adjudicated separately, on grounds of the individual's capacity to understand and consent to the procedure and its implications. As Allen Buchanan and Dan Brock put it, competence is thus "decision-relative."[11] Moreover, legal competence and moral competence are not equivalent. Children, for example, are not legally competent to make medical decisions for themselves, but many children are competent to make moral (and immoral) decisions. (I discuss this topic further in Chapter 11.) Similarly, many retarded adults are capable of making their own moral (and immoral) decisions, even though others may legally decide in their behalf.[12] To preempt this possibility on paternalistic grounds constitutes discrimination against the disabled. Coincidentally, it violates at least two of the guidelines elaborated in Chapter 1: the obligation not to thwart the choices of others (4), and the obligation to treat others as who or what they are (5).

To understand how competence applies to the issue of sterilization, we need to distinguish among three types of competence: intellectual, volitional, and practical. Intellectual competence is limited but not necessarily eliminated through mental retardation. Volitional competence has to do with the capacity to make a free or voluntary decision. Many people of normal intelligence make decisions that are not really free because the stress of the situation, social pressures, or other factors such as ignorance impede the voluntariness of a particular decision. If a person is practically competent, then impediments such as limited time for deciding, or limited ability (temporarily or permanently) to communicate a decision, are absent. When we realize that retarded people are sometimes competent intellectually to make moral decisions, and that all of us may be impeded volitionally or practically in making moral decisions, then the role of competence regarding the morality of sterilization of retarded persons takes on a different caste. Competence in all three ways is possible for some of the retarded population with regard to this issue.

Because sterilization is generally intended to insure permanent prevention of reproduction,[13] a person who freely consents to sterilization should have some understanding of what "permanent" and "reproduction" mean. While the young or the retarded often lack understanding of long-term implications, they usually understand what "permanent" means. For example, most retarded persons would not expect a dead animal to move again; they thus exhibit a sense of death's permanence.

Similarly, the meaning of reproduction is often understood through observation of reproductive practices, pregnancy, and birth of humans and other animals. If a retarded person understands the basic meaning and implications of sterilization, he or she may consent to the procedure voluntarily so long as there are no impeding factors such as pressures from parents or others. If that same person refuses consent, sterilization can only be performed involuntarily, and (on egalitarian grounds) this is legally and morally as unjustified for the retarded person as for the nonretarded person. For those who cannot understand what sterilization involves, however, neither voluntary nor involuntary sterilization can occur. What may then occur, and may be morally justified, is nonvoluntary sterilization. When an individual is wholly incapable of moral decisions, sterilization may be looked on as morally wrong if it opposes the person's own best interests, or as morally right if it is consistent with those interests. In other words, voluntariness is a critical factor in determining whether a retarded person may or may not be sterilized; when this is not determinable the situation calls for further principles or reasons in support of either alternative.

While it is important to recognize that our obligations to persons apply to them as individuals rather than as categories, what we know about different degrees of mental retardation helps us to deal with the individuals who embody those characteristics in a way that respects individual differences. At one end of the scale, we may be dealing with a profoundly retarded person who is incapable of understanding or fulfilling the demands of parenthood but capable of expressing and receiving affection. In such a situation, sterilization is morally justified insofar as it fulfills the principle of beneficence towards the patient or client, without violating autonomy because the person is not capable of autonomy. The benefit of the procedure must outweigh its harm, with recognition that this is the least harmful way of obtaining that benefit for the patient. Benefits of sterilization for the retarded include the elimination of menses and the enhancement of other relationships.[14]

At the other end of the spectrum is the mildly retarded person who may be capable of autonomous decisions and competent parenting.[15] The potential presence of a capacity for competent parenting (which appears lacking in some persons who are not retarded) argues for respecting the individual's choice regarding sterilization. Less intrusive and less permanent means of curtailing fertility should be encouraged so long as the patient is capable of exercising these options. In general, however, respect for autonomy is the decisive principle in dealing with mildly retarded persons. When we consider moderately and seriously retarded persons, beneficence towards them becomes more significant to the extent that capacity for autonomy is reduced; and beneficence towards others, including possible offspring, is also relevant. Less permanent or less intrusive means of contraception need to be considered to the extent that these are realistic options.

What we really have, then, is a continuum between the two principles of beneficence and respect for autonomy, requiring that beneficence be maximized to the extent that an individual is not autonomous, and vice versa. The continuum extends across degrees of retardation. A complicating but unavoidable feature of our obligation regarding beneficence is concern for a possible child, and the needs of others besides the retarded person. A complicating but unavoidable feature of our obligation to respect autonomy is that this extends to the autonomy of family members and caregivers, even where it does not apply to the patient or client. An egalitarian perspective requires equal distribution of harms and benefits, and respect for conflicting decisions in proportion to the degree to which individuals are affected by those decisions (see guidelines 6 and 7). The complex issue of sterilization can thus only be addressed in a morally justifiable way by considering the interests and preferences of all of those involved and prioritizing these according to the effects on each. To avoid discrimination against the disabled, such considerations are as applicable to nonretarded as to retarded persons.[16] In the next section I argue that they are also applicable to the issue of sex selection.

Sex Selection: Before and After Conception

The practice of infanticide has a long history.[17] In general the practice has applied to disabled or anomalous newborns and to female infants who are left to die or directly killed. From an egalitarian perspective, arguments against infanticide for either group of infants are compelling.[18] In a less compelling way, the same arguments apply to fetuses. Arguments against genetic selection or sex selection prior to conception are less compelling still, and may in fact fail to support the claim that such selection is immoral. In what follows, therefore, I want to distinguish between sex selection before and after conception.

Prior to conception, sex selection basically involves efforts to ensure that fertilization results in an embryo or pre-embryo of the desired sex. As many studies show, the desired sex is usually male.[19] Historically, techniques for preconceptive sex selection have been undertaken solely by the potential parents; they mainly involve the timing and frequency of intercourse, coital position, use of douches, and diet.[20] Scientifically, none of these "home remedies" has proved effective. Techniques requiring medical assistance or intervention have shown more promise. These involve different methods of separating gynosperm (sperm that bear the chromosome that produces daughters) from androsperm (sperm bearing the chromosome that produces sons). By separating these two types from each other and then inseminating the woman with the appropriate fraction during the most fertile period of her cycle, the probability of achieving a fertilized ovum of the desired sex is increased.[21]

Methods of sex selection following conception involve two parts: determination of the sex of the conceptus and elimination of conceptuses that are not of the desired sex. Many genetic counselors are unwilling to provide their services solely for the sake of sex selection unrelated to sex-linked disease. However, in the course of prenatal diagnoses (through ultrasound, chorionic villus biopsy, or amniocentesis) undertaken for health reasons, clinicians routinely ascertain the sex of the fetus, and communicate this information unless the patient asks not to be told. Having received the information, some patients choose abortion when a healthy fetus is not of the desired sex. Practitioners may feel manipulated by such patients.

Although methods of sex selection following conception are more effective than preconceptive methods, they are more problematic because they are inevitably associated with abortion as the means through which fetuses or embryos of the undesired sex are eliminated. If human fetuses or embryos have no value in their own right, or if their value is no greater than that of ova or sperm, then sex selection is no more problematic before conception than after conception. As we have already argued, however, there is a morally relevant difference between gametes and embryos, or between possible and actual embryos, and the difference suggests that obligations to the latter are greater than those towards the former. Accordingly, even if sex selection is morally acceptable, it is more acceptable prior to conception.

Reasons for sex selection are not necessarily sexist. The strongest argument supporting it occurs in situations where an embryo is likely to be affected by a serious genetic disease that primarily or exclusively afflicts males (for example, hemophilia). Although Bayles regards as inherently sexist all nonmedical reasons for sex selection,[22] Christine Overall argues that reasons of "sexual similarity" or "sexual complementarity" are morally acceptable reasons for wanting a child of a certain sex.[23] She bases her argument on the significance of sexuality, both heterosexuality or homosexuality, to interpersonal relationships. In sexual similarity, what is sought

> is a likeness, an affinity, of experience and capacities—the groundedness of being with one's own kind. The notion of sexual complementarity, on the other hand,... is not merely a matter of dissimilarity,...[but also] the desire for the new, for what will change and enlarge one's own experience.[24]

In arguing for the legitimacy of sex preselection on this basis, Overall exhibits a sensitivity to the uniqueness of human relationships, consistent with a care-based ethic.

Although Overall generally favors additional involvement of men in various aspects of child care, she stops short of advocating equal involvement of heterosexual partners in sex selection of their children.[25] Through its emphasis on attention to relevant differences, an egalitarian view would not only stop short, but insist that one partner, the woman, has a stronger claim in this regard. Since it is the woman who carries the main burden of bringing a child of either sex into the world, and usually also the main burden of nurturance, the woman's preference regarding the sex of her offspring is more compelling than a comparable preference on the part of her male partner. This reasoning would also apply to a lesbian couple in whom one partner has a different preference regarding the sex of a future child than the other: the woman who would undergo insemination, gestation, and birth has the stronger claim regarding the sex of the child.

If a policy permitting sex selection led to numerical predominance of one sex over another, such a result would not necessarily be inegalitarian. It is at least possible for relationships between members of a minority and members of a majority to be fair and equal. It is also possible for a minority group to be politically stronger and to form an economic majority despite its numerical minority. So even if sex selection led to a majority of males in the population, it does not follow that subjugation of women to men would thereby be supported or intensified. Nonetheless, the threat of increased sexism remains a matter of concern to those interested in promoting an egalitarian (less inegalitarian) society.

In an analogy with decisions made in behalf of disabled fetuses, another argument regarding the morality of sex selection arises. This argument starts from the basic feminist claim that women have historically been oppressed and that such oppression is morally wrong. Both claims may be relevant to the issue of sex determination. If we consider them in the context of decision making by potential parents, a potential parent who wants what is best for her future child might reason as follows prior to conception:

> I know that girls do not generally get a "fair shake" in society. Despite an "egalitarian" upbringing, a female child that I might bear is likely to have fewer advantages in life than a male child. Even with natural talents equal to a son, she probably would not reach the same income or prestige level. Parenthood and gender stereotypes would reduce her chances of success, as they would not comparably affect a man. If I choose to have a daughter despite these drawbacks, I am choosing a future that is less than optimal for my child. Perhaps, therefore, I am morally bound to choose a son.[26]

Obviously, if every potential parent thought and acted in keeping with the preceding rationale, society would in time arrive at an overwhelming preponderance of males. Although such a situation is not necessarily sexist, it is likely to reinforce the sexism that already prevails even if the rationale for sex selection is the promotion of the future child's best interests. Accordingly, while the rationale is explicative of decisions by individual parents, it does not merit universalization, or enactment into social policy.

Benefits of the availability of techniques for sex selection should be acknowledged. Among those cited are the virtual elimination of sex-linked diseases, reduced likelihood that children will be unwanted because of sex, reduction of the birthrate, and possibly a better balance of males and females in the elderly population.[27] The last result would occur because of the probability that the number of men in the general population would substantially increase.[28] The greater longevity of women might be matched by the greater proportion of the larger population of men who survive to more advanced age. Of course, achievement of this end would also mean that the ratio of men to women would be greater earlier in life. So it is only in advanced age that numerical equality may be achieved, and it is even possible then that men would predominate. The anticipated numerical predominance of men is predicated on the well-supported preference of both men and women for sons, especially as firstborns.[29]

What are the potential harms of general availability of sex selection techniques?[30] As already suggested, these have mainly to do with the reinforcement of sex-role stereotypes and sexist practices. They may also intensify the burden, if the technique fails, of having or being an unwanted child. Regardless of their sex, firstborns are generally more ambitious and successful than second-born children, who are generally more sociable than their older siblings. If firstborns are predominantly sons and second-born children are predominantly daughters, these differences parallel the sex-role stereotypes discussed in Chapter 2. With fewer women in the world, Amitai Etzioni maintains that the more venal traits of men would prevail—for example, their tendency to violence and criminality.[31] Women would more prevalently be treated as objects, valued for their worth to men rather than for themselves, with their freedom suppressed to increase their availability. Men who least exemplify male stereotypes might also be exploited. Alternatively but improbably, women might be treated as queen ants, dominating the mostly male world as supreme sovereigns. Either scenario is sexist and inegalitarian, and therefore morally unacceptable.

Clearly, the morality of sex selection following conception is tied to the morality of abortion. But the meanings of abortion and of pregnancy are also crucial to determining whether elimination or destruction of an unwanted conceptus is equivalent to abortion. It is now possible, for example, to determine the sex of a pre-embryo in vitro, and to choose or decline to implant it in a woman's uterus. A woman whose pre-embryo exists in vitro is not physiologically pregnant. In fact, a woman who is not genetically related to her may become pregnant through implantation of her pre-embryo. According to guideline 2, whether or not elective termination of a conceptus is technically defined as abortion, the act is not morally neutral. However, the fact that the conceptus exists ex utero means that a woman's body need not be invaded to affect it and that others can and do affect the outcome. While it is doubtful that the entirety of human gesta-

tion will ever be technically sustainable in vitro, that development would increase the responsibility of those who could provide the technology.

Genetic Counseling, Eugenics, and Fetal Reduction

Charges of eugenics have sometimes been levelled against clinicians who provide genetic counseling. To some extent the charges are valid because such individuals are usually interested in reducing the incidence of defective genes in the
general population. Their primary concern, however, is communication rather
than prevention. Genetic counselors are sometimes assumed to advocate abortion in cases of genetic abnormality; yet their practice tends to reduce rather than
increase the incidence of abortions. Preconception counseling may lead to fewer
abortions than would otherwise occur. But even among patients who undergo
prenatal diagnosis, only about 5 percent are discovered to be carrying fetuses
with genetic defects.[32] Although most of these select abortion when the anomaly
is disclosed, at least some of the other 95 percent would otherwise have terminated pregnancies because of fear of fetal defect. Consider, for example, the
number of pregnant women over 35 who are aware that the incidence of Down
syndrome increases with maternal age. The vast majority of these women dis-
cover after prenatal diagnosis that their fetuses are chromosomally normal. If
they were unable through prenatal diagnosis to insure that their fetuses were
chromosomally normal, some would abort in order to negate the possibility of
giving birth to a genetically defective child.

Typically, genetic counselors attempt to provide information in an impartial
fashion, recognizing that this is an ideal that can only be approximated. On the
whole, however, most are open to abortion in cases where fetal abnormalities have
been detected, even while they prefer to conduct genetic testing prior to pregnancy
in order to prevent that occurrence. As already suggested, one exception to a generally liberal view on the right of a woman or couple to obtain prenatal diagnosis
is when the motive is sex selection unrelated to medical need.[33] The appropriate
concern here is avoidance of social sexism. Yet, termination of fetuses because of
their anomalies may be another kind of chauvinism or social prejudice. Although
the rationale of fetal euthanasia may be offered in instances where the genetic defect forecasts unavoidable suffering for the newborn, the usual basis for abortion
of an anomalous fetus is avoidance of suffering and costs to others.

Another ethical issue associated with genetic counseling involves the use and
interpretation of statistical data. Consider, for example, an obstetrician counseling
a 36-year-old woman who is pregnant for the first time. On grounds that the risk
of chromosomal abnormalities increases with maternal age, the physician recommends prenatal diagnosis to ensure that the fetus is normal. In his explanation of

two available diagnostic procedures, chorionic villus sampling and amniocentesis, he acknowledges the risks of each but describes them as "very low." The reason for recommending prenatal diagnosis, however, is that the woman's risk of delivering a child with a chromosomal abnormality is "high" in comparison with, say, women in their twenties. In fact, the risk of chromosomal abnormality at birth for a woman at 36 years of age is 1 in 164, while the risk associated with amniocentesis (chiefly through spontaneous abortion) is 1 in 200.[34] To the woman, of course, giving birth to a seriously disabled child may be a greater burden than aborting a healthy fetus, which means that mere comparison of numbers is not an adequate means of assessing risk. Rather than use the terms "high risk" and "low risk," which are inevitably value laden, the clinician might avoid biases of interpretation by citing the relevant numbers, explaining the mathematical probability or improbability they denote, along with the fact that most genetic anomalies are unpredictable on the basis of prenatal diagnosis. This provides a sense of context that allows maximal fulfillment of egalitarian guidelines, particularly guideline 4.

Regarding the charge of eugenics, consider the distinction between (a) the natural moral inclination of parents to promote the health of their offspring, and of others to promote the health of future generations, and (b) individual or social practices that demean and sometimes destroy imperfect human beings. Immoral eugenics occurs in the context of (b) rather than (a), that is, when suboptimal traits, which are appropriately recognized as such, are taken to define rather than describe the worth of individuals. It also occurs, as the past so tragically illustrates, when one set of traits that are comparable in value to another is promoted as superior, while the other (comparable) traits are treated as inferior. Eugenic practice may take the form of positive or negative eugenics. Positive eugenics attempts to promote better or superior genetic traits in the human population; negative eugenics aims at decreasing undesirable traits. Because negative eugenics is associated with elimination of severe and debilitating diseases, it is perceived as less problematic than positive eugenics. However, neither form is morally acceptable if it entails social prejudice, that is, judgment based on morally irrelevant criteria.

How do we define relevant criteria for decisions regarding fertility curtailment? An egalitarian perspective provides only general guidelines. Clearly, the autonomy and welfare of those most affected by the decisions are paramount. Prior to conception, the couple are most affected. After conception, sentiency or viability of the fetus is relevant. Future persons (that is, the persons the fetuses may become) and all of society, including members of future generations, are influenced by population decisions on the individual level and on the policy level. In general, however, the needs and interests of actual persons supersede those of potential persons. The right to have children is not inconsistent with responsibility to avoid conception in certain circumstances; for example, a population explosion that threatens the welfare of those already living, when serious genetic abnormality is likely for potential offspring, or when obligations to others are more compelling. Even if there is no duty to practice contraception, the right of a woman to fertility curtailment is still defensible. But the right may be immorally exercised if its purpose involves a possible social harm (such as reinforcement of sexism) or if the means of curtailment entails greater harm to others.

As with amniocentesis and chorionic villus sampling, there are risks associated with different types of fertility curtailment, and these are obviously relevant to decisions regarding their use. The term risk may also apply to the possibility of producing an unwanted outcome or a tragic outcome even in cases where parenthood is much desired. Consider the following cases as examples of both kinds of risk.

CASE I: FETAL REDUCTION TO AVOID TWINS.

A woman who had been undergoing infertility treatment for two years became pregnant after taking perganol. She had been told that this drug might cause multiple gestation. At eight weeks' gestation, ultrasound confirmed the presence in utero of two fetuses, both of which appeared healthy. One week later, the woman asked her physician to reduce the pregnancy to a single fetus. Although the patient was informed that the procedure for doing this involved some risk for both fetuses, she persisted in her request for fetal reduction, indicating that if this could not be done, she would seek abortion of both fetuses and "try again" for another pregnancy.

CASE II: FETAL REDUCTION TO PROMOTE LIVE BIRTH.

After four years of infertility treatment, a 35-year-old married patient became pregnant after in vitro fertilization and embryo transfer. In order to optimize the chance of a successful pregnancy, that is, a pregnancy producing one or more live births, it is preferable to transfer no more than three or four embryos to the woman's uterus. In this case, seven ova had been fertilized in vitro, and the couple, who were opposed to the alternatives of discarding or freezing the extra embryos, asked that all seven be transferred to the woman's reproductive tract. Nine weeks after this was done, ultrasound confirmed that all seven were developing in utero. It was extremely improbable that any would survive unless the number was reduced. The couple wondered whether they should request fetal reduction in order to preserve the lives of at least one or several fetuses.

The ethical dilemmas posed by these cases could not have occurred fifteen years ago. Although we then had fertility drugs, we could not have diagnosed multiple gestation as early as now, and in vitro fertilization with embryo transfer was unheard of. Nor would we have had available the technical possibility of reducing a multiple gestation in utero. The risk involved in the first case is mainly one of denying to the pregnant woman her wish of having a singleton rather than twins. In most cases, a twin pregnancy introduces relatively little added medical risk to the woman or her fetuses. Some infertile women are especially anxious about their ability to be the "perfect mother" they aspire to be, and this anxiety may well be exacerbated when the woman has conceived twins. Sensitivity to such anxiety is morally appropriate, but it does not imply that the selective abortion is morally justified.

In the second case, however, the moral argument for requesting fetal reduction[35] is to save lives rather than to lose them. If the only way of having any fetuses survive is to remove some of them from the uterus, the moral reasons for doing so are more compelling than in the first case. In fact, one could say that it is more pro-life to reduce the multiple gestation in such a situation than to permit the continuation of a pregnancy in which all the fetuses would otherwise be destined to die in utero.[36]

Guideline 2 suggests that life, human as well as nonhuman, is an a priori value to be considered in moral decision making. To promote that value while risking its loss is defensible from an egalitarian standpoint, regardless of whether human fetuses or embryos count as persons. In a developmental view of the world and human nature, the burden of overriding a priori values becomes greater as an organism develops, and is always greater for potential than for actual lives.

As the last two chapters illustrate, the ethical issues raised by humankind's increasing ability to curtail fertility are unavoidably complex and controversial. In this chapter I have also examined issues that arise because of techniques for prenatal diagnosis and fertility enhancement. Next I focus on the enhancement aspect, which raises questions beyond those considered here.

NOTES

1. For example, see Pope Pius XII, "Ordinary Means to Prolong Life," in *Medical Ethics,* ed. Kevin D. O'Rourke and Philip Boyle (St. Louis: The Catholic Health Association, 1989), 207–08. The President's Commission (for the Study of Ethical Problems in Medicine and Biomedical and Behavioral Research) has rejected this distinction because it is "both difficult and controversial and can lead to inconsistent results." Most biomedical ethicists concur with this assessment. See President's Commission, *Deciding to Forego Life-Sustaining Treatment* (Washington, D.C.: U.S. Government Printing Office, March 1983), 82–90.

2. See O'Rourke and Boyle, eds., 103; and President's Commission, 80.

3. Michael Bayles, *Reproductive Ethics* (Englewood Cliffs, New Jersey: Prentice Hall, Inc. 1984), 8. Bayles elaborates his objections to the natural law argument on pp. 8–9.

4. See Pope Paul VI, *Humanae Vitae,* 1968; and Joseph Cardinal Ratzinger, Congregation for the Doctrine of the Faith, *Instruction on Respect for Human Life in Its Origin and on the Dignity of Procreation,* 1987.

5. About one-half of all conceptions are not clinically recognized because they are sponta-neously aborted before pregnancy is detectable. See Steven G. Gabbe, Jennifer R. Niebyl, and Joe Leigh Simpson, eds., *Obstetrics: Normal and Problem Pregnancies* (New York: Churchill Living-stone, 1986), 651. After clinical recognition, the loss rate is 12–15%, declining to 1% by the six-teenth week of gestation. See Joe Leigh Simpson, "Incidence and Timing of Pregnancy Losses," *American Journal of Medical Genetics* 35 (1990): 165–73.

6. Benedict M. Ashley and Kevin D. O'Rourke, *Health Care Ethics* (St. Louis: Catholic Health Association, 1982), 272–76.

7. Note that the existent "conditions" to which I refer include actual living beings, whether these be gametes, embryos, or fetuses. Thus, guideline 2 is also involved. Some authors seem to treat these entities as nonexistent and/or nonliving, arguing that we have no obligations to merely possible beings. See, for example, Mary Anne Warren, "Do Potential People Have Moral Rights?" in *Obliga-tions to Future Generations* ed. R. I. Sikora and Brian Barry (Philadelphia: Temple University Press, 1978), 14–30.

8. See Thomas Aquinas, *Summa Theologiae,* Prima Pars (Matriti: Biblioteca de Autores Cris-tianos, 1955), Ques. 96, art. 3. Referring to whether "men" would have been "equal" in the state of innocence, Aquinas uses the term "aequales." He quotes Augustine's use of the terms "parium" and "disparium" for "equal" and "unequal," and also uses "disparitas" for "inequality."

9. See Aquinas, Ques. 92, art. 1, which concludes the passage as follows: "Nor is inequality ('inaequalitas') among men excluded by the state of innocence."

10. Herbert J. Grossman, ed., *Classification in Mental Retardation* (Washington, D.C.: American Association on Mental Deficiency, 1983), 11.

11. Allen E. Buchanan and Dan W. Brock, *Deciding for Others* (Cambridge, England: Cam-bridge University Press, 1989), 18.

12. Consider, for example, a pregnant schizophrenic adult whose father obtained guardianship with the intent of securing an abortion for her. The young woman had declared her desire to continue the pregnancy because she considered abortion "immoral." Although a psychiatrist maintained that the woman's statement was rational, and that she was competent to make it, a judge ruled that the guardian's decision should be upheld. See articles based on this incident, with some details changed, in Mary B. Mahowald and Virginia Abernethy, "When a Mentally Ill Woman Refuses Abortion," *Hastings Center Report* 15, no. 2 (April 1985): 22–23.

13. *Williams Obstetrics* defines surgical sterilization as a form of contraception (18th ed., ed. E. Gary Cunningham, Paul C. MacDonald, and Norman F. Gant [Norwalk, Connecticut: Appleton & Lange, 1989], 936). *Stedman's Medical Dictionary* defines *sterility* as "the inability to produce prog-eny" (23d ed., 1334).

14. Jane C. Perrin, Carolyn Sands, Dorris E. Tinker, Bernadette C. Dominguez, Janet T. Dingle, and Marianna J. Thomas, "A Considered Approach to Sterilization of Mentally Retarded Youth," *American Journal of Diseases of Children* 130 (March 1976): 288–90.

15. The term *retardation* connotes a slower rate of development, which suggests that at least some retarded persons may *in time* achieve a level of development that allows for competent parenting.

16. Two other categories of persons may be discriminated against regarding sterilization: those who are quite young (e.g., those in their early twenties or younger) and those who have never had children. Physicians are sometimes loathe to perform sterilizations in these cases on the grounds that patients may later wish to have children. An egalitarian perspective argues for providing sterilization to such individuals so long as their request is adequately informed and free.

17. See, for example, Robert Weir, *Selective Nontreatment of Handicapped Newborns: Moral Dilemmas in Neonatal Medicine* (New York: Oxford University Press, 1984), 5–16; and Laila Wil-

liamson, "Infanticide: An Anthropological Analysis," in *Infanticide and the Value of Life,* ed. Marvin Kohl (Buffalo, New York: Prometheus Books, Inc., 1978), 61–73.

18. Among contemporary ethicists, even those who support infanticide in some circumstances tend to oppose it in these circumstances (e.g., Weir, 273), or their rationale for its permissibility prescinds from discriminatory criteria and applies to all newborns (e.g., Michael Tooley, "In Defense of Abortion and Infanticide," in *The Problem of Abortion,* ed. Joel Feinberg [Belmont, California: Wadsworth Publishing Company, 1984], 120–34).

19. For example, Nancy E. Williamson, *Sons or Daughters: A Cross-Cultural Survey of Parental Preferences* (Newbury Park, California: Sage Publications, 1976). In the United States and in Europe, the preference for sons is weaker than in other parts of the world, but it is still predominant. See Mary Anne Warren, *Gendercide* (Totowa, New Jersey: Rowman and Allanheld, 1985), 17–18.

20. Christine Overall, *Ethics and Human Reproduction* (Boston: Allen and Unwin, 1987), 17–19; and Warren, *Gendercide,* 6–12.

21. See Warren, *Gendercide,* 11.

22. Bayles, 34–37.

23. Overall, 26–27.

24. Overall, 27.

25. Overall, personal communication to the author, September 25, 1989.

26. Warren suggests more compelling circumstances than I have here portrayed—for example, societies that are extremely oppressive to women, where inheritance or other economic rights and privileges are limited to males. See Warren, *Gendercide,* 85.

27. Warren, *Gendercide,* 160–76; and Overall, 29–33.

28. Other ways of increasing the number of men in the population include more medical and social attention to the most frequent causes of death among men, for example, homicide and stress-related work roles.

29. William D. Althus, "Birth Order and Its Sequelae," *Science* 151 (Jan. 7, 1966): 44; and Overall, 29.

30. Warren, *Gendercide,* 108–58; and Overall, 29–33.

31. Amitai Etzioni, "Sex Control, Science and Society," *Science* 161 (Sept. 13, 1968): 1109.

32. Aubrey Milunsky, *Genetic Disorders and the Fetus,* 2d ed. (New York: Plenum Press, 1986), 2.

33. Laurie Abraham, "Sex Selection through Technology? Geneticists Weigh Ethics, Patient Rights," *American Medical News* (Feb. 24, 1989): 2, 22.

34. See Ernest B. Hook, "Differences between Rates of Trisomy 21 (Down Syndrome) and Other Chromosomal Abnormalities Diagnosed in Live Births and in Cells Cultured after Second-Trimester Amniocentesis—Suggested Explanations and Implications for Genetic Counseling and Program Planning," in *Sex Differentiation and Chromosomal Abnormalities,* ed. R. L. Summit and D. Bergsma, Birth Defects Original Article Series, vol. 14, no. 6C (New York: Alan R. Liss, 1978), 247–49. Concerning risk of the procedure, see Joe L. Simpson and James L. Mills "Methodologic Problems in Determining Fetal Loss Rates," in *Chorionic Villus Sampling: Fetal Diagnosis of Genetic Diseases in the First Trimester,* ed. Bruno Brambati, Guiseppe Simone, and Sergio Fabro, Clinical and Biochemical Analysis Series (New York: Marcel Dekker, 1986), 227–52.

35. The term *fetal reduction* more accurately reflects the practice discussed here than other terms that have been used, such as "selective abortion," "pregnancy reduction," or "selective termination of pregnancy." In the cases described, the women wanted to remain pregnant, and they were not selecting particular fetuses to die or survive, but only that the number of fetuses gestating be reduced. "Pregnancy enhancement" has also been proposed, but this seems to exaggerate the positive intent of the procedure. For an excellent account of clinical as well as ethical aspects of this issue, see Mark I. Evans, John C. Fletcher, Ivan E. Zador, Burritt W. Newton, Mary Helen Quigg, and Curtis D. Struyk, "Selective First-Trimester Termination in Octuplet and Quadruplet Pregnancies: Clinical and Ethical Issues," *Obstetrics and Gynecology* 71 (March 1988): 289–96.

36. However, the usual method through which fetal reduction is performed is not removal of some fetuses from the uterus, but ultrasound-guided injection of a lethal substance (potassium chloride) into the fetal heart. Pro-life advocates are unlikely to support so direct a means of terminating some fetuses, even when the procedure is intended to save the lives of other fetuses.

6

Fertility Enhancement and the Right to Have a Baby

Discussions of reproductive issues are replete with references to rights: rights of women, men, children, fetuses or embryos, and rights of others in society. The "right to have a baby" is often invoked as grounds for providing women or couples with technical or social means of fertility enhancement or assisted reproduction. In this chapter, therefore, I begin with an examination of the concept of rights in order to set the context for addressing issues that arise with regard to artificial insemination, egg donor programs, and surrogate gestation.

Rights and Rights Language

Rights language is especially pervasive in American society because of its stress on individual liberty. Although deontological theories rely on rights more than other ethical theories do, utilitarian, contractarian, and even natural law theories may also be defended or explained in terms of rights. In addition, feminist arguments on reproductive issues are sometimes based on rights. As pointed out in Chapter 1, however, rights language may not be congenial to a "feminine" model of moral reasoning. In an effort to develop a concept of equality compatible with an ethic of care as well as justice, I avoided the term in developing guidelines that stress responsibilities rather than rights. Discussions of rights are so imbedded in our culture, however, that it seems impossible to address issues without that language. As Joan Callahan and Patricia Smith observe, it is not "rights" as such that present a problem, but a tendency to view the protection of rights as "the whole of the moral story." While insisting on the moral importance of rights, they propose that a *"preoccupation* with rights [be] abandoned in favor of a preoccupation with responsiveness to others, that urges the provision of care, preventing of harm, and maintaining relationships, including creating and maintaining the communities and social arrangements that make meaningful and morally appropriate relationships possible and natural for *all* persons."[1]

Such an interpretation is supported by explanations of rights as necessarily involving mutual or relation-based responsibilities. Stanley Benn, for example, defines a right as "a duty looked at from the standpoint of the other term in the same relationship."[2] Although the terms *duty, obligation,* and *responsibility* are often used interchangeably, I prefer the term responsibility because it is more suggestive (through its association with "response") of interpersonal relationships as grounds for moral judgments. Responsibilities arise not only because of impartially defined duties or obligations, but also because of ties formed by relationships or attachments among persons.

Rights may broadly be defined as justified or justifiable claims or entitlements. In applying the concept to concrete circumstances, the immediate questions that arise are: Claims or entitlements to what, for whom, and from whom? In the context of fertility enhancement, the first question is often answered as "the right to have a baby." Generally, the answer to the second question (a right for whom?) is a woman or couple who wish to have a child related to them genetically or gestationally. The answer to the third question (a right from whom?) may be anyone whose involvement is necessary in order to exercise the right to have a baby—for example, a sexual partner, gamete donor, technician or clinician, or possibly an insurance company or government program that funds infertility treatment. At a more basic level, rights may be derived from the "law of nature," "consent of the governed," or from God.[3]

Unfortunately, when rights language is used in common parlance, and often when it is used in formal argument, different types of rights are not distinguished. Among the possible distinctions are human, natural, moral and legal, absolute or prima facie, and positive and negative rights. For each of these, there may be correlative responsibilities or duties.[4]

Human rights involve the basic needs or interests of all human beings.[5] They are thus associated with the principle of justice or a concept of equality: all human beings equally deserve to have their fundamental needs or interests fulfilled. In light of their applicability to all humans, human rights may also be construed as natural rights.[6] Within the context of natural law theory, both are identified with moral rights because what is natural to human beings is defined as moral. But moral rights generally include rights based on voluntary action as well as those based on human need or interest.

Moral rights are also broader than legal rights. Jeremy Bentham and John

Austin view moral rights as based on public opinion rather than law or statute.[7] In contrast, legal rights assume the existence of a legal system whose rules govern their exercise, often restricting their application to certain groups. Although they often articulate moral or human rights, legal rights are formally stipulated by social conventions enacted through legislative or constitutional processes. Moral rights extend to informal as well as formal agreements or commitments among persons.

Whether or not the right to have a baby is a human or natural right depends on whether having a child is a basic need or interest of those who desire parenthood. Although social stigmas and personal disappointment sometimes accompany childlessness, becoming a parent is not necessary to survival, and therefore not a fundamental human need. Parenthood is in the *interest* of some individuals or couples, but whether this interest is so fundamental that it counts as a human or natural right remains questionable. Other interests, such as professional or economic success, clearly do not count as human or natural rights of individuals. Nonetheless, the right to have a baby is a claim that social convention both informally and formally supports, so long as its exercise does not impugn the rights of others. As a moral right, this right may appeal to obligations of beneficence or charity as its correlate.[8]

The distinction between absolute and prima facie duties has a parallel in discussions of rights. An absolute right is exceptionless in the demand it imposes on others. Although opposite sides of the abortion debate seem to regard their values in this way, it is difficult to see how either the right to life or the right to choose is truly absolute. A right to life does not obligate others to risk their lives in my behalf, and a right to choose is surely limited by the obligation of nonmaleficence. Yet either of these rights seems more compelling than a right to have a baby,[9] and a right to reproduce without another's assistance seems more compelling than a right to fertility enhancement or assisted reproduction. If these rights are not absolute, but may be overridden while recognizing their validity, they are prima facie rights.

The distinction between positive and negative rights, corresponding with a distinction between positive and negative duties, is particularly pertinent to issues of fertility enhancement. A negative right implies another's responsibility not to interfere with the expression of that right. A positive right implies the other's responsibility to support or facilitate its expression. If an individual's or couple's right to have a baby is a positive right, practitioners have an obligation to provide infertility treatment to those who require and desire it; if the right to have a baby is a negative right, such assistance is not obligatory. It thus seems clear that the right to have a baby is at most a prima facie negative right of individuals. Health professionals are not obligated to respond affirmatively to every individual or couple requesting reproductive assistance.

A Right to Have a Baby?

What does it mean, then, to assert one's right to have a baby? Ordinarily, when the claim is made by or on behalf of an infertile person or couple, it means the right to procreate by becoming a biological (genetic or gestational) parent. It may also mean the moral or legal right to become a social parent, through adoption or through one's committed relationship to a biological parent. Using Michael Bayles' s terminology, we may therefore distinguish between the right to beget, the right to bear, and the right to rear a child. Presumably, the exercise of these rights is based on a desire on the part of a prospective parent. Bayles argues that the desire "to beget for its own sake" is irrational because fulfillment of the desire "will not contribute to one's life experiences."[10] The desire may become rational if begetting is seen as a means to fulfillment of other (rational) desires, such as the desire to rear a child. But the "right to have a baby" deserves analysis on another level, namely, critique of its key terms, *have* and *baby*.

To have has a different meaning when it applies to persons or subjects rather than to objects. To have an object means to own or possess it, and this implies a prima facie right to dispose of the object as one wishes. To own something is to treat it as property, which has no rights of its own. When we use the verb "have" in referring to persons (as in "I have a friend," or "I have a husband"), we refer to the relationship itself rather than to a person. We "have," for example, a relationship of friendship or marriage to a particular person. The "having" of such interpersonal relationships does not imply ownership. Typically, it implies responsibility more than rights. As Antoine de Saint-Exupéry observed in *The Little Prince,* you become "responsible for what you have tamed,"[11] that is, for those with whom you have established ties. If having a baby means having a parental relationship with someone, it fulfills this view of responsibility toward another.

The right to *have* a baby may not be literally equivalent to the right *to* a son or daughter, but rather a right to "the having" or reproducing of one. In that context the claim to a biological tie with one's progeny is affirmed, and this may be expressed through gamete donation or surrogacy as well as through ordinary means of reproduction. One can then "have" a baby without a commitment to its nurturance. Obviously, this raises problems regarding the baby's rights. But what do we mean by the term *baby* in discussing the right to have one? The simplest and most direct answer is to say we mean a future baby or child, that is, a newborn. Because human gametes, embryos, and fetuses are not yet babies and may never be, the assertion of a right to have any of them is not equivalent to the right to have a baby. Moreover, if embryos and fetuses are not persons, the relationship signified by "having" is probably closer to a property relationship than an interpersonal relationship. Gametes are more like property than embryos or fetuses because they are formed entirely from within the "owner's" body, and are, in fact, a part of that body. If I may donate a kidney or blood because it is my property, I might also donate ova for the same reason. Legal restrictions on the selling of body parts do not imply that they are not owned by the person whose body contains them.

So long as they do not harm others, human beings have a right to do what they will with their own bodies. Nonetheless, American society has been loathe to treat human tissue or organs as mere property, that is, as subject to commerce. When the tissue involved is human ova or sperm, the term *donation* is used to characterize the exchange of money for tissue. As we will see subsequently, however, that characterization is misleading.

Artificial Insemination by "Donors"

Natural law theory opposes artificial insemination of women even when the semen donor is the woman's husband. In Roman Catholic teaching, this prohibition is based on the requisite of conformity with the natural law paradigm of sexual intercourse between married partners as crucial to the morality of reproduction.[12] Those who subscribe to a psychosomatic rather than physicalist view of human nature reject this interpretation of natural law ethics. Catholic theologian Lisa Soule Cahill, for example, argues in support of homologous but against heterologous reproductive technologies.[13] Homologous techniques are those that involve only the married couple as biological parents; heterologous methods involve third parties. "Artificial" insemination of a woman with her husband's sperm (AIH) is homologous; artificial insemination with another man's sperm (AID or TID) is heterologous.[14] Within the broader interpretation of natural law, the rationale for permitting artificial insemination of a woman with her husband's sperm is its fulfillment of the couple's commitment to have children within the context of the marital relationship as a whole, and not in the context of specific acts of sexual intercourse.

Cahill and others argue against artificial insemination with donor sperm on grounds that it violates the commitment to procreate only with one's spouse, which is essential to marriage.[15] In general, however, donor insemination presents more controversial issues with regard to its anonymity, and by the secrecy generally observed by couples who practice it. More recently, controversy has centered on the question of whether single women should be inseminated with donor sperm.[16] Each of these issues is assessable from an egalitarian perspective.

"Protection of donors" is the reason often given in support of their right to anonymity.[17] Donor recruitment, it is thought, would be extremely difficult if donors' identities were known or knowable to the individuals in whose behalf their semen is utilized. Assurance of anonymity, however, along with the possible satisfactions of benefitting an infertile couple, or of passing on one's genetic endowment, has apparently not been considered an adequate incentive for semen donation in the United States. Financial remuneration undoubtedly provides persuasive incentive to those who might not otherwise "donate."[18]

Supposedly, semen donors have a right to anonymity as protection from possible legal, financial, and emotional claims on the part of future children to whom they are genetically related. Clinicians are generally so supportive of this prima facie right that they themselves remain unaware of donors' identities, keep incomplete records regarding them, and sometimes insist that the infertile husband's name be recorded on birth certificates of children born by donor insemination.[19]

The "protection of donors" rationale, with its utilitarian argument about maintaining a sufficient number of donors, ignores the interests of others who deserve protection when donor insemination is employed. Foremost among those whose interests may be compromised by secrecy are future offspring, who, on analogy with the rights of adopted children, have a right to know the facts regarding their origin and genetic parents.[20] The argument for the child's right to know is strongest when the information is medically relevant to her interests. As genetic knowledge escalates, so do the possibilities for genetic therapies. The relevant information can usually be provided without identifying the donor. In some cases, however, the information cannot be obtained without tracking down the donor, thus jeopardizing his anonymity.

In an age of AIDS (acquired immune deficiency syndrome), the argument regarding medically relevant information also applies to the woman who may be impregnated with donor semen, and to her husband. Both might be infected by viruses undetectable at the time of semen donation. Careful screening mechanisms have been set up at infertility clinics to eliminate that possibility and to test for other diseases also.[21]

Even if donor anonymity imposed no risk on the couple or their potential offspring, the question of whether it is truly supportive of the donor's own interests and of society's interests remains relevant. Anonymity is intended to assure the donor that he has no responsibility for the potential offspring or its parents. Minimally, however, he is responsible for honest and full disclosure of risk factors associated with his "donation." He is also responsible for facilitating the reproduction of another human being, whose existence is not always a moral desideratum. Donor insemination implies that genetic fatherhood is separable from social fatherhood, and anonymity suggests the appropriateness of maintaining that separation. It thus parallels the questionable view that procreation is not importantly related to sexual expression.[22] Society's interests may not be well served by reinforcing the separation between the two.

An egalitarian framework calls for balancing the rights and responsibilities of all of those affected by decisions regarding donor insemination: women, their partners, semen donors, potential children, and even the larger community. Possibly, laws could be introduced that would ensure some protection of donors' rights while also protecting others. But clearly the position that donor anonymity is an absolute requirement for donor insemination must be rejected. Interestingly, few have been as concerned about anonymity of ova donors as about that of semen donors. To treat men and women equally, however, the same standard of anonymity or confidentiality should apply to gamete donors of either sex.

Beyond the issue of identifying donors, couples prevalently tend to practice secrecy regarding donor insemination, and this distinguishes it from other methods of assisted reproduction. Judith Lasker and Susan Borg suggest that the main reason for this tendency is that donor insemination is used exclusively to solve the problem of male infertility, and "male infertility is a condition with greater stigma attached to it than female infertility."[23] While this observation may be correct with regard to couples, it obviously does not apply to situations where single women seek donor insemination. (I deal with this point shortly.) The secrecy practiced by those involved with donor insemination extends beyond donor anonymity to withholding information so as to mislead others regarding the fact of donor insemination. The others are relatives, friends, offspring, and sometimes even the couple themselves.[24] Some physicians continue to advocate not telling husbands of their infertility or that donor semen is going to be used, and most encourage couples to keep donor insemination a secret from everyone. Sometimes the semen of the infertile husband is mixed with that of the donor so as to provide the couple with some basis for believing that the child produced through the artificial insemination is genetically related to the husband.[25]

Parental reasons for secrecy regarding donor insemination are commonly framed in terms of the interests of the child, as well as those of the sperm donor and the legal father, usually the husband of the woman who is inseminated.[26] It is assumed that children might be teased by their peers if the latter know they were conceived through donor insemination. By itself this possibility is surely not an adequate reason for foregoing donor insemination, because children may be teased by peers for many inappropriate reasons (and may be complimented for inappropriate reasons also). Nonetheless, parents fear disapproval on the part of others, including family members and the children themselves, if conception by donor insemination is revealed.

Parents seldom offer the explanation of secrecy that may figure centrally in their own motivation—protection of the legal father from the embarrassment of infertility.[27] Although infertility is undoubtedly stressful for women, Lasker and Borg claim that men "are more likely than women to want to keep donor insemination secret, and women are likely to cover up their husband's infertility by taking on the blame for reproductive problems."[28] The cover-up achievable with donor insemination is not achievable when other reproductive interventions or adoption is practiced, but it inevitably exacts a greater psychological toll. For some at least, there is the strain of "living the lie" that they are biological parents who have reproduced their offspring in the "normal" way. Despite such stresses, the cover-up is maintained to protect the man from what is apparently the greater harm of "a social stigma."[29]

Whether men in fact suffer more from the "stigma" of infertility remains an open question. Reasons offered in support of this point are the following: (1) unlike women, men tend to judge themselves by criteria of performance at work, in bed, and in producing offspring; (2) since the woman's ongoing role is crucial to produce offspring while the husband's is not, the traditional view of male dominance over their wives is thereby threatened; (3) donor insemination poses an additional threat to male dominance in society by barring the possibility of "patriarchal descent."[30] On examination, each of these reasons can be rejected as inegalitarian. Criteria of performance, after all, are no more (or less) legitimate for men than for women, and the reproductive performance exacted of women is surely a tougher requirement than that exacted of men. Male dominance, or for that matter female dominance, whether in families or society is clearly opposed to an egalitarian approach to these institutions. Sex differences thus provide justification for different but not unequal treatment. Should differences in marital status be treated similarly?

Possibly the issue of donor insemination for single women is the most controversial aspect of donor insemination. Consider, for example, the following case:

A 35-year-old physician at a leading university hospital was not interested in marriage and had not dated for several years. Attempts to adopt a child had been unsuccessful. Although she was aware that health problems of her own would be exacerbated by pregnancy, she requested donor insemination from her gynecologist. "More than anything else, I want to be a mother," she maintained, "and time is running out on my biological clock."

This case illustrates the fact that single women who request donor insemination tend to be well educated and economically independent.[31] They are often professional women, with ready access to the medical establishment. Given the circumstances of this case, it is surprising that the physician had not been able to arrange for a private adoption. However, the fact that she had tried to adopt a child suggests that she was not fixated on having a genetically related offspring.

If donor insemination is morally justified in cases such as the above, all of those affected by the decision should be treated fairly. Oddly enough, the one who might have been least fairly treated in this case was the patient herself, for whom pregnancy involved a special health risk. Nonetheless, her recognition and acceptance of the risk as necessary means to achievement of her goal made it possible for the physician to support her decision. If the medical risk the woman was ready to assume was so high that physician assistance would be tantamount to malpractice, it would have been wrong to perform the donor insemination.[32] Although the risk was not that severe, the physician still had a right to refuse the service because it was not medically necessary for the patient, or for reasons relevant to the possible offspring. If, for example, the physician considered it wrong to deliberately produce a child who would have only one social parent, he could deny the service on that basis. Many doctors and some institutions routinely deny reproductive technologies to unmarried individuals.[33] However, if their denial is based solely on *prejudice* against single parenthood, the refusal of treatment is morally unjustified.

The exercise of the right to deny medical services is based on the autonomy of the practitioner rather than his own reasons for the denial. Obviously, autonomy and the reasons for its exercise are related, such that irrational or counterfactual reasons may render autonomy impossible or invalid. A parallel issue for refusal to provide donor insemination is that of abortion, which practitioners are neither legally nor morally bound to perform. Although some feminists disagree, so long as reproductive technologies or abortions are not necessary for women's health, physicians are not professionally obligated to provide them. The same standard applies to other medical treatments.

Whether or not infertility should be treated as a disease is a question that merits fuller consideration than I can give here.[34] The very fact that the question is raised suggests that conditions that are unquestionably diseases are more deserving of medical interventions. As medical technology introduces more and better means of satisfying people's desires to improve their lot in life, requests for interventions unrelated to health multiply, mainly from those who are affluent enough to pay for them.[35] It is ironic that infertility be considered a disease subject to treatment by medical personnel, while fertility or pregnancy is also a medicalized event. Both of the latter conditions, after all, are normal and healthy for most women, for part of their lives. As the stigma of single parenthood has lessened in society, requests for donor insemination by single women have increased.[36] To some the issue becomes especially troublesome if the woman requesting donor insemination is an acknowledged lesbian. Among the questions raised for the endocrinologist who has elected to treat cases of infertility is the following: What right or competence do I have to refuse medical assistance to someone who wants to be a parent?

The endocrinologist who was involved in the case previously cited has a defensible but controversial practical response to this issue.[37] First of all, he acknowledges that he is responsible for making it possible for a woman to bear a child when she would not otherwise do so. He therefore agrees, albeit reluctantly, that he has some responsibility for the outcome, and should not provide assistance without considering the effect on the potential child. He is aware that children born to married couples are often raised for much of their lives by only one parent, and that one parent may be as effective as two. Reasoning that his own training has not equipped him to assess competence for parenthood, he notes that others in society are credited with such expertise, for example, adoption specialists. Accordingly, he asks persons who request donor insemination, in circumstances where society tends to question the appropriateness of parenthood, to consult such a specialist to obtain an evaluation of competence for parenthood. If the person refuses or the evaluation is negative, he declines to perform the procedure. If the evaluation is positive, he proceeds with donor insemination.

It may be objected that the logic of the endocrinologist's argument extends to requests for AIH as well as AID, and in general to married couples who seek access to reproductive technologies. The physician claims he has no right to deny infertility treatment to married couples because society generally supports their right to reproduce. However, so long as infertility is not a disease demanding treatment, he is not obligated to provide artificial insemination to anyone, married or single, homosexual or heterosexual. Women may inseminate themselves, do so with the assistance of nonmedical friends, or achieve pregnancy through casual sexual intercourse. Obviously, physicians have no control over such circumstances; they are responsible only for cases where their involvement makes a difference. The endocrinologist's practice fulfills that responsibility.

Suppose, however, the same endocrinologist were approached by a single man who wanted very much to have his own genetically related child. Suppose, further, that the man had a friend who was willing to be artificially inseminated with his sperm, to carry a resultant pregnancy to term, and to deliver a baby that she would then give to the man to nurture and raise. Would competence for parenthood be an adequate criterion for determining whether to proceed with the insemination? Would egalitarian considerations require a parallel policy for the endocrinologist? In addressing these questions, we need to examine whether in fact the two situations are parallel, whether egg donation or surrogate gestation by women for men who wish to be genetic fathers is comparable to sperm donation for women who wish to be biological mothers. Although the two types of donation by women often occur together, I treat them separately because the differences between them are morally relevant.

Egg "Donation"

The process through which eggs are retrieved from women is clearly more difficult, costly, and riskier than the process through which sperm are obtained for artificial insemination. Sperm are provided through the nonclinical (and normally pleasurable) practice of masturbation, whereas eggs are obtained after drug induced hyperstimulation of the ovaries, followed by aspiration through the vagina.[38] The disparity in contribution of egg and sperm "donors" is acknowledged in the different "compensations" they receive: about $50 for sperm "donors," and $1500 or more for egg "donors."[39]

Both types of donation provide a means of overcoming infertility, or of avoiding transmission of a genetic defect. In egg donation, however, the recipient of the egg may also be the gestational mother. Unlike the infertile husband in donor insemination, the nongenetic (gestational) parent is biologically related to her offspring. In the ordinary process of reproduction, women contribute more, at greater risk and cost to themselves, for a much longer period than do men. Through egg donation and surrogate gestation, two aspects of biological motherhood are divided, but either aspect taken singly is still more demanding than biological (genetic) fatherhood. If equality of rights regarding offspring is based on the extent of contribution to their reproduction, the gestational mother has the most compelling claim, followed by the genetic mother, followed by the genetic father. As already indicated, however, having rights does not imply that they must be exercised. In fact, the exercise of rights relevant to reproduction may be freely forfeited, donated or sold, as occurs in donor insemination, egg donation, and surrogate gestation. Neither do the rights of gamete donors or gestational mothers alone suffice for determining the fate of embryos, fetuses, or infants when their fate is disputed by biological and social parents. I return to this point in the next section, but the concept of "donation" of gametes needs to be examined first.

A donation is a gift, and a gift is something of worth that is freely offered to another. Ordinarily, a gift is worth something to the giver also, or the process of obtaining the gift costs the giver something. It thus represents the giver. A gift is not equivalent to what is bought or sold, or exchanged for something of agreed value. When we discard items that are of no use to us, those who receive them are hardly receiving a gift. In fact, they may be giving us something by relieving us of objects we do not want. When human gametes are taken from would-be donors, the gametes themselves are of no intended "use" to the "donors" who are paid or compensated for inconvenience, pain, or risk that their "donation" entails. These aspects make the relationship one of exchange or commerce rather than gift. Accordingly, the term *donation* is misleading in that context.

In the context of ova donation, the term *compensation* is also misleading. Compensation is an attempt to reduce inequality between individuals or groups. The rationale for compensation of ovum donors thus involves recognition of the disparity between the recipient—who lacks ova, and the donor—who has ova and must undergo medical procedures in order to provide them for another. When (extra) ova are retrieved in conjunction with in vitro fertilization, there is no added risk for the donor. Ova may also be retrieved in conjunction with medical procedures such as tubal ligation, with little risk or inconvenience to the donor. Compensation has not generally been provided in these cases, but it has been practiced when the donor would not otherwise undergo a medical procedure. At least one center in the United States has elicited egg donation through public announcement.[40] Allegedly, the announcement appealed to altruistic motives on the part of donors, that is, their desire to help an infertile couple. But provision of compensation while still calling the program egg donation presents a confusing message. The socioeconomic status of possible donors clearly influences whether participation in the program is undertaken. For example, a college student who had enrolled in an egg donor program made the following remark: "I would never go through this (drug-induced hyperstimulation and ova retrieval) if I weren't being paid $1000. After all, I'm a poor student."[41] Her comment illustrates the obvious point that compensation for some is enticement for others. Although society generally rejects the sale of human tissue or organs, the amount offered for ova retrieval apart from other medical procedures is clearly more than compensation. Thus "ova donation," as we know it, plainly amounts to "eggs for sale."

While the concept of commercial trafficking in human organs or tissue is worrisome to many, it is not without valid supportive arguments. Most of these appeal to the saliency of mutual informed consent or agreement between buyers and sellers, and consistency between this practice and other means of capitalizing on one's body in a capitalistic society. When the concept extends beyond sperm or ova to embryos and fetuses, however, commercialization becomes more troublesome still. At that point, the exchange has been labelled baby-selling. Whether the label is appropriate is considered in the next section.

Surrogate Gestation and the Case of Baby M

Ordinarily, a woman becomes a mother when her baby is born, that is, when her pregnancy is terminated through delivery (surgically or vaginally) of an infant. If we were to fix on a moment for the initiation of motherhood, it would be that moment at which the umbilical cord is cut so that the child is in fact separate (and not simply separable) from the mother. The man who impregnated the woman is generally regarded as becoming a father at the same moment. Both parents are regarded as parents even if the child is stillborn or later adopted. In the case of donor insemination, however, the fatherhood of the sperm donor is generally not acknowledged.

Women can and do become mothers in other ways as well. Such ways have long been available because social motherhood is separable from biological motherhood. Although there is a parallel distinction between social and biological fatherhood, a patriarchal pattern of the marital relationship has resulted in the tendency to attribute more importance to genetic than to social aspects of fatherhood. Men, in general, are defined by others and themselves in terms of their external roles in society rather than by the familial roles that predominate in perceptions of women.

More recently, biological motherhood has been separated into genetic and gestational components. Because women are capable of nursing children to whom they are neither genetically nor gestationally related, lactational motherhood provides an additional option that is rarely exercised apart from gestational motherhood. There are thus four possible forms of "surrogacy" for women: genetic, gestational, lactational, and social. Most women fulfill all four functions in their expression of motherhood.[42] When the functions are separated, the question of whose maternal rights and responsibilities have priority over others is inescapable. A second unavoidable question follows when conflicts regarding parental rights and responsibilities extend to genetic and social fathers. Before addressing these questions, however, we need to consider the different forms of surrogacy in more detail.

Literally, a *surrogate* is a substitute, that is, someone who stands in place of another. But whom does a surrogate mother stand in place of? If we define the paradigm case of "real" motherhood as one whose ovum is fertilized, who becomes pregnant, gives birth, nurses and raises her child, then a surrogate may fulfill any but not all of these roles. Yet I would not want to argue that "real" motherhood requires fulfillment of all of these functions. Many women do not nurse their children, and others assist in their nurturance, especially fathers. They are nonetheless "really" mothers of their offspring. Adoptive mothers are also "really" mothers, despite the fact that they are not biologically related to their children. They are not mothers in every sense in which they might be, but they are truly mothers. Similarly, genetic and social fathers are really fathers, although neither considered separately is a father in every possible sense.

What makes a woman a surrogate rather than a real mother? Or is the term surrogate inappropriate for multiple maternal roles? Comparable questions may be asked regarding sperm donation or genetic fatherhood and social fatherhood. Are sperm donors really fathers to their genetically related offspring, and are genetically unrelated social fathers real fathers?

On a very basic level, it must be admitted that all of these individuals are parents as well as surrogate parents. To the extent that any of us stands in place of parents toward others' children (for example, as teachers or health care providers), we serve as surrogates, and to the extent that others act parentally toward our children, they serve as surrogates. Thus the parentalism described in Chapter 2 as a moral paradigm for interpersonal and social relations exemplifies a kind of generalized surrogacy. What we are trying to do when we distinguish between surrogate and real parenthood is to determine a priority of parental rights and responsibilities. In keeping with an egalitarian framework, the priority is determinable by considering differences among those who participate in parental arrangements. The differences involve two types of criteria, one subjective and the other objective. The subjective criterion is the intention of the participants; the objective criteria include the duration, cost, necessity, and effectiveness of their involvement in parenting.

If an egalitarian perspective entails respect for the choices of individuals (as in guideline 4), the intention to be a social parent is clearly relevant to its determination of whose parental tie has priority. Ironically, the intention not to be a parent while participating in the process of reproduction has generally characterized those considered surrogate mothers.[43] In contrast, the intention along with the desire to be a parent has characterized those who cannot participate in this process, who depute the surrogate to provide them with a child. In biblical times, the slave girl Agar was deputed by Sarah, the barren wife of Abraham, to bear them a child.[44] Similar "natural" arrangements are possible nowadays,[45] but artificial insemination rather than sexual intercourse is the usual means of achieving conception. According to Lori Andrews, one of the earliest cases of modern surrogacy involved insemination of a "surrogate" by one of the friends (a couple) for whom she had volunteered to bear a child. When the doctors they consulted were unwilling to perform donor insemination,

> the threesome finally handled the matter themselves. Debbie purchased a diabetic syringe at the drug store and filled it with her husband's sperm. Following the directions in a family medical guide, she successfully injected the sperm into her friend Sue, a twenty-four-year-old virgin, and their child was born nine months later.[46]

Because no money was exchanged, and because the surrogate was a friend of the infertile couple, the surrogate could be described as giving rather than selling the baby she bore to the infertile couple. Empirically, she was both the genetic and gestational mother, but her intent excluded her from being the social mother. Apparently, when she chose to be a surrogate she equated social motherhood with real motherhood. In contrast, some surrogates refer to the child they bear as solely the child of the men whose sperm are used for their insemination. Such men, however, also intend to be social fathers. Judge Harvey Sorkow affirmed this sentiment in the Baby M case by comparing the surrogate's position with that of a paid sperm donor who has no right to the offspring produced through his sperm.[47] Because the biological mother had agreed to accept payment for her surrogacy, the biological father alone had a right to Baby M. As Sorkow put it, "At birth the father does not purchase the child. It is his own biological genetically related child. He cannot purchase what is already his."[48]

Sorkow's decision was eventually overturned by the New Jersey Supreme Court, which acknowledged the full maternal rights of the woman who was both genetically and gestationally the child's mother.[49] Given the context in which this decision was made, intention was still crucial. Although the pregnancy had been initiated in fulfillment of a contract, the surrogate's intention had changed during the course of her pregnancy. By the time the baby was born, she regretted the commercial agreement and wanted to be the child's social mother.

Intention also seems to be crucial in cases where the surrogate is not genetically related to the child she gestates. The first reported case of this type occurred in Cleveland in 1985 when a woman requested in vitro fertilization of her egg with her husband's sperm, and embryo transfer of the resultant embryos to the uterus of a friend.[50] When this attempt failed, a paid surrogate was recruited, who eventually gave birth to a child who was genetically related to each of the partners who intended to raise the child. Before the baby was born, a court ruled that she should be considered the offspring of both genetic parents. Contradicting the traditional definition of mother as "a woman who has borne a child,"[51] the motherhood of the woman who gave birth was not acknowledged.

Like most surrogate arrangements, the preceding case did not result in conflict when the time came for the surrogate to give the baby to the infertile couple. In some cases, however, conflict arises because either of the contracting parties reneges on the original agreement. Usually it is the surrogate.[52] When she changes her mind and wants to be a social as well as biological mother, whose right to have the baby should prevail? From a deontological perspective, the contract between the surrogate and the infertile couple should be upheld no matter what consequences may ensue. From an egalitarian perspective, however, the objective differences among competing "parents" should be respected. Mary Beth Whitehead, who had conceived, gestated, given birth, and nursed her newborn might have remained a surrogate if she had completed the terms of her agreement with William and Elizabeth Stern to give them the child she bore in exchange for $10,000. Instead, Whitehead refused the money and took the child, claiming that her biological motherhood justified her doing so. The New Jersey Supreme Court supported her claim, describing the commercial surrogacy arrangement as babyselling.[53]

Gestation obviously involves a duration of commitment greater than that of gamete donation; it also involves risk and pain greater than that experienced through sperm donation. Accordingly, when weighing the parental right of Whitehead against that of William Stern, Whitehead's claim was more compelling so long as she also intended to be the child's social parent. This would be true even if Whitehead had not been both genetic and gestational mother. If the Sterns had *both* been the genetic parents, and Whitehead was only the gestational parent, Whitehead's maternal right would still be more compelling because the risk, duration, necessity and effectiveness of her parental role was clearly more substantial than either or both of the Sterns. I disagree, therefore, with a ruling by the California Court of Appeals, which held that the genetic relationship should take precedence over the gestational tie.[54] The justification for denying maternal rights to a gestator who is not genetically related to the child she bears is dependent on the gestator's intention not to be a social mother, and on her decision not to exercise her right to have a baby to rear. She is then a biological parent who has chosen not to be a social mother.

The right to have a baby need not be exercised. When it is exercised, however, the expression of the right is not without moral limits. It would be wrong, for example, for a man to impregnate a woman against her wishes, or force her to maintain a pregnancy, solely because he wants to be a father. Similarly, a woman's right to have a baby does not impose an obligation on clinicians to provide technology necessary to facilitate that, for example, by prescribing infertility drugs or by performing in vitro fertilization. Neither does the right to have a baby imply the right to sell the baby.

In the Baby M Case, Judge Sorkow rejected the allegation of babyselling by claiming that Whitehead sold her services but not her child. Like men who sell their sperm so that an infertile couple might have a child, Whitehead agreed to accept money for providing an environment in which William Stern's child could develop. This rationale only makes sense if one negates or trivializes Whitehead's biological tie to her offspring. Surrogacy provides a conceptual framework for doing that. The very term surrogacy misleadingly suggests that a woman who conceives, gestates and gives birth to a child is not a mother.[55]

In general, a care model of moral reasoning stresses both social and biological ties between parents and their offspring. Because surrogacy sunders those relationships, it can only be supported by this model if its practice reinforces other basic ties, such as those between the surrogate and other family members. It is possible that the intention harbored by a surrogate who gives her child to an infertile couple epitomizes the nurturant responsibility of motherhood. Just as a woman who allows her child to be adopted may be acting heroically for the child's best interests, a surrogate may be acting similarly. Payment for the act suggests that it is based on self-interest rather than the child's interest, but this is not necessarily the case. It is possible that a woman who accepts $10,000 (or more) for undergoing pregnancy and childbirth in behalf of an infertile couple perceives herself as giving inestimably more to them than she is receiving. If human life is of inestimable value, the gift motif may still be valid in commercial surrogacy.

Nonetheless, most feminists are wary of the possibilities for exploitation of women that surrogacy, even more than egg donation programs, involves.[56] Socioeconomic differences between infertile persons who utilize reproductive "services" and those whose bodies provide such "services" were dramatically illustrated in the Baby M case.[57] Elizabeth Stern was a pediatrician, her husband a biochemist; Whitehead was an unemployed housewife, her husband a sanitation worker. The Sterns had postponed having children while pursuing their doctoral degrees; neither of the Whiteheads was educated beyond high school, they had married young and had children early, their financial status as well as their relationship had been "rocky" from time to time. Mary Beth Whitehead represented an opportunity for the Sterns to have a child genetically related to one of them. Apparently, that prospect was especially important to William Stern because he was an only child whose parents had experienced the holocaust. Although Elizabeth Stern was not infertile, the contract signed by Whitehead indicated that she was.[58] Elizabeth had a mild form of multiple sclerosis, and there had been concern that the disease could be exacerbated by pregnancy. The Sterns might also have been concerned that Elizabeth's age (40) increased the risk of genetic abnormality.

Exploitation occurs in an inegalitarian milieu, that is, one in which differences among individuals prompt evaluative judgments based on irrelevant criteria. Clearly, socioeconomic status is not an adequate guide for settling competing claims about the right to have a baby. Most feminists oppose surrogacy because of its implications regarding exploitation of women, but some liberal feminists support the practice as consistent with the principle of reproductive choice.[59] Those who oppose surrogacy sometimes compare it with prostitution: women's bodies are thus used in the interests of others.[60] The target of their critique is the set of social conditions that make surrogacy a genuine and unavoidable option for individual women. A free-market model may also use the analogy with prostitution, arguing in defense of surrogacy that reproductive choice entails women's right to do as they wish with their bodies and the products of their bodies. Unless infants have a moral status equivalent to that of more mature humans, the baby-selling that surrogacy may involve presents no problem for a free enterprise ethic. From an egalitarian perspective, however, it violates guideline 5, which requires that individuals not be "treated as other than who or what they are." If other human beings may not be bought or sold as commodities, then neither should newborn humans.

If competing rights to have the same baby are settled on the basis of objective differences among claimants, all of whom intend to be social parents, the woman who gestates and gives birth has the most compelling claim. Those who have no biological tie to the child have the least compelling claim. Although genetic mothers and fathers are equally related from a biological point of view, they differ regarding the prospect of child rearing. Even if a genetic mother does not nurse her child, she is more likely to be the crucial parental figure in the child's early development. This circumstance gives greater weight to her parental claim vis-à-vis the genetic father.

But competing rights of intentional social parents are not the only morally relevant factors in determining who should rear the baby. Clearly, the child has rights also, and these may override those of would-be parents, regardless of their biological tie to the child. Minimally the child has a prima facie right to receive the care necessary to survive and grow, but this does not imply that the child has a right to be placed in the parental situation that will provide the most advantages. Although the best interests of the child is the standard commonly invoked in settling custodial disputes between parents, that standard is vague and subject to biased interpretations.[61] By that standard, for example, financial advantages alone would have made the Sterns preferable parents for Baby M because they could buy her opportunities such as a large house, music lessons, and private school education. By that standard those of us who cannot do or be all that is best for our children might lose our custody of them to those who can provide greater material advantages.

To be invoked in an egalitarian framework, the principle of the child's best interests must not be regarded as an absolute criterion for settling questions of parental rights. So long as the Whiteheads had demonstrated adequate parenting skill, they might have been preferable parents for Baby M despite their lower income and educational level (as compared with the Sterns). As it was, however, two factors raised concerns about that adequacy: the possibility that Mary Beth Whitehead had become emotionally unstable,[62] and the fact that the child had already spent most of her infancy in the custody of the Sterns. It was not unreasonable, therefore, for the courts to give custody to the Sterns as a means of protecting the child from further instability. As a negative right, protection from possible harm is more compelling than the positive right to have one's best interests promoted. Had the child spent most of her infancy with the Whiteheads, and had that household provided a stable, although modest, environment, custody could reasonably have been awarded to the Whiteheads.

Summarily, the issue of surrogate motherhood, as the case of Baby M well illustrates, raises a plethora of thorny ethical questions about rights to have a baby. In this chapter I have mainly considered the competing rights of parents, and I have suggested an egalitarian framework for those rights according to the intentions of, and objective differences among, the rights holders. I have only touched on the rights of children, but I deal more with this topic in Chapter 11.

NOTES

1. Joan C. Callahan and Patricia G. Smith, "Liberalism, Communitarianism, and Feminism," in *Liberalism and Community,* ed. Noel Reynolds, Cornelius Murphy, and Robert Moffat (Lewiston, New York: Edwin Mellen Press), forthcoming.

2. Stanley I. Benn, "Rights," in *Encyclopedia of Philosophy,* ed. Paul Edwards (New York: Macmillan Publishing Company, 1967), vol. 7, 196. Benn attributes this view to Wesley Hohfeld. That he endorses it himself is evident in the following statement from Stanley I. Benn and R. S.

Peters, *The Principles of Political Thought* (New York: The Free Press, 1959), 102: "Rights and duty are different names for the same normative relation, according to the point of view from which it is regarded." He insists, nonetheless, that the correlation between rights and duties "is a logical, not a moral or legal relation."

3. Natural law theory supports the view that rights are based on the nature of human beings. This rationale may reflect the religious belief that nature is defined or ordered by God (Eternal Law). Social contract theorists such as Rousseau and Locke construe rights as derived from consent of the governed.

4. See Benn, *Encyclopedia,* 195–99; H. Tristram Engelhardt, *The Foundations of Bioethics* (New York: Oxford University Press, 1986), 94–97; Tom L. Beauchamp and James F. Childress, *Principles of Biomedical Ethics,* 3d ed. (New York: Oxford University Press, 1989), 56–60; and John Finnis, *Natural Law and Natural Rights* (Oxford: Clarendon Press, 1980). Beauchamp also distinguishes between fundamental and derivative rights. See his *Philosophical Ethics* (New York: McGraw-Hill Book Company, 1982), 194–95. Concerning the correlativity thesis, see Joel Feinberg, "The Nature and Value of Rights," *Journal of Value Inquiry* 4 (1970): 243–57. Feinberg rejects the correlativity thesis for some positive rights.

5. As Alan Gewirth states: "We may assume as true by definition, that human rights are rights that all persons have simply insofar as they are human" (in *Human Rights: Essays on Justification and Application* [Chicago: University of Chicago Press, 1983], 41). But human rights may also be possessed by humans who are not persons, or whose personhood is questionable.

6. See Abraham Irving Melden, *Rights and Persons* (Berkeley: University of California Press, 1977), 1.

7. Jeremy Bentham and John Austin in Benn, *Encyclopedia,* 197.

8. Some philosophers (e.g., libertarians) deny that there are any obligations of charity. Even if beneficence and charity are considered obligatory, however, their obligatoriness is less compelling than the obligation of nonmaleficence. As Hippocrates suggested, not harming is more important than doing good. See "Selections from the Hippocratic Corpus," in *Ethics in Medicine,* ed. Stanley Rieser, Arthur J. Dyck, and William Curran (Cambridge, Massachusetts: MIT Press, 1977), 7.

9. Although the two are often equated, the right to reproduce is not equivalent to the right to have a baby. For a woman, the right to reproduce includes the right to conceive, gestate, and give birth. The right to *have* a baby may be superseded by the child's right not to be harmed, even by a parent.

10. Michael Bayles, *Reproductive Ethics* (Englewood Cliffs, New Jersey: Prentice Hall, Inc., 1984), 13.

11. Antoine de Saint-Exupéry, *The Little Prince,* trans. Katherine Woods (New York: Harcourt, Brace and World, 1943), 71.

12. See Pope Pius XI, "Encyclical Letter on Christian Marriage" (Dec. 31, 1930), as cited in *Medical Ethics,* ed. Kevin D. O'Rourke and Philip Boyle (St. Louis: Catholic Health Association, 1989), 116.

13. Lisa Soule Cahill, "Women, Marriage, Parenthood: What Are Their 'Natures'?" *Logos* 9 (1988): 11–35.

14. AIH means "artificial insemination by husband"; AID means "artificial insemination by donor"; TID means "therapeutic insemination by donor." AID and TID are interchangeable terms. While not prevalently used, the acronym TID avoids the close association between AID and AIDS. My use of quotation marks in the first but not the second clause is intended to call attention to the difference between insemination from husband's sperm (artificial?) and from anonymous donor sperm.

15. Cahill, 22–26; and Benedict M. Ashley and Kevin D. O'Rourke, *Health Care Ethics: A Theological Analysis* (St. Louis: Catholic Health Association of the United States, 1982), 287–88.

16. Carson Strong and Jay S. Schinfield, "The Single Woman and Artificial Insemination by Donor," *Journal of Reproductive Medicine* 29 (1984): 293–99.

17. Kamran S. Moghissi, "The Technology of AID," in *New Approaches to Human Reproduction,* ed. Linda M. Whiteford and Marilyn L. Poland (Boulder, Colorado: Westview Press, 1989), 128. As Judith N. Lasker and Susan Borg suggest, however, protection of social fathers may be the main reason for secrecy regarding AID. See their "Secrecy and the New Reproductive Technologies," in Whiteford and Poland, 138–41.

18. According to Moghissi, remuneration currently ranges from $20 to $35 per ejaculate. See Moghissi, 122. But Lori Andrews says that the usual payment is $50 a "donation," as cited in Nadine Brozan, "Babies from Donated Eggs: Growing Use Stirs Questions," *New York Times* (Jan. 18, 1988), 9.

19. Moghissi, 128

20. George J. Annas, "Fathers Anonymous: Beyond the Best Interests of the Sperm Donor," in *Genetics and the Law II,* ed. Aubrey Milunsky and George J. Annas (New York: Plenum Press, 1980), 331–40.

21. Moghissi, 120–21.

22. Natural law theology probably provides the strongest support for an essential relation between procreation and sexuality, but the relation between the two can also be supported on utilitarian and deontological grounds, without arguing against contraception or Victorian mores.

23. Lasker and Borg, 133. AID is also used by couples who are carriers for debilitating genetic diseases. Observing secrecy about AID then obscures the fact that the male is a carrier for the disease.

24. In 1884, when the first reported instance of successful AID occurred in Britain, the procedure was performed without revealing it to the couple. See Lasker and Borg, 134.

25. Ronald Munson calls this practice CAI (confused artificial insemination). See his *Intervention and Reflection,* 3d ed. (Belmont, California: Wadsworth Publishing Company, 1988), 415.

26. Lasker and Borg, 136–42.

27. Lasker and Borg, 138.

28. Lasker and Borg, 139.

29. Lasker and Borg, 140.

30. Lasker and Borg, 140–41.

31. Cheryl F. McCartney, "Decision by Single Women to Conceive by Artificial Insemination," *Journal of Psychosomatic Obstetrics and Gynecology* 4 (1985): 321; Maureen McGuire and Nancy J. Alexander, "Artificial Insemination of Single Women," *Fertility and Sterility* 43 (Feb. 1985): 183; and Miriam B. Rosenthal, "Single Women Requesting Artificial Insemination by Donor" in *Psychiatric Aspects of Reproductive Technology,* ed. Nada L. Stotland (Chicago: American Psychiatric Press, Inc., 1990), 113–21.

32. The Hippocratic imperative, "to help, or at least to do no harm," would thus have been violated. See Reiser, Dyck, and Curran, 7.

33. Moghissi, 129. For example, Mount Sinai Hospital in Cleveland, Ohio, offered reproductive technologies to married couples only.

34. Leon Kass, *Toward a More Natural Science* (New York: Free Press, 1985), 159–64; and Christine Overall, *Ethics and Human Reproduction* (Boston: Allen and Unwin, 1987), 139–51.

35. Growth hormone therapy is a particularly apt example in this regard. I treat this further in Chapter 12.

36. Possible evidence of the increase of requests for artificial insemination by single women is the increase of articles on the topic, see note 31.

37. I wish to thank the reproductive endocrinologist involved in this case, Paul Schnatz, who is now at Providence Hospital in Southfield, Michigan, for sharing with me his rationale and practice regarding requests for artificial insemination from single women.

38. See John A. Robertson, "Ethical and Legal Issues in Human Egg Donation," *Fertility and Sterility* 52 (Sept. 1989): 358. Nonsurgical removal through the vagina is now the preferred means of ova retrieval from donors who are not undergoing medical procedures for themselves.

39. Brozan, 9; and Paula Monarez, "Halfway There," *Chicago Tribune* (Feb. 2, 1992), sect. 6, 4.

40. "Clinic in Ohio Starts Egg Donor Plan," *New York Times* (Sept. 15, 1987), 10.

41. The student took my course on "Moral Problems in Medicine" at Case Western Reserve University in 1987. After undergoing drug-induced hyperstimulation of her ovaries, this student underwent laparoscopic removal of her ova.

42. Although "wet nurses" have not socially been regarded as mothers, they fit the meaning of motherhood as one who fulfills an essential nurturant role.

43. See Phillip J. Parker, "Motivations of Surrogate Mothers: Initial Findings," *American Journal of Psychiatry* 140 (Jan. 1983): 117–18.

44. Genesis 16: 1–4. This early case of "surrogacy" had its problems. Once pregnant by Abraham, Agar despised Sarah, who in turn treated her so badly that she ran away from the household (16: 5–6).

45. One such "natural arrangement" is recounted by Juliette Zipper and Selina Sevenhuijsen in "Surrogacy: Feminist Notions of Motherhood Reconsidered," in *Reproductive Technologies,* ed. Michelle Stanworth (Minneapolis: University of Minnesota Press, 1987), 118.

46. Lori B. Andrews, *New Conceptions* (New York: St. Martin's Press, 1984), 202.

47. *In re Baby M,* New Jersey Superior court, No. FM-25314-86F (March 31, 1987).

48. "To Serve 'the Best Interests of a Child'," *New York Times* (April 1,1987), B2.

49. *In re matter of Baby M,* 1988 New Jersey Lexis 1, 79; New Jersey Supreme Court No. A-39, (February 1988).

50. Wulf Utian, Leon Sheeham, and James Goldfarb, "Successful Pregnancy after *In Vitro* Fertilization and Embryo Transfer from an Infertile Woman to a Surrogate," Letter, *New England Journal of Medicine* 313, no. 21 (Nov. 21, 1985): 1351–52.

51. *Webster's New World Dictionary,* 2d college ed. (New York: Simon and Schuster, 1982), 928.

52. Anna Johnson, for example, had agreed to be a gestational surrogate for Crispina Culvert, who with her husband had provided the gametes that produced a child. Although Johnson changed her mind and wanted to raise her son, a California court ruled that giving birth did not make her the child's mother. See *Anna J. vs. Mark C.,* 234 Cal. Ap. 3rd., 1557 (1991); and Susan Peterson and Susan Kelleher, "Genes Settle Legal Argument," *Chicago Tribune* (Oct. 10, 1991), sect. 1, 6.

53. *In re matter of Baby M,* February 1988. See Leonard Fleck, "Surrogate Motherhood: Is It Morally Equivalent to Selling Babies?" *Logos* 9 (1988): 135–45.

54. Peterson and Kelleher, 6.

55. LeRoy Walters reinforces this misleading interpretation by distinguishing between full and partial surrogacy. Full surrogacy occurs when the "surrogate" is both gestational and genetic mother. Partial surrogacy occurs when the "surrogate" is only the gestational mother. Moreover, despite the fact that neither adoptive fathers nor sperm donors are "surrogate fathers," he treats the issue of surrogate motherhood under the heading "Surrogate Parenthood." See LeRoy Walters, "Genetics and Reproductive Technologies," in *Medical Ethics,* ed. Robert Veatch (Boston: Jones and Bartlett, Publishers, 1989), 210–12.

56. For example, Overall, 111–31. The following feminists joined a brief opposing surrogacy on behalf of Amici Curiae (*Foundation on Economic Trends,* In the Matter of Baby M, New Jersey Supreme Court, No. FM-25314-86E): Betty Friedan, Gloria Steinem, Gena Corea, Barbara Katz Rothman, Lois Gould, Michelle Harrison, Phyllis Chesler, and Letty Cottin Pogrebin.

57. Lisa H. Newton, "Surrogate Motherhood and the Limits of Rational Ethics," *Logos* 9 (1988): 113–14.

58. "The Baby M Contract: Is it Enforceable?" *New Jersey Law Journal* 119 (Feb. 26, 1987): 1.

59. Lori B. Andrews, "Feminism Revisited: Fallacies and Policies in the Surrogacy Debate," *Logos* 9 (1988): 81–96; also, Newton, 113–34.

60. Overall, 116–19.

61. Robert Hanley, "Baby M's 'Best Interests' May Decide Case without Firm Legal Precedent," *New York Times* (Feb. 2, 1987), 15. See my "Ethical Decisions in Neonatal Intensive Care," in *Human Values in Critical Care Medicine,* ed. Stuart J. Youngner (New York: Praeger Press, 1986), 74–82. John J. Arras provides an excellent critique of this principle in his "Toward an Ethic of Ambiguity," *Hastings Center Report* 14, no. 2 (April 1984): 25–33.

62. As suggested, for example, in Whitehead's statement: "I'd rather see me and her [i.e., her baby] dead before you get her." George J. Annas, "Baby M: Babies (and Justice) for Sale," *Hastings Center Report* 17, no. 3 (June 1987): 13.

7

In Vitro Development and Childbirth

The birth of a child has traditionally signalled entry into the human community. Birthdays celebrate the beginning of a person's life span as if the individual did not exist before the moment in which the newborn was propelled from womb to world with a mighty push. Census figures do not count fetuses, even if they are viable and thus capable of independent existence. So long as the fetus remains within a woman's body, its existence is generally unacknowledged.[1]

Why is there such a difference between extrauterine and intrauterine human life? Several probable reasons come to mind. First is the fact that the fetus cannot be seen or heard, and can only be felt by the pregnant woman or others who feel its movement through her abdomen. Second, the fact that the fetus is contained within the woman's body suggests that it is part of her rather than a separate or even separable entity. Third, unlike a newborn, a fetus in utero does not depend on outside assistance in order to develop. Although the pregnant woman may cultivate a life-style conducive to fetal welfare, the fetus is likely to survive even without her caring for it in utero.

Ordinarily, the fetus is more immature than a newborn. This is not always the case because the magic moment of birth is not necessarily identical with full-term gestation, that is, about 40 weeks. A substantial number of births are preterm (less than 38 weeks), and some are postterm (greater than 42 weeks).[2] Accordingly, an individual newborn conceived seven months earlier is developmentally younger than an unborn counterpart conceived nine months previously. While the preterm infant is legally a person, as well as morally a person by most accounts, the moral status of the nine-month fetus remains controversial.

If externalization of the fetus through birth has such significant social, legal, and moral ramifications, what may be said of externalization prior to that event?

114

In vitro fertilization (IVF) has externalized the very early development of the human embryo, so that it can now be observed and directly affected just as a newborn can, but with several important differences. In vitro fertilization produces a one-cell fertilized ovum, called a zygote. Until about six weeks after fertilization the developing organism is called an embryo; from then until birth it is called a fetus. Some infertility specialists refer to the zygote or embryo until the time of its implantation in the uterus as a pre-embryo.[3] To the naked human eye, the in vitro embryo appears as a pin-point; under a microscope it resembles a group of undifferentiated cells such as one sees in micrographs of early embryonic development.[4] Because of its lack of differentiation, the embryo has no ability to produce sounds; for that matter, neither does the early stage fetus. The strong healthy wail of the newborn is many months ahead in the developmental sequence. Micromanipulators may be used to capture one- to eight-cell embryos in a nutrient solution, but if one were to touch an embryo directly it would most probably be destroyed. The experience of feeling, hearing, and seeing externalized zygotes or early embryos is thus drastically different from experiencing an infant just delivered from its mother's womb.

Externalization is nonetheless common to both an in vitro embryo and a newborn, but not to a viable fetus, which in other respects is much more like the newborn than the embryo. In exceptional cases, externalization also occurs for purposes of fetal treatment.[5] While externalization is morally relevant, it is not an adequate criterion for determining responsibility to the developing human organism. In what follows I wish to consider further criteria, focusing on events that initiate and conclude the period of human development that ordinarily takes place within the body of a woman.

In Vitro Fertilization

In the summer of 1978, the birth of Louise Brown in London was greeted with more excitement and publicity than previous reproductive advances such as fertility drugs and techniques for artificial insemination had been afforded.[6] For some, the event shook their life-long belief that human beings cannot create life. Patrick Steptoe and Robert Edwards had in fact taken human sperm and ova, mixed them in a petri dish where they observed signs of fertilization and multiple cell cleavage, and transferred the embryos to Ms. Brown's uterus through the cervix by means of a small plastic catheter.[7] Pregnancy was initiated when the embryo that was to become Louise Brown implanted successfully and began to gestate in an in vivo environment. Nine months later the previously infertile couple, along with the team of scientists and clinicians who cared for them, rejoiced in the birth of a healthy baby girl.

Since it takes at least two individuals to achieve a pregnancy, I have referred to the couple rather than to the woman who gave birth to Louise Brown as (previously) infertile. Male infertility in fact accounts for about 30 percent to 40 percent of infertility cases, in a population where about 15 percent of those in the reproductive age group are infertile.[8] At the inception of IVF, the procedure was mainly intended as a fallopian tube by-pass for cases where a problem with the fallopian tubes was the cause of infertility. Fertilization usually takes place within the fallopian tubes. If a woman's tubes are irreparably obstructed or absent but she has ovaries and a uterus, IVF provides her with the possibility of achieving gestational as well as genetic motherhood. As the practice of IVF has become more widespread, other medical reasons such as male infertility have been provided for its use.[9]

From an egalitarian perspective, the use of IVF to solve problems of male infertility is more troubling than the use of artificial insemination. In both situations, the fertile woman sustains the risk and discomfort of a procedure that only benefits her to the extent that she wants to bear a child that she cannot conceive through intercourse with her infertile partner. Because she is fertile, she could conceive through intercourse with another man or by being artificially inseminated with a donor's sperm. To conceive through artificial insemination is a relatively simple procedure, whereas IVF requires that the woman take superovulatory drugs, and undergo ova retrieval and embryo transfer.[10] In a sense, the fertile woman's body is thus used to solve someone else's problems. Only through an emphasis on the woman's autonomy in the context of her relationship to her partner can this problem be adequately resolved. Although couples usually regard infertility as a mutual problem, an egalitarian perspective calls for consideration of disproportionate risks to either partner.

Use of IVF for cases of male infertility (instead of AID) suggests the importance of a man's genetic tie to his offspring. Women may place less weight on the genetic tie because they also relate to their offspring through pregnancy, birth, and lactation. I once observed anecdotal support for this view in the responses of four women to the following query: If you could only be a gestational mother or a genetic mother, which would you choose? Without hesitation, each said "gestational."[11] Presumably, if the question were put to more women, some would answer differently. Social parenthood is probably more important than either genetic or gestational parenthood to many men and women, and one's desire to be gestationally or genetically related to offspring is usually connected with the desire to be a social parent as well.

In the case of Louise Brown, the use of IVF raised few ethical concerns because the situation reflected the conventional norm for becoming parents: a married couple wished to have and raise their genetically and gestationally related offspring. The principal ethical questions introduced through IVF involve departures from this norm by combining techniques of gamete donation and surrogate gestation, manipulation of in vitro gametes or embryos, and storage or disposal of extra embryos. Like surrogacy, IVF also represents an option that is financially beyond the reach of most infertile women or couples. A single in vitro fertilization cycle can cost more than $8000,[12] and the procedure fails about 85 percent of the time.[13]

Since its first success was reported in Britain in 1978, IVF has become widely practiced elsewhere as well. The initial shock of having fertilization, which to some signifies the beginning of new human life, occur in a petri dish rather than within the body of a woman, subsided as thousands of IVF infants were born throughout the world. One factor influencing general acceptance of IVF is the realization of fertile women and men that infertility is an immense source of anguish for many infertile persons, and that the right to have children in an age of reproductive technologies may extend to others besides themselves. Appropriately, those who have become parents without need for technological assistance may feel self-conscious about addressing the issue of infertility without ever having experienced it. As one among that number, I am acutely aware that my views about the rights of infertile persons may fail to reflect adequate sensitivity to their experience.[14]

At least one aspect of IVF places responsibility on others besides the infertile: the in vitro status of zygotes or embryos, which sets up possibilities for their manipulation, treatment, and disposal. Some of the most controversial cases regarding use of the new technology relate to these possibilities.[15]

Responsibilities for In Vitro Zygotes or Embryos

According to the egalitarian framework elaborated in Chapter 1, we should not destroy what already exists, or kill what is living unless there is a suitable overriding reason for doing so. Overriding reasons are easier to identify the more they relate to other values in addition to life or mere existence, values such as the capacity for pain or autonomy (see guidelines 3 and 4).

After fertilization in a petri dish, zygotes grow by cell division so long as they are in fact living rather than dead. However, embryos are far from sufficiently developed to experience pain, as may a newborn or a fetus in which the nervous system has begun to develop. Gametes may also be described as living or dead, but responsibilities towards them are surely not greater than those toward embryos for a number of reasons. First is the huge wastage of gametes that occurs naturally in the course of human development. Except for periods of pregnancy, normal females waste ova each month from puberty to menopause, and normal males waste millions of sperm in each ejaculation.[16] Although a substantial number of zygotes and embryos are also wasted in the usual course of events, many survive, grow, and develop into fetuses and newborns. Such development occurs naturally within the body of a woman who sustains their growth, even when she is unaware of their presence.

In contrast, gametes do not develop toward independent existence unless an intervention occurs through a specific act or process of fertilization, whether in vivo or in vitro. Fertilization does not occur inadvertently; it can only occur through a deliberate act.[17] When it occurs in a petri dish, the continuation of life and development of the zygote does not occur spontaneously as it would within a woman's uterus. Here too, deliberate intervention is required to prevent the otherwise certain demise of the embryo: it must be transferred to a receptive uterine environment. If one could know in advance that the survival of the in vitro embryo was likely to result in a painful existence for the sentient entity it would eventually become, the intervention necessary to ensure its survival should probably not be undertaken. At such times guideline 3 supersedes guideline 2: the lives of those whose existence is mainly fraught with suffering should not be prolonged.

A second reason why responsibilities toward embryos override those towards gametes is that the embryo contains the full and unique genetic endowment of a possible future person, whereas gametes considered separately only constitute half of that endowment. Ova and sperm are like two kinds of building material, both of which are essential for any building; a zygote represents a plan for the building, as well as the onset of its construction. Unlike other building materials, ova and sperm are no longer separable and reusable once construction is initiated through fertilization. In the rudimentary stage of early embryonic development, the building plan may be radically altered: instead of one, two identical buildings may be planned, or one may replace a previous plan for two. Possibilities for multiplication and fusion of embryos occur for about two weeks after conception.[18] Nonetheless, the development of at least one new building or person has been initiated even if it might not complete its development. It is irreversibly distinct from the building materials from which it was formed, and also from the older buildings (builders?) who contributed their different types of genetic material to the project. However, the in vivo embryo is not clearly distinct from the environment in which it develops, namely, the body of the pregnant woman. In light of her legal right to abortion, the fact that the in vivo embryo is dependent on her, and cannot be separated from her without invading her body, gives the pregnant woman control over the fate of the embryo. Her exercise of such control involves a moral decision.

Taken together, the factors previously described (the excess of sperm and ova available through normal men and women, the extensive wastage of gametes that typically occurs, the fact that gametes cannot develop into human beings without specific acts or interventions, and the fact that they cannot experience pain or suffering) suggest that it is hardly necessary to stipulate a priori reasons for destroying extra or defective gametes. There may in fact be compelling reasons for destroying them. For example, if sperm are collected from a person later discovered to be HIV positive, the sperm may be discarded in order to prevent harm to others. Because of their superabundance, healthy sperm are routinely discarded after fertility testing is performed on samples of semen.

Helga Kuhse and Peter Singer argue that disposal of embryos is comparable to disposal of gametes.[19] Consider, they suggest, three somewhat similar situations. In the first, ova and semen have been taken from a man and woman enrolled in an IVF program. Just before the gametes are placed together in the petri dish, it is learned that the woman has a medical condition that makes pregnancy impossible. Although the ova could be fertilized and returned to the woman's uterus, the resultant embryos would soon die anyway. So the ovum and sperm are flushed separately down the sink. According to Singer and Kuhse, this is clearly justified from a moral point of view.

The second situation resembles the first except that the information about the woman's inability to sustain a pregnancy arrives just after in vitro fertilization has occurred. Some would question the appropriateness of disposing of the embryo in the same manner as the separated gametes. But consider a third situation in which ova and sperm are disposed of separately, but remain lodged together in the sink because the drain is blocked.

> A nurse is about to clear the blockage and flush them both away when a thought occurs to her: perhaps the egg has been fertilized by the semen that was thrown on top of it! If that has happened, those who believe that the embryo has a special moral status which makes it wrong to destroy it must now believe that it would be wrong to clear the blockage; instead the egg must now be rescued from the sink, checked to see if fertilization has occurred, and if it has, efforts should presumably be made to keep it alive.[20]

Singer and Kuhse believe that the third situation illustrates the inadequacy of arguments supporting "the moral status" of the human embryo, in contrast with the lack of "moral status" of human gametes. I prefer to consider responsibilities derived from relationships rather than independent moral status, partly because the term *moral status* prescinds from the social context in which decisions about disposition of embryos, zygotes, or gametes are made. Moral-status language purports to attribute rights to individuals regardless of their social context. A care-based ethic emphasizes responsibilities that arise because of different contexts. An egalitarian perspective insists on attention to these differences.

If gametes were capable of experiencing pain, we would have moral responsibilities toward them regardless of their superfluity and lack of potential for becoming persons without intervention. If gametes were extremely scarce, we might then have responsibility to preserve them for the sake of the infertile, or to replenish the world's population. Singer and Kuhse do not deny that there are empirical differences between gametes and embryos, but they insist that the differences are not adequate to assign moral status in either case. While I eschew the moral status issue, I believe the differences are morally relevant.[21] As William James puts it, "there is no difference anywhere that doesn't make a difference everywhere."[22] The embryo is undeniably closer to full humanhood than separate gametes, and, unlike them, constitutes the genetic endowment of at least one new human organism. Although these differences do not provide a basis for imputing moral status to either, they suggest, in accordance with guideline 5, that gametes and embryos ought to be treated differently.[23] Prima facie, this means not interfering with their respective tendencies toward development or wastage. Thus, removal of gametes for IVF, or termination of developing embryos, needs to be justified on other grounds.

Interventions in the natural course of human development may be moral, immoral, or amoral.[24] What Singer and Kuhse as well as other writers on the topic do not address, however, is that embryos in a petri dish are not in their "natural" environment anyway. Left to themselves, they will *all* "naturally" die. This fact goes beyond Richard McCormick's claim that "as long as artificial procedures do not result in many more lost embryos than the large numbers lost in the natural process, this is not a key argument against IVF."[25] McCormick would allow the loss, apparently ignoring the intervention necessary to avoid it. But responsibilities toward in vitro embryos are different from those we have toward in vivo embryos. It is one thing to argue that a developing embryo should be left alone in order to die or survive, and another to claim that it should be manipulated for either reason. Even if an in vitro embryo has a "right" to survival assistance, it has no right, nor do others have a right, to coerce a woman to become pregnant in order to provide that assistance. We are left then with different levels of responsibilities towards (a) gametes, (b) in vitro embryos, and (c) in vivo embryos. Other things being equal, our responsibilities toward gametes (a) are least compelling, and those towards zygotes or embryos capable of development without intervention (c) are most compelling. If, as I have argued earlier, the views of those most affected by ethical decisions should have the greatest weight, a woman in whose body an embryo resides should be the paramount decision maker regarding intervention. By the same criterion, the sperm provider's views should count more when the embryo is externalized than when it is developing within a woman's body.

Recently, the role of the sperm provider has been addressed in terms of rights rather than responsibilities.[26] In the ova retrieval that precedes IVF, extra eggs are often obtained, sometimes producing more embryos than optimal for transfer to the woman's uterus. The extra embryos may then be frozen or discarded. Because fewer embryos survive freezing, the freezing procedure does not totally avoid the issue of responsibilities to embryos. Do gamete donors have a right to freeze or flush down a drain a living embryo? If they do so, are such actions equivalent to abortions?

Freezing and flushing both involve the arrest of vital functions, but the arrest is intended to be temporary in the case of freezing. Death means the permanent loss of vital function, and this is apparently the intent when an embryo is discarded. If letting a person die is sometimes morally justified, so also is letting an embryo die in the laboratory. For example, in situations of limited resources, one person's life may not be maintained so that others may live. So extra embryos may be allowed to die rather than transferred to a woman's uterus in order to optimize the possibility that the other embryos will implant, gestate, and develop to term. Allowing them to die is not equivalent to abortion because abortion means the termination of a pregnancy, and pregnancy is not established until there is a physiological tie between a woman and the embryo.[27] A woman whose ova have been retrieved and fertilized in vitro cannot become pregnant until the embryo is transferred to her body. Even when fertilization occurs in vivo, pregnancy is not established until the embryo has implanted in a woman's uterus.[28]

When pregnancy has been established in vivo, the integral tie or relationship between embryo or fetus and pregnant woman is crucial to moral decisions regarding either. When the relationship is involuntary, as occurs when pregnancy is due to rape or failed contraception, the moral and emotional tie is obviously more fragile than in intentional or voluntary pregnancies. To paraphrase *The Little Prince,* we are less responsible for what we have not tamed, and least responsible for relationships to which we have not consented, especially where those relationships jeopardize other relationships.[29] Carol Gilligan's study of women considering abortion shows that women generally base their decisions on a desire to preserve relationships already established or accepted.[30] Such prioritizing is consistent with an ethic based on respect for differences in relationships.

Because voluntariness plays so significant a role in relationships, it is possible that the tie between a woman and an in vitro embryo developed from her ova makes her "more responsible for what she has tamed" than the woman whose in vivo embryo is the result of an involuntary pregnancy. The man who has provided sperm with the intent of becoming a parent may also be more responsible for the embryo. In either case, however, taming connotes commitment to the future of the one tamed, that is, to social parenthood if the embryo develops to term. Anonymous sperm or ova donors do not establish ties with the embryos that their gametes produce.

Moral responsibilities for others, while based on relationships and differences among them, do not bind us to them in a way that impedes either our or their development. The parental relationship particularly illustrates the need to let go so that one's child may become increasingly independent, and thus able to form new relationships. As I have already suggested (Chapter 2), the moment in which a woman becomes a mother epitomizes this meaning: literally, the tie between her and the fetus is cut so that the newborn can breathe and grow in the larger world beyond her body.

Midwifery and Childbirth

This entry into world from womb is an exciting and somewhat perilous event for both of the principal participants. Throughout history the birthing process has been facilitated by those whose desire to assist was first sparked by their own birthings. Midwifery arose as a profession of those who further learned the lore through years of practice and attunement to the nuances of each delivery. The practice belonged entirely to women until the latter part of the seventeenth century, when the practice of employing a male midwife or *accoucheur* spread among the French upperclass.[31] According to Adrienne Rich, the "fad" was precipitated by the attendance of a male physician on Louise de la Valliere, the favorite mistress of Louis XIV. However, the rise of obstetrics as a predominantly male specialty was mainly associated with difficult births at which special maneuvers or instruments were used. Unlike the midwives, whose care for birthing women entailed their presence throughout labor, birth, and the immediate postpartum period, obstetricians' involvement with their patients was episodic. The differences between the two professions illustrate the perennial contrast between the atomistic character of "masculist" reasoning and the emphasis on continuity and relationship within the "feminine" model.

Connections between midwifery and the art of healing are chronicled by Barbara Ehrenreich and Deirdre English, who further relate the two to witchcraft.[32] Partly because of their wondrous capacity to give birth, and their skills in facilitating the birthings of other women, women throughout history have extended their life-giving talents to others through the art of healing. For centuries,

> they were the unlicensed doctors and anatomists,... abortionists, nurses and counselors. They were pharmacists, cultivating healing herbs and exchanging the secrets of their uses. They were midwives, travelling from home to home and village to village.... They were called "wise women" by the people, witches or charlatans by the authorities.[33]

To label midwives witches was clearly a means of suppressing them so that a new, predominantly male, medical profession might develop. Initially, the new profession was protected by the ruling classes, and midwives and female healers continued to treat the poor and peasants. As witch hunts spread, thousands of women were executed in their wake. Gradually the medical profession assumed predominance over the midwives.

A horrendous example of how some men replaced women as "experts" regarding childbirth involves the development of forceps by the Chamberlens in England in the late sixteenth century. Three generations of men preserved the "Secret" through which they were able to assist successfully in difficult deliveries. The "Secret ... consisted of a kit of three instruments: a pair of obstetric forceps, a vectis or lever to be used in grasping the back of the head of the fetus, and a fillet or cord used to help in drawing the fetus, once disengaged from an abnormal position, out through the birth canal."[34]

When the Chamberlens were called to a birth, they arrived carrying a huge chest containing those items. No one was permitted to observe the contents of the chest, and the women giving birth were blindfolded. In 1721 a Belgian barber surgeon named Jean Palfyne developed an instrument comparable to the Chamberlens' forceps without their help, and presented it to the Paris Academy of Science. The actual design of the forceps developed by the Chamberlens was withheld until Edwin Chapman, a surgeon and male midwife, revealed it in his *Essays for the Improvement of Midwifery* in 1773. From then on, according to Rich, "the forceps was available to all male — and to almost no female — practitioners of the obstetric art."[35]

The early history of obstetrics thus supports feminist criticism that it wrested control of reproduction from the women most invested in the process both personally and professionally. Contemporarily, there has been a resurgence of midwifery practiced by specially trained nurses, but its legitimization has mainly occurred through their attachment to the profession of obstetrics. As in the days of the Chamberlens, the obstetricians deal with the difficult cases, where their episodic involvement draws greater prestige and income than midwives whose long hours of assistance at "ordinary" births is less materially rewarding. In practice, then, the relationship between nurse midwives and male obstetricians reinforces sex-role stereotypes.

Concomitantly, it must be acknowledged that advances in obstetrics have not only led to increased survival rates for infants, but also to reduction in the risks of pregnancy to women themselves. As one whose health was seriously threatened by a well-known complication of pregnancy (toxemia), I am grateful for these advances. As the mother of two premature infants, I am doubly grateful for the technological supports that allowed them to survive and develop into healthy children. In my particular circumstances, I needed and wanted the "medicalization" of childbirth. For most women, however, pregnancy is a normal, healthy condition that produces normal, healthy children without the necessity of medical technology. That it has become a doctor-directed experience for so many, regardless of their need for its medicalization, seems unfortunate. It is pregnant women, after all, who deliver their babies from womb to world, and midwives or obstetricians receive them or, at most, assist in their delivery. To say that the obstetrician delivers the infant is to suggest a passive role for the woman who in fact carried the "package" from conception to completion.

In the medicalization of childbirth as in other areas of medicine, the emphasis has been on the product rather than the process. Generally, the desired product is a healthy newborn and a healthy mother. In the majority of cases, the interests of both are coincident.[36] In fact, women willingly undertake inconveniences and alter their life-styles so as to maximize the probability of producing a healthy infant.[37] Sometimes this extends to prepregnancy planning through genetic counseling, timing, and spacing of children; typically it involves regular prenatal visits (with their concomitant expense), rest and nutrition supplements, and curtailment of smoking and of caffeine and alcohol consumption. None of these is an extreme inconvenience, and obviously such practices promote the health of pregnant women as well as their newborns, but our foremothers were hardly burdened with such concerns. The more medical science has learned about negative effects of certain maternal behaviors on fetal development, the more pregnant women have voluntarily abstained from those behaviors. To avoid premature birth, with its threat to fetal welfare, some women undergo extra medical procedures and prolonged hospitalization or bed rest. To optimize newborn welfare, they forego pain medication for themselves, and choose methods of delivery that are riskier for them than the alternatives.[38]

Because of this general pattern of behavior by pregnant women, it is legitimate to assume, in cases where their wishes are not discoverable, that so long as they have chosen to continue their pregnancies they would want what is best for their fetuses, even at some cost to themselves. A questionable application of this assumption occurs in cases where a woman who is comatose or "brain dead" is kept "alive" for the sake of her fetus.[39] The assumption is questionable here because the pregnant woman's altruism in behalf of the fetus may arise from an expectation that she would not only give birth but also raise her child. The impossibility of her becoming the child's social mother might radically alter her willingness to continue the pregnancy.

Rosalind Hursthouse argues that women's capability for bearing children places special moral demands on them.[40] In an egalitarian society, however, moral burdens as well as other burdens and benefits are shared. That women are generally inclined to sacrifice themselves for the sake of others, especially those with whom they maintain special relationships, is an indication of their generosity or virtue, as superogatory rather than obligatory morality. The following cases illustrate this tendency of women to promote the interests of newborns even at the expense of their own.

CASE I: MATERNAL ALTRUISM AND A VIABLE FETUS.

Lee March was a 26-year-old pregnant woman who had Eisenminger's syndrome from birth. The disease involves a cardiac defect that may be repaired during infancy but cannot successfully be repaired later in life. Because Ms. March had not had the problem corrected, her condition had deteriorated, and she was not expected to survive more than two years.

Prior to her marriage Ms. March was warned that pregnancy would be life-threatening, especially during the time of parturition. When she became pregnant, she rejected the option of abortion and indicated that she wanted to live, but also that she wanted to have the baby. As she approached term, a decision was needed regarding the method of delivery. On learning that the method best for the fetus might not be best for her, Ms. March stated that she wanted the baby to survive even if she might not. Her family disagreed with this view, but supported her decision. Hoping to limit the time of labor and to relieve fetal distress, the obstetrician performed a cesarean section. After delivery, Ms. March was rushed to the intensive care unit, where she died five days later despite maximal support. The newborn was noted to have Eisenminger's syndrome, but the condition was repaired during early infancy.

CASE II: OBSTETRIC INTERVENTION AND A NONVIABLE FETUS.

Nan Ost was approaching term during her third pregnancy. Ultrasonographs taken early and later in gestation had indicated that the fetus had not developed calcification around the brain. In other words, there was no protective skull. In order to optimize the chance of fetal survival, Ms. Ost wanted to undergo operative delivery. After thoroughly exploring possible postdelivery treatment, a medical team concluded that survival of the newborn was virtually impossible. The obstetrician did not wish to perform the cesarean section because it imposed a risk on the woman with no realistic expectation of benefit to the fetus. He considered surgery in such circumstances medically irresponsible.

When Ms. Ost was presented with the medical opinion that the fetus was nonviable, she asked for a cesarean section delivery. Her rationale was twofold: (1) the medical opinion might be incorrect, and (2) vaginal delivery would be more painful for the fetus regardless of its viability status. The obstetrician explained his reasons for confidence in the medical consensus. He also indicated that there was no way of knowing which method of delivery would be more painful for the fetus, or even if the fetus was capable of experiencing pain. Although Ms. Ost was informed that another physician might perform a cesarean section delivery, she opted to follow the advice of her original obstetrician. At 39 weeks' gestation she delivered vaginally a still-born infant.

Obviously, both of these cases involve conflicts among those affecting and affected by the outcome. Although the two "canonical" principles of contemporary biomedical ethics, respect for autonomy and beneficence, are inadequate for resolving the questions raised, the principle of distributive justice may be invoked as a means of adjudicating the conflicts. But, as we noted earlier, different theories of justice imply different concepts of equality. A libertarian theory affords no resolution of the contradictory wishes of those involved, for example, Lee March and her family (Case I). A utilitarian view might support a decision made solely in behalf of the fetus in Case I and solely in behalf of the pregnant woman in Case II; it could also support opposite positions. A Marxist account based on need seems to by-pass the moral relevance of respect for autonomy of all of those affected.

An egalitarian perspective provides a means of analyzing both cases. Guideline 2, for example, states that individual lives should neither be shortened nor terminated. In Case II, cesarean section delivery is not expected to extend the life span of the fetus, and it involves greater risk than vaginal delivery for the pregnant woman. In Case I, the woman might have lived for several years if she survived the postpartum period. The guideline says nothing, however, about comparing two years of survival for Ms. March with a full life span for her fetus. Indeed, it seems ageist to argue against saving the woman on that basis. Two years more with her in whom a complex set of relationships, ties of affection and commitment had developed over the years were more compelling to family members than a full life span for someone (the fetus) in whom no such relationship or ties had yet developed. Such relationships were also relevant to clinicians who had consciously established professional and, quite possibly, personal bonds with Ms. March.

Guideline 3, which stipulates that those capable of suffering should not be caused to suffer is applicable to both cases and to other situations in which different methods of delivery and treatment during delivery are likely to cause pain. Ordinarily, cesarean section delivery not only means greater risk of complications to the pregnant woman, but greater discomfort and a longer period of recovery for her than vaginal delivery. In Case I, however, the physician reasoned that quicker delivery by cesarean section would shorten the duration of labor; in Case II, he believed that deliberately inflicting pain through surgical delivery was unjustified because it offered no benefit to either Ms. Ost or her fetus.

Guideline 4 affirms that choices should not as a rule be ignored or impeded. In Case I, Ms. March's wishes were in conflict with those of her family, but she would be more affected than they by a decision regarding the fetus within her.

Thus her autonomy should prevail. If Lee March had asked that her life be saved, even though only for a severely limited life span, and even though her survival might involve death for her fetus, her autonomy should still prevail. In the case described, respect for Ms. March's autonomy is supported by the best interests of the fetus, as well as the reluctant preference of the physician, and together these goods outweigh the opposing wishes of family members.

In Case II, Nan Ost and her obstetrician finally agreed to vaginal delivery of the apparently nonviable fetus. Had she insisted on surgical delivery, the physician would not have been obligated to act counter to his professional judgment that this meant unjustifiable risk to the patient. It would have been appropriate at that juncture, however, to offer her transfer of care to someone else. Another physician might have considered the risk negligible, and performed the cesarean section solely on the basis of respect for Ms. Ost's autonomy.

Guideline 5 asserts that individuals should not be treated as other than who or what they are. If Lee March (Case I) were defined solely as a receptacle for the fetus developing within her, this guideline would have been violated. What transformed the situation from one in which Ms. March might have been used in such a way is the fact that she chose to maintain her pregnancy for the sake of the fetus. Respecting her choice could not constitute misuse or abuse. Neither would Nan Ost (Case II) have been misused or abused by undergoing cesarean section delivery if she had chosen the procedure with full knowledge of its risks. However, had the physician been required to act contrary to his professional judgment, guideline 5 would have been violated in his regard.

Guideline 6 requires that individuals be given an equal share of pertinent resources. For the fetus in Case II, unfortunately, the mobilization of treatment resources for management of a newborn with no protective skull around her brain would have been a futile, and therefore nonpertinent, expenditure. In Case I, however, Ms. March had a 50/50 chance of surviving for several more years, and this would clearly have been a benefit to all concerned. Accordingly, the mobilization of hospital and medical resources (maximal treatment in an intensive care unit) constituted her equal share of pertinent available resources. Maximal treatment of her fetus also constituted an equal share of pertinent resources.

Guideline 7 says that human beings have priority over other sentient beings in distribution of limited resources. Limited resources were indirectly a feature of decision making in both cases. In declining to perform the cesarean section, Ms. Ost's obstetrician prevented the huge allotment of hospital personnel and technology that might have been necessary to attempt to sustain the life of a newborn who would inevitably die soon anyway. Presumably, such expensive futile treatment would have been used more effectively for those with a chance of survival, even if the survival were limited, as in Lee March's case.

A distinction between human beings and nonhuman sentient beings was not relevant in either case because mature fetuses, that is, those ready to be born, are surely sentient. At early stages of development, embryos and fetuses are not sentient, but their humanness remains morally relevant. Although fetuses as well as pregnant women in both cases were living, sentient, and human, an egalitarian approach did not demand that they be treated in the same way. If all cannot be saved, only an impoverished notion of equality would argue for letting all die instead of saving some. In such tragic circumstances, other differences noted in the guidelines are determinative. Respect for the autonomy of those most affected was a critical factor in both of these cases.

Once children are born, those who are not their mothers may assume control over their welfare. If a birth mother wishes to allow others to adopt her newborn, she is free to do so, and if she is unable to adequately care for the child herself, the child may be given over to others for care even without her consent. As with in vitro embryos, therefore, the extrauterine status of newborns imputes a greater range of responsibilities to those who can and do affect their fate. Recent court battles illustrate that custody disputes occur at either point of extrauterine existence. However, so long as women are most affected by resolution of the disputes, and so long as the welfare of their offspring will not thereby be threatened, their autonomy should be respected.

Since the decline of midwifery and the predominance of male obstetricians, women's experience of childbirth has been altered to reflect more episodic and impersonal treatment than a care model involves. Hospital birth and procedures have prevalently separated birthing women from their homes, from their families, and even from their newborns, introducing barriers rather than reinforcement to the very relationships that produce and nourish offspring. In some cases the technology is a great boon to infants as well as their mothers; in other instances, the medicalization of childbirth is unnecessary, and negative in its effects on the principal participants. Criticism of the excesses of medical involvement and control over a process that is unique to women's history and experience is thus deserved. As we shall see in the next chapter, however, violations of the rights of pregnant women have also been wrought by the decisions of American judges.

NOTES

1. Exceptions to the general lack of acknowledgment include "wrongful birth" and "wrongful life" suits, liability claims for damages to pregnant women, and conflicts such as those described in Chapter 8. See, for example, James Coplan, "Wrongful Life and Wrongful Birth: New Concepts for the Pediatrician," *Pediatrics* 95 (Jan. 1985): 65–72; and "Wrongful Death Suit Seeks $550,000 for Fetus," *The Cleveland Plain Dealer* (July 20, 1984), 13B.

2. See Richard E. Behrman, *Nelson Textbook of Pediatrics,* 13th ed. (Philadelphia: W. B. Saunders Company, 1987), 375–76, 383. According to Behrman, the term "premature" is often used to

denote immaturity. Since both fetuses and infants are immature, "preterm" more accurately describes the status of a fetus or infant at less than 38 weeks' gestation.

3. This term was also used by the Warnock Committee, which distinguished between the pre-embryo and the "definitive" embryo that begins to form at 15 or 16 days after fertilization when the "primitive streak" appears. According to the Warnock report, research should be permitted on pre-embryos but not embryos. See Patricia Spallone, "Reproductive Technology and the State: The Warnock Report and Its Clones," in *Made to Order,* ed. Patricia Spallone and Deborah Steinberg (New York: Pergamon Press, 1987), 177–78.

4. For example, in Keith L. Moore, *The Developing Human,* 3d ed. (Philadelphia: W. B. Saunders Company, 1982), 24–36.

5. Such procedures involve surgical removal of the fetus from the woman's uterus, performance of the intended procedures, return of the fetus to the woman's uterus, and surgical closure of the uterine wall and abdomen. See Gina Kolata, "Lifesaving Surgery on a Fetus Works for the First Time," *New York Times* (May 31, 1990), A1, B8.

6. Ronald Munson, *Intervention and Reflection,* 3d ed. (Belmont, California: Wadsworth Publishing Company, 1988), 403–04, 407–08.

7. Patricia M. McShane, "*In Vitro* Fertilization, GIFT and Related Technologies — Hope in a Test Tube," in *Embryos, Ethics and Women's Rights,* ed. Elaine Baruch, Amadeo D'Adamo, and Joni Seager (New York: The Haworth Press, 1987), 37.

8. McShane, 32; and Christine Overall, *Ethics and Human Reproduction* (Boston: Allen and Unwin, 1987), 139.

9. Judith Lorber, "*In Vitro* Fertilization and Gender Politics," in Baruch, D'Adamo, and Seager, 120.

10. Munson, 408; McShane, 32–38; Lorber, 121–22.

11. The question was posed at an informal gathering of participants in a meeting on reproductive genetics sponsored by the National Institutes of Health, Bethesda, in November, 1991.

12. American Fertility Society, *IVF & GIFT — A Patient's Guide to Assisted Reproductive Technologies* (Birmingham, Alabama: American Fertility Society, 1989), 7. See also Philip Elmer-Dewitt, "Making Babies," *Time* (Sept. 30, 1991); and Michael D. Lemonick, "Trying to Fool the Infertile," *Time* (March 13, 1989), 53.

13. This rate is based on live births. See Medical Research International and Society for Assisted Reproduction, "In Vitro Fertilization-Embryo Transfer (IVF-ET) in the United States: 1990 Results from the IVF-ET Registry," *Fertility and Sterility* 57, no. 1 (Jan. 1992): 15. Success rates are sometimes misleadingly reported on the basis of fertilizations or pregnancies achieved rather than live births. See also Lemonick, 53; and McShane, 39–42.

14. For an account of the emotional impact of infertility and its ethical relevance, see Jay S. Schinfield, Thomas E. Elkins, and Carson Strong, "Ethical Considerations in the Management of Infertility," *Journal of Reproductive Medicine* 31, no. 11 (Nov. 1986): 1038–42. Such considerations are particularly pertinent to a care-based ethic.

15. Custody disputes about "ownership" of extra frozen embryos have emerged because of the deaths of gamete providers and because of divorce and disagreement between them. See Munson, "The Orphaned Embryos of Mario and Elsa Rios," 404–05; and Carole Ashkinaze, "Divorce Court to Decide Who Gets Embryos," *The Cleveland Plain Dealer* (April 9, 1989), 4-C.

16. Helga Kuhse and Peter Singer, "The Moral Status of the Embryos," in *Test-Tube Babies,* ed. William A. W. Walters and Peter Singer (New York: Oxford University Press, 1984), 58.

17. Although the act that causes fertilization is deliberate (e.g., intercourse or IVF), fertilization itself is often indeliberate. Moreover, intercourse may be a deliberate act for a man but not the woman with whom he copulates, as occurs in rape situations.

18. The possibility of twinning and recombining of the early embryo (or pre-embryo) is crucial to Norman Ford's argument regarding responsibilities to new human individuals. See his *When Did I Begin?* (Cambridge: Cambridge University Press, 1989). According to Ford, each of us begins when our individuality is settled.

19. Kuhse and Singer, 58.

20. Kuhse and Singer, 59.

21. As James Knight and Joan Callahan put it, the human embryo or fetus is "something of value" even if it lacks moral status or personhood. See their *Preventing Birth* (Salt Lake City: University of Utah Press, 1989), 222. This view is similar to L. W. Sumner's concept of "moral standing," but not to his conception of "moral status," which is attributed to "every physical object," regardless of whether it has moral standing. See his *Abortion and Moral Theory* (Princeton, New Jersey: Princeton University Press, 1981), 26. An anonymous reviewer of this chapter has claimed that "a being that has moral status is one to whom we have moral responsibilities." I concur with this statement, but not with its converse. Although we have moral responsibilities to entities that are neither human nor persons, this does not imply "moral status" for those entities. It does imply moral status for those who have such responsibilities. In other words, my concept of moral status is narrower than Sumner's and Sumner's concept is narrower than that of the anonymous reviewer.

22. William James, *Pragmatism* (New York: The World Publishing Company, 1965), 45.

23. This position is commonly defended by those who hold that contraception is morally as well as medically preferable to abortion. See Knight and Callahan, 222–23. A similar point may be made for newborns, namely, that even if they are not persons, we have moral responsibilities to them that are greater than those we have towards fetuses. Infanticide is thus more difficult to justify than abortion.

24. See Knight and Callahan, 186, for an excellent account of problems raised by reasoning "from the way nature is to the way persons may or may not function."

25. Richard A. McCormick, "Ethical Questions: A Look at the Issues," *Contemporary OB/GYN* 20 (Nov. 1982): 229.

26. Karen Farkas, "Frozen Embryos Become Point of Dispute in Divorce," *The Cleveland Plain Dealer* (Sept. 27, 1989), 1; and John A. Robertson, "Ethical and Legal Issues in Cryopreservation of Human Embryos," *Fertility and Sterility* 47, no. 3 (March 1987): 372–73.

27. Although disposal of in vitro embryos is not equivalent to abortion, the disposal is not morally neutral. Guideline 2 opposes termination of individual lives unless other guidelines are overriding.

28. Until then, pregnancy is not scientifically detectable. It is also possible for a woman to become pregnant without ever having *her* ova fertilized. This occurs in cases of "genuine surrogacy," that is, where genetic and gestational mothers are different persons.

29. Antoine de Saint-Exupéry, *The Little Prince,* trans. Katherine Woods (New York: Harcourt, Brace and World, 1943), 71.

30. Carol Gilligan, *In a Different Voice* (Cambridge, Massachusetts: Harvard University Press, 1982).

31. Adrienne Rich, *Of Woman Born* (New York: Bantam Books, 1977), 130.

32. Barbara Ehrenreich and Deirdre English, *For Her Own Good* (New York: Doubleday, 1978), 33–98.

33. Barbara Ehrenreich and Deirdre English, *Witches, Midwives, and Nurses: A History of Women Healers* (Old Westbury, New York: The Feminist Press, 1973), 3.

34. Rich, 134.

35. Rich, 136. Rich cites Harvey Graham on this; see his *Eternal Eve: The Mysteries of Birth and the Customs that Surround It* (London: Hutchinson, 1960), 106–22.

36. Michael Bayles, *Reproductive Ethics* (Englewood Cliffs, New Jersey: Prentice Hall, Inc., 1984), 74.

37. Overall, 95.

38. Some, for example, have requested cesarean section delivery when it was not medically indicated, to optimize the outcome for the fetus. See George Feldman and Jennie Freiman, "Prophylactic Cesarean Section at Term?" *New England Journal of Medicine* 312, no. 19 (May 9, 1985): 1264–67. However, the authors contend that cesarean sections in such circumstances are not optimal for the fetus anyway.

39. See William P. Dillon, Richard V. Lee, Michael J. Tronolone, Sharon Buckwald, and Ronald J. Foote, "Life Support and Maternal Brain Death during Pregnancy," *Journal of the American Medical Association* 248, no. 9 (Sept. 3, 1982): 1089–91.

40. Rosalind Hursthouse, *Beginning Lives* (New York: Basil Blackwell, 1987).

8

Coercive Treatment After Fetal Viability

Although the United States Supreme Court has affirmed the right of women to terminate pregnancy, the choices of women who decide against abortion have increasingly been subject to social pressures and legal restrictions. Examples of possible pressures or restrictions include signs in bars advising pregnant women against alcohol consumption; special warnings against drugs and smoking; exclusion from certain occupations;[1] and standard medical advice regarding diet, sexual intercourse, bed rest, hospitalization, and method of delivery. Probably the most coercive among these interferences is what George Annas calls the "most unkindest cut of all": court ordered cesarean sections.[2]

Situations in which pregnant women refuse treatment recommended for the sake of the fetus have been characterized as "maternal-fetal conflicts."[3] Ellen Stein has suggested that this characterization is misleading because such situations primarily involve conflicts between women and "the medical or legal establishment" rather than conflicts between women and their fetuses.[4] The following case poignantly illustrates this point.

Angela Carder was a 28-year-old woman from Washington D.C., whose recurrence of cancer was discovered in June 1987, during her twenty-fifth week of a wanted pregnancy.[5] Her prognosis was grim: she was expected to die within weeks. Ms. Carder agreed to treatment that would prolong her life to improve the fetus' chance of healthy survival, but insisted that her own care and comfort be given priority. One week later her death was imminent, and a cesarean section was recommended for the sake of the fetus. Ms. Carder declined to give consent for the surgery, and family members and clinicians supported this decision. A neonatologist estimated the probability of fetal viability as 50 to 60 percent, with a less than 20 percent risk of serious impairment.

Legal counsel asked the court to apprise the institution of its responsibilities in the situation. After a hearing at the hospital at which different lawyers represented Ms. Carder, the fetus, and the institution, Judge Emmett Sullivan ruled that the cesarean section should be performed for the sake of the fetus. When request for a stay of the order was denied, Ms. Carder underwent surgical delivery. Her previable infant died about two hours later; she succumbed after two more days.

Subsequent to these deaths, the decision to allow the surgery was legally challenged through the Court of Appeals for the District of Columbia. More than 100 briefs from individuals and organizations, including the American Medical Association, were filed on behalf of the family. On April 26, 1990, the court ruled in support of the pregnant woman's refusal of treatment. According to Judge John A. Terry, writing for the majority, "A fetus cannot have rights in this respect superior to those of a person who has already been born."[6]

This chapter provides an egalitarian critique of the issues that the preceding case epitomizes.[7] Several writers have discussed legal aspects such as the possible relevance of child neglect laws and the U.S. Supreme Court decision in *Roe*.[8] Substantive positions have been developed both supporting and opposing coercive intervention for the sake of the fetus.[9] I wish here to consider the right to privacy and the concept of fetal viability on which legal rulings about pregnancy and refusal of treatment during pregnancy have been based. My aim is to illustrate the complexity of these decisions by analyzing different types of cases involving women who do not consent to cesarean delivery. After arguing that court orders for cesarean sections are inconsistent with refusal by the courts to require individuals to undergo less invasive procedures for the sake of a family member, I propose alternative remedies for the inconsistency of that practice. Consistency would promote equality in current practice.

The Right to Privacy

Although the United States Constitution does not explicitly mention a right to privacy, the U.S. Supreme Court has acknowledged that "a right of personal privacy, or a guarantee of certain areas or zones of privacy, does exist under the Constitution."[10] The roots of that right have been found in the United States Bill of Rights, and in the first, fourth, fifth, ninth, and fourteenth amendments.[11] *Griswold v. Connecticut* (1965), a case involving contraception, was the first to deal with privacy as a right that extends not only to personal information but also to personal activities.[12] In content, the right to privacy is a negative right, that is, a right to be let alone, implying an obligation on the part of others not to interfere with those activities.

In health care decisions, the right to privacy has been associated with the requirement of informed consent, which is intended to protect the self-determination of the patient. Intrusions on privacy may be seen as impeding informed consent, thus violating the principle of respect for autonomy. Justice William O. Douglas enumerates the following range of interests encompassed by the privacy right:

> First is the autonomous control over the development and expression of one's intellect, interests, tastes, and personality.... Second is freedom of choice in the basic decisions of one's life respecting marriage, divorce, procreation, contraception, and the education and upbringing of children.... Third is the freedom from bodily restraint or compulsion, freedom to walk, stroll, or loaf.[13]

Each of these interests links privacy with autonomy. As Judith Thomson suggests, the right to privacy may in fact be subsumed by other concepts and rights, particularly that of autonomy.[14] It may also be subsumed by guideline 4, which stipulates prima facie that the thoughts and choices of persons should not be impeded.

Even where the right to privacy is strongly affirmed, it is not held to be an absolute right. For example, while court decisions have emphasized the right to privacy as the basis of a woman's right to terminate a pregnancy, they have also maintained that the right to privacy may be subordinated to the state's interests in a viable fetus and in promoting the health of pregnant women. After viability, termination of pregnancy with a living survivor is clinically and legally defined as premature or preterm birth rather than abortion. Prior to viability, a physician must agree to the procedure. Even for a first trimester abortion, *Roe* asserts that "the abortion decision and its effectuation must be left to the medical judgment of the pregnant woman's attending physician."[15]

The fact that medical assistance is required for safe and effective performance of elective abortions or terminations of pregnancy points to a significant problem in the right to privacy rationale. Consider the following alternative implications: (a) the right to privacy refers only to the existential, solitary choice of an individual regardless of whether the choice can be exercised, or (b) it refers also to the social context in which the pregnant woman's choice can be practically implemented. The former alternative leaves us with an unacceptably ineffectual concept of "choice" or decision making, one which permits horrendous intrusions on free actions. On this basis, for example, one could claim that the pregnant woman who refuses cesarean section delivery has, by virtue of her privately determined refusal, adequately exercised her autonomy. The refusal of others to respect her decision does not then violate either her right to privacy or her right to choose differently than those others.

The second alternative exposes an erroneous assumption in the right to privacy rationale. Clearly, it is an error to assume that pregnant women can make these decisions in private, if "making a decision" involves power to implement one's choice. If and when the drug RU 486 or comparable means to terminate their own pregnancies safely and effectively are available to pregnant women without requiring medical approval, supervision, or assistance, then the privacy rationale will make more sense.[16] Meanwhile, for the vast majority of women, privacy is inevitably compromised through dependence on others for the performance of abortion or later (induced) termination of pregnancy.

The right to privacy of others, especially clinicians, also limits the choices of pregnant women. A study of obstetrician-gynecologists in a northeastern community in the United States showed that the majority of physicians in private practice, including those not opposed to abortion on moral grounds, elect not to perform abortions.[17] The right to determine the nature of their private practice had the effect of reducing the privacy of those patients who sought abortions, who were referred to clinics in another city to obtain the procedure. For women whose physicians request court orders to perform cesarean sections, the right to privacy is compromised not only by their dependence on the physician for medical assistance but by their situation becoming a matter of public record. In a tragic case such as that of Angela Carder, media access to court records may deprive the affected parties of privacy at a time when they most deserve and need it.

The points just made regarding the right to privacy can be made with regard to other health care decisions as well. For example, the right to refuse or terminate treatment, even lifesaving treatment, has often been defended on the basis of the patient's right to privacy, control over his own body, life, and so on; [18] yet the solitary decisions of individuals are seldom respected solely on that basis. An appeal to the right to privacy has not provided an adequate basis for resolving troublesome ethical questions that emerge in the clinical setting, particularly those regarding conflict between women's autonomy and fetal interests.

Viability

Curtailment of the right to privacy has generally been associated with obligations of beneficence and nonmaleficence, as exemplified in guidelines 2 and 3. For example, patients may be involuntarily hospitalized for a limited period of time if they pose a serious threat to themselves or others. The presumption in such cases is that the patient has not competently declined recommended treatment. Competent patients who are parents of dependent minors have been ordered to undergo lifesaving blood transfusions despite their opposition on religious grounds. Situations of refused cesarean section typically involve competent patients who decline treatment recommended for the sake of a viable or possibly viable fetus, for the sake of the pregnant woman, or for the sake of both. Such situations may include any of the following conflicts: (a) the wishes of the pregnant woman versus fetal interests where the woman's welfare is not at stake; (b) the pregnant woman's welfare versus fetal welfare where the woman's wishes are unknown; (c) the pregnant woman's welfare versus fetal welfare where the woman herself gives priority to the fetus; (d) the welfare of both the fetus and the pregnant woman versus the woman's wishes. Subsequent case descriptions develop these scenarios.

The *Roe* decision of the U.S. Supreme Court is instructive but not determinative regarding the role of viability in resolving such conflicts. Viability, which usually occurs at the end of the second trimester of gestation, determines whether the state's interests in the fetus *may* override a pregnant woman's decision to terminate a pregnancy. Prior to fetal viability, abortions are legally permissible for women of any age. Legislation may be enacted to ensure that abortions are performed in a medically responsible manner, but no state is required to fund abortions.[19] After viability has been achieved, however, the laws of many states preclude elective termination of pregnancy unless it is necessary for the health of the pregnant woman. The woman's right to privacy is then outweighed by the state's "interest" in the fetus. Supporting that interest, the U.S. Supreme Court affirmed in *Webster v. Reproductive Health Services* (1989) that the states may require physicians to conduct viability tests for fetuses at more than 20 weeks' gestation.[20]

The term *viable* is defined in *Roe* as "potentially able to live outside the mother's womb, albeit with artificial aid."[21] Although viability was estimated to occur at about seven months' gestation (twenty-eight weeks), *Roe* acknowledged that it could occur earlier, even at twenty-four weeks. Accordingly, the ruling stipulated both the numerical criterion (twenty-four weeks) and the achievement-of-viability criterion for determining when abortions might be proscribed by the states. The numerical criterion thus depends on the viability criterion. Since the *Roe* ruling, advances in neonatology and techniques for terminating pregnancy have moved viability to an earlier point in gestation, at least for those fetuses for which the new technology is available. These advances in part explain the willingness of the majority in the *Webster* decision to support the requirement of viability testing earlier than twenty-four weeks. As Justice Sandra Day O'Connor has suggested, *Roe* may be on a collision course with itself because of modern technology.[22] If "artificial aid" could support fetal life from conception onward, the viability criterion might allow the states to legally preclude elective termination of pregnancy at any stage of gestation. Termination would then be tantamount to induced preterm birth rather than abortion.

Although clinical texts define viability as "ability to survive ex utero," they also stipulate more conservative figures than *Roe* uses as the cutoff point between abortion and preterm birth: usually twenty weeks' gestation or less than 500 grams for fetal weight.[23] Similarly, the National Commission for the Protection of Human Subjects, while acknowledging that it was "not aware of any well-documented instances of survival of infants of less than 24 weeks...gestational age and weighing less than 600 grams," has set the boundary between nonviability and possible viability at 500 grams and twenty weeks' gestation.[24] The crucial concern involves recognition of different or greater obligations to fetuses or infants that are viable than to those that are nonviable. While acknowledging other possible cut-off points for determining obligations to the fetus (for example, "quickening," live birth), *Roe* establishes viability as the "compelling point" that permits the states to proscribe abortion. The rationale for selecting viability as the criterion is that "the fetus then presumably has the *capability of meaningful life* outside the mother's womb"[25] (my italics). *Capability* implies "not yet" with regard to "meaningful life", suggesting a responsibility toward those that are clearly not yet persons. But by introducing the term "meaningful", the court goes beyond its own definition of viability as "ability to survive." Fetuses stricken with severe neurological devastation are able in some cases to survive with assistance, but whether their lives are "meaningful" to them or others remains a legitimate question. As Peter Singer observes, the ruling offers no explanation of why "life outside the womb should be more 'meaningful' than life inside the womb."[26]

As already suggested, a better argument for the legal relevance of viability may be based on the ability of others besides the pregnant woman to ensure the survival of the newborn that the fetus may become. "Ability to survive ex utero" means that even if the pregnancy is terminated before term, the fetus can survive with the assistance of others. The term viability thus refers to individuals that are already alive either in utero or ex utero, and are capable of continuing to live (meaningfully?) for an indefinite period of time. Nonviability, in contrast, refers to living individuals that are incapable of continuing to live (meaningfully or not) for an extended period; in other words, they are dying. There is, of course, a trivial, extremely broad sense of "dying," by which living is itself construed as dying, because death is inevitable, and each moment of life brings us closer to the inevitable moment of death. On such an account, there is no distinction between viable and nonviable; nonviability characterizes all of life. Needless to say, those who posit the legal or moral relevance of viability invoke the narrower sense of the term, usually viewing it as survival with or without a "meaningful" level of existence. Whether or not this is a morally adequate criterion, it has apparently been pivotal to court orders for cesarean sections.

Although both *Roe* and *Webster* emphasize the importance of viability as a criterion by which the states may proscribe "abortion," neither denies the right of a woman to terminate pregnancy after the fetus is viable if the termination is considered necessary for her health.[27] Cesarean section delivery represents a risk to pregnant women's health that even the states' interest in the fetus may not override. Accordingly, while viability provides legal grounds for the states' "interest" in the fetus, it does not imply that the states may order a pregnant woman to undergo coercive surgery for the sake of the fetus.[28] Nonetheless, in cases such as the following, as in the case of Angela Carder, court orders for cesarean section have been requested and obtained, and the surgery performed, without the pregnant woman's consent.

Illustrative Cases Involving Cesarean Section without Consent

CASE I: REFUSAL OF SURGICAL DELIVERY RECOMMENDED FOR THE FETUS.

A woman who has been pregnant for thirty-four weeks experiences preterm labor after rupture of her membranes. The fetus is in breech position, and there are signs of fetal distress. The physician recommends cesarean section in order to optimize fetal outcome. The woman, who had successful outcomes from both vaginal and cesarean deliveries in previous pregnancies, refuses permission for the procedure.

CASE II: REFUSAL OF SURGICAL DELIVERY RECOMMENDED FOR THE PREGNANT WOMAN.

A 38-year-old woman, pregnant for the first time, experiences severe preeclampsia during the thirtieth week of her pregnancy. This condition, which threatens both the pregnant woman and her fetus, is usually improved dramatically after delivery. In this situation, however, fetal tracings suggest that the fetus is still well. Medication has failed to halt the woman's rising blood pressure, and she shows signs of imminent seizures and possible liver and renal failure. Delivery at this point would involve risk to the fetus because of lung immaturity. When cesarean delivery is recommended for the sake of the woman's health and life, she declines consent to the procedure, saying she prefers to risk her own health rather than jeopardize that of the fetus.

CASE III: UNCONSCIOUS PREGNANT PATIENT AT TERM.

An 18-year-old pregnant woman is brought to the emergency ward, apparently in diabetic coma. The woman's condition is stabilized, but she remains unconscious. Clinical examination indicates that this is a full term pregnancy, but the fetal heart rate is rapidly decelerating. The physician prepares to do an emergency cesarean delivery, without the woman's consent to the procedure.

CASE IV: REFUSAL OF SURGICAL DELIVERY RECOMMENDED FOR BOTH PREGNANT WOMAN AND FETUS.

During her last month of pregnancy, a woman suffers complete placenta previa, a condition that entails serious risk both to her and to the fetus until and unless delivery occurs. The physician opts to deliver by cesarean section, but the woman refuses permission on religious grounds. She realizes, she says, that her refusal places her own health in jeopardy.

The preceding scenarios illustrate different circumstances and reasons for refusal of treatment. In each case, the outcome could be positive or negative for either the woman or the fetus, regardless of whether the cesarean section is done. Case III differs from the others in three important respects: the woman is unconscious; she has not refused permission for the surgery; and the procedure must be done immediately if it is to save the fetus. According to a study published in 1987, in the majority (88 percent) of reported requests for court orders, the order was obtained in less than six hours, 19 percent were obtained in less than one hour, and at least one order was granted by telephone.[29]

The decision to seek a court order is in theory at least separable from the decision that a cesarean section should be done. For example, the order may be sought as a means of resolving the moral ambiguity of the parties involved. However, the opinions of most obstetrician-gynecologists are apparently divided rather than ambiguous on this issue. In the study just cited, twenty-seven of fifty-seven (47 percent) physicians who head fellowship programs in maternal-fetal medicine indicated their support of court ordered cesarean sections for the sake of the fetus, and their belief that the precedent should be extended to other procedures that might be lifesaving for the fetus, such as intrauterine transfusion. Moreover, the study suggested the improbability that requests for court orders would be denied. In all but one of fifteen reported requests, the orders to perform cesarean sections were obtained.[30] If the Appeals Court ruling in the Angela Carder case has impact in other jurisdictions, fewer court orders will be sought, and where they are, they will be less likely to be granted.

In particular cases, knowledge of the outcome tends to provoke the fallacy of Monday morning quarterbacking in reasoning about the morality of the decision to perform or not to perform a cesarean section. If the fetus dies or is seriously and irreparably damaged, then either decision is likely to be faulted, as also occurs if the woman is seriously harmed or dies. If both the woman and fetus do well, regardless of the decision, then either decision is likely to be judged appropriate. In none of these situations, however, is the outcome an adequate basis for assessing the decision. Positive outcomes have occurred for both the woman and the fetus where court ordered sections have been performed, but they have also occurred in cases of vaginal delivery after refusal of permission for cesarean surgery. Negative outcomes also occur, for both fetus and woman, in either circumstance.

Arguments Supporting Court Ordered Cesarean Sections

Egalitarian guidelines may be invoked by those who support as well as those who oppose coercive treatment after viability. In support of the procedure, consider first the right of the fetus to the life it has begun (guideline 2), particularly at a point where that life may be sustained ex utero. If fetal life is threatened by failure to perform a cesarean section (Cases I, III, and IV), this right to life needs to be weighed against the woman's right to refuse treatment (guideline 4). The right to life is more basic because it represents an essential condition for the autonomy that allows refusal of treatment. A similar argument can be made even if the fetus has no rights because it is not a person. We have responsibilities to living entities not to end their lives or cause them suffering (guidelines 2 and 3) without sufficient reason. The woman's refusal to undergo cesarean section (Cases I and IV) is not a sufficient reason if the life or health of a viable fetus outweighs respect for her autonomy.

A second reason in favor of court ordered cesarean sections in the interest of the fetus is that we have obligations to future persons that may be as great or greater than those toward present persons. These obligations relate both to the anticipated choices of future persons and to future persons' welfare. In both respects, for Cases I, III, and IV, it may be reasonably presumed that the future person that is presently a fetus would wish to be provided with optimal treatment, even though this might involve some risk to the pregnant woman who would be his mother. In Case IV, it may be reasonably assumed that the future person would also want the cesarean for the sake of the pregnant woman. In Case II, however, it should not be assumed that the future person would risk his own life or health (at fetal stage) for the sake of the pregnant woman; that presumption stretches beyond reasonable behavior to moral heroism. Moreover, if the fetus dies, consideration of obligations to future persons is no longer relevant.

Cases II and IV suggest another reason for court ordered cesarean section, namely, an obligation to promote the health or to save the life of a pregnant woman. This reason becomes more persuasive to the extent that the risk to the fetus of preterm delivery is reduced. A woman could, for example, seriously jeopardize her own health in order to optimize fetal outcome. A crucial point in both cases, however, is that the woman has indicated her preference, knowing that this is at odds with her own welfare. Accordingly, if we are to override her refusal of surgery for the sake of the fetus, it must be argued that beneficence or nonmaleficence toward her is the stronger claim. This line of argument is paternalistic, and therefore subject to justification or refutation on that basis.[31] The fact that the interests of both fetus and pregnant woman are thus promoted makes the argument for cesarean section stronger here than in the other cases.

A fourth reason could strengthen the argument for performing a cesarean in Case II, namely, the family's desire to save the woman's life, or reduce risk to her, where these interests are threatened by those of the fetus. Their motivation is paternalistic if it is based on the desire to promote her interests by overriding her autonomy; or self-interested if it is based on the desire to avoid their loss of her through death. Regarding others' interests, it should be acknowledged that we have moral obligations to others besides the pregnant woman and fetus, particularly to those most affected by the decision. The role of family members would be relevant in situations I, II, and IV. In Case III, I have assumed that no family member or proxy is available when the physician must decide about the cesarean delivery. In such cases, however, the primary responsibility of the proxy or family member is to inform caregivers regarding the patient's own views about the issue.

In Cases I, II, and IV, the input of others, especially family members, would ordinarily be helpful in assessing whether the patient's refusal of treatment is consistent with her general pattern of decision making. If the refusal is inconsistent with that pattern, it is possible that the patient is not making a truly autonomous decision. If it is honestly anticipated that the patient would later acknowledge that her refusal was not a rational, informed, and free decision, and that she would have agreed to the cesarean had she been truly autonomous, then we have a fifth reason supporting the procedure: the anticipation of subsequent consent or presumed consent. This rationale is more clearly applicable to Case III, where the woman herself is not capable of providing consent to cesarean section. Since we are unaware of the reason for her coma, it may be reasonably presumed that a woman who had apparently chosen to continue her pregnancy to term would agree to measures thought necessary to deliver a healthy infant. This presumption is tenuous, however, because it is also possible that the woman would not want the child to survive without her.

Several further reasons reflect the state's interests in conflicts between pregnant women and their fetuses: the obligation to protect the lives of viable individuals (regardless of their personhood status), and a concern about avoiding costs that might be incurred through the birth of a disabled infant. The government has a special responsibility toward those who cannot protect themselves. As already indicated, the *Roe* and *Webster* decisions extend that sense of responsibility to viable fetuses. For Cases I, III, and IV, these rulings argue for the court ordered cesarean, but Case II remains problematic because the pregnant woman may be viewed as needing protection also. The state's interests in limiting costs would argue for whatever interferences are least likely to produce morbidity, either for the woman or the fetus. A problem here is that, by monetary measure, death is the least expensive alternative.

One last reason in support of court ordered cesarean section is the claim that practitioners should do all that they can for the health of both fetus and pregnant woman. Their primary professional and moral obligation is to fulfill the basic imperative of helping or at least avoiding harm to patients.[32] Exceptions to that commitment are bound to impugn confidence in the physician–patient relationship, which can and should be therapeutic. This rationale argues for performing the cesarean without the pregnant woman's consent in Cases I, III, and IV, but not for II. It does not shed light on how a physician is to deal with the competing needs of equally needy patients. It does suggest, however, that if either the fetus or the woman is in greater need, the obligation to that individual is the greater.

Arguments Opposing Court Ordered Cesarean Sections

Among reasons that oppose court ordered cesarean sections, consider first the primacy of the informed consent requirement for treatment of any patient. Although the legal doctrine of informed consent is a comparatively recent development, it accords with a long-standing emphasis in American medicine on the freedom of the individual to choose or reject treatment.[33] Since cesarean section constitutes a surgical procedure, it should be noted that this emphasis has been particularly evident in surgery. Even when the treatment is lifesaving, the courts have generally upheld the refusal of treatment by a competent patient. In fact, treatment provided in the absence of informed consent (Case III) has been regarded as malpractice unless consent is legitimately presumed. Treatments for which these considerations apply include hospitalization. Every day competent patients sign out of hospitals against medical advice (AMA). Practitioners regretfully allow them to do so, aware of their legal obligation to respect the wishes of these patients, often despite disagreement with their decision. They thus recognize the limits of paternalism and the validity of guideline 4.

Second, consider the individual's right to bodily integrity. We are familiar with this claim in connection with abortion, and we generally recognize a responsibility to respect bodily integrity even with regard to the dead.[34] In Cases I, II, and III, we are violating that right by forcibly removing what is naturally within the woman's body. The fact that this type of coercion is expressed mainly by men (the doctors who request the court orders and the judges who grant them) and directed only toward women suggests an unequal (sex-based) standard for decision making.

A third reason for supporting a woman's refusal of cesarean section is that the personhood of the fetus has not been legally or morally settled. Even beyond viability, the U.S. Supreme Court does not assert that the fetus is a person but only that the state "has an interest" in the viable fetus. Different authors assign the onset of personhood to different points in the developmental process, including conception, brain activity, human resemblance, birth, and milestones achieved after birth.[35] So long as such ambiguity and disagreement remains, it is reasonable to maintain that the interests of the individual who is indisputably a person should prevail. Where there is conflict between the pregnant woman's welfare and autonomy, the priority to be observed if we are consistent with policy regarding refusal of treatment in other health care situations is clear: our first obligation is to respect autonomy. This priority implies that a woman is permitted to subject herself to serious risk by refusing a cesarean section in cases where fetal interests are promoted (Case II), as well as where the fetus is also subjected to serious risk (Case IV). In both cases, support for her decision is consistent with society's overall support of a competent person's informed refusal of lifesaving treatment. It is also consistent with the egalitarian perspective elaborated in earlier chapters.

Fourth, cesarean section should not be performed on a pregnant woman against her wishes because of the risk that surgery and use of anesthesia impose on the woman, and also because of the lengthier and more uncomfortable period of recovery that surgery involves as compared to vaginal birth. In Case I, the patient, who has already delivered healthy babies both vaginally and surgically, is more experientially aware of the pain and inconvenience associated with both procedures than is an obstetrician who has undergone neither of them. Surely, the woman should not be ordered to undergo the known burden and risk of surgery involuntarily.

Fifth, the state has an obligation to protect the individual against bodily harm. This reason may support a court order on paternalistic grounds for Cases II and IV, since in those instances the greater harm to the pregnant woman would probably be done by not performing the section. In Cases I and III, however, it argues against the cesarean, as subjecting the woman to greater harm. Consider the parallel on this point with cases of possible organ or tissue donation. Even where the possible donor can provide the organ or tissue least likely to be rejected, where the recipient is in lifesaving need of the donation, and where the donation is a relatively small risk to the donor (for example, in bone marrow or blood donation), no one has been forced to serve as a donor. It is inconsistent, therefore, to seek court orders for cesarean sections for viable fetuses, while we do not obtain court orders for organ or tissue donations for living persons. It is this inconsistency that grounds the charge of sexism with regard to the practice. So long as men's bodies may not also be violated for the sake of another life, the charge remains legitimate. As Annas puts it, women are thus treated as "unequal citizens," mere "fetal containers."[36]

Another argument against court ordered cesarean sections is that the obstetrician–gynecologist's primary commitment is to the patient with whom the physician–patient relationship is begun, that is, the pregnant woman. As indicated in Chapter 3, confidence in doctors has waned in recent years, especially among women, and especially with regard to obstetrician-gynecologists.[37] Increased expectation of coercive interventions during pregnancy is likely to further undermine trust, prompting some women to forego appropriate prenatal care entirely. This avoidance of care could in turn cause an increase in prenatal and perinatal loss or damage to women as well as to fetuses or newborns. Because abortion is a legal option for women at least until fetal viability is achieved, women's and fetal interests cannot always be promoted to the same degree throughout pregnancy. Moreover, there appears to be an appropriate complementarity between the responsibilities of the neonatologist and the obstetrician, a complementarity that is effective precisely because of the different roles of the two specialists. The physician may serve as an advocate for the woman in either or both of two ways: by respecting her autonomy and by promoting her health or welfare. If the physician primarily affirms the woman's autonomy, court orders will not be obtained in any of the cases described except Case III. If the physician gives priority to the woman's health, this decision supports cesarean sections in Cases II, III and IV, but probably not in Case I.

A final reason for opposing court ordered cesarean sections is the uncertainty of prenatal diagnosis. In one of the first reported cases where a court order was obtained, the woman delivered a healthy baby vaginally while the ruling was appealed.[38] The situation was comparable to that described for Case IV: the lives of both the woman and the fetus were at stake. In a report similar to Case I, although a cesarean was done over the woman's objection, the newborn was considerably healthier than had been anticipated, suggesting that the section might not have been needed. With regard to fetal viability, the case of Angela Carder dramatically illustrates the fallibility of prenatal assessment. In contrast to these outcomes, a case I encountered in 1987 resulted in the demise of the newborn when the woman's informed refusal of cesarean section was honored. The uncertainty thus evident in prenatal diagnosis, despite impressive improvements during the last decade, makes an argument for respecting the wishes of the pregnant woman still more compelling.

Consistency in the Face of Complexity

I have tried to elaborate a full and clear range of reasons both in support of and in opposition to court ordered cesarean sections to illustrate the complexity of the issue. However, certain assumptions could dispel the complexity and impel us toward a clear conclusion. For example, we could assume that no one ever has any responsibilities to a fetus, or we could assume that responsibilities to a fetus are even greater than those towards human beings already born. With the former assumption it would follow that court ordered sections are always morally wrong and should be prohibited. With the latter assumption, it follows that the pregnant woman's rights should always be superseded by the interests of the fetus, and that, therefore, not only cesarean sections but other interventions such as hospitalization, intrauterine transfusion, and fetal surgeries should be mandatory.

While both of these assumptions can be defended, each is obviously contestable. So long as neither wins the contest, the standoff serves only to further illustrate the dilemmatic nature of the issue. The dilemma may be irresolvable at a policy level because agreement about basic assumptions continues to elude us, as occurs with abortion. But the way in which American society has addressed the abortion issue is by prohibiting government interference in the decisions made by individuals. A similar response seems appropriate for refusal of cesarean section.

As already suggested, there are inconsistencies between our manner of dealing with refusal of requests for organ or tissue donation and our manner of dealing with refusal of cesarean delivery. As a judge in the state of Washington remarked, "I would not have the right to require the woman to donate an organ to one of her other children, if that child were dying.... I cannot require her to undergo that major surgical procedure for this child."[39] The analogy is of course imperfect. For example, the judge's use of the term *child* rather than *fetus* obscures an important difference between the situations, namely, that a prospective organ recipient is acknowledged to be a person, while the personhood of the fetus remains disputed. Moreover, although the judge referred to organ donation (kidney), an even stronger claim can be made with respect to tissue donation (for example, bone marrow). Cesarean section is a more invasive procedure than bone marrow donation, but the latter procedure has never been legally mandated, even for one's child or spouse. These differences between the procedures argue that the opposite situation should prevail, that is, that refusal of cesarean section should be managed less coercively than refusal of bone marrow donation. On egalitarian grounds, we have more stringent obligations to those who are clearly persons than those who are not, and to respect refusals when consent involves the greater risk.

The analogy with kidney donation is also less persuasive than the analogy with bone marrow donation because those afflicted with end-stage renal disease have other options: transplantation with an organ from a cadaver or continuation of treatment by dialysis. The fetus, on the other hand, is totally dependent on what is done or not done to the pregnant woman. Although the success rate of survival for transplanted bone marrow recipients has been improving,[40] the fetus has a chance of survival whether or not a cesarean section is performed. The uncertainty of prenatal diagnosis weakens arguments for coercive intervention.

Any of three alternatives would remedy the inconsistency of current practice regarding refusal of bone marrow donation as well as refusal of cesarean section for the sake of the fetus: (a) allow both coercive bone marrow donation and coercive cesarean section for the health of the fetus, (b) allow coercive bone marrow donation, but do not allow cesarean section if the pregnant woman does not consent to the procedure; (c) allow neither. Alternative (c) is consistent with an overall insistence on the right of any patient to refuse treatment or participation in another's treatment. Alternative (b) is consistent with an obligation to curtail an individual's choice only if that choice jeopardizes another person's welfare; so long as the fetus is not regarded as a person, this obligation does not extend to the fetus. Alternative (a) is consistent with an obligation to limit any choices that impugn other lives, including fetal life; this alternative would only apply to viable fetuses. If the other lives were to include nonviable fetuses, then yet more radical, liberty-limiting laws would have to be enacted.

Consistency would promote a more egalitarian situation than currently exists by applying the same standard to men as to women. In applying that standard, the difference between an ethic of virtue and an ethic of obligation is again introduced. Although most pregnant women act superogatorily in behalf of their fetuses, this surely does not imply the obligation of every woman to do so. Presumably, legislation imposes obligations that are equally applicable in relevant respects to all citizens. Until and unless others in society undergo comparable risks and coercion, pregnant women should not be required to undergo treatment involuntarily.

NOTES

1. The U.S. Supreme Court in *Automobile Workers v. Johnson Control* (Mar. 20, 1991) states that it is illegal to deny employment to women on the basis of their reproductive capacity. Despite the unanimity of that decision, it is possible that a more narrowly defined exclusion would be upheld, as suggested by the separate opinion filed by Justices Byron White, William Rehnquist, and Anthony Kennedy, and another filed by Justice Antonin Scalia. See Linda Greenhouse, "Court Backs Right of Women to Jobs with Health Risks," *New York Times* (March 21, 1991), A1, A12.

2. See George J. Annas, "Forced Cesareans: The Most Unkindest Cut," *Hastings Center Report* 12, no. 3 (June 1982): 16.

3. For example, American College of Obstetrics and Gynecology Committee on Ethics, "Patient Choice: Maternal-Fetal Conflict," *ACOG Committee Opinion* 55 (Oct. 1987): 1; Lawrence J. Nelson and Nancy Milliken, "Compelled Medical Treatment of Pregnant Women," *Journal of the American Medical Association* 259, 7 (Feb. 19, 1988): 1060; Nancy K. Rhoden, "The Judge in the Delivery Room: The Emergence of Court Ordered Cesareans," *California Law Review* 74 (1986): 1951; and Carson Strong, "Ethical Conflicts Between Mother and Fetus in Obstetrics," *Clinics in Perinatalogy* 14, no. 2 (June 1987): 313.

4. Ellen J. Stein, "Maternal-Fetal Conflict: Reformulating the Equation," in *Challenges in Medical Care,* ed. Andrew Grubb (Chichester: John Wiley and Sons, Ltd., 1992), 91–92.

5. George J. Annas, "She's Going to Die: The Case of Angela C.," *Hastings Center Report* 18, no. 1 (Feb./March 1988): 23–25.

6. See Linda Greenhouse, "Forced Surgery to Save Fetus Is Rejected by Court in Capital," *New York Times* (April 27, 1990), A1, A8.

7. Most of this chapter is a revised and updated version of my "Beyond Abortion: Refusal of Caesarean Section," in *Bioethics* 3, no. 2 (April 1989): 106–21.

8. Janet Gallagher, "Prenatal Invasions and Interventions," *Harvard Women's Law Journal* 10 (Spring 1987): 9–58; Susan Goldberg, "Medical Choices During Pregnancy: Whose Decision Is It Anyway?" *Rutgers Law Review* 41, no. 2 (Winter 1989): 591–623; Dawn Johnsen, "The Creation of Fetal Rights: Conflicts with Women's Constitutional Rights to Liberty, Privacy and Equal Protection," *Yale Law Journal* 95 (1986): 599–625; Nelson and Milliken, 1061–62; Alice Noble-Allgire, "Court Ordered Cesarean Sections," *Journal of Legal Medicine* 10, no. 1 (1989): 211–49; Rhoden, 1951–2030.

9. Except for Noble-Allgire, the authors mentioned in the preceding note are all opposed to coercive interventions, as are Annas ("Forced Cesareans") and Jeremy Wilde, "Caesarean Section: Whose Choice and for Whom?" in *Doctor's Decisions: Ethical Conflicts in Medical Practice,* ed. Gordon Reginald Dunstan and Elliot A. Shinebourne (New York: Oxford University Press, 1989), 45–50. In addition to Noble-Allgire, the following argue in support of coercive interventions, at least in certain circumstances: Eike-Henner W. Kluge, "When Caesarean Section Operations Imposed by a Court Are Justified," *Journal of Medical Ethics* 14 (1988): 206–11, and Strong, 313–28. While Strong acknowledges that treatment of the fetus requires invading the pregnant woman's body, he claims there are "no morally relevant differences between newborns and near-term fetuses" (p. 316). Frank Chervenak and Laurence McCullough maintain that coercive cesarean sections are permissible in situations of well-documented complete placenta previa, when the patient does not withdraw from the role of patient. See their "Justified Limits on Refusing Intervention," *Hastings Center Report* 21, no. 2 (March/April 1991): 12–17.

10. *Roe v. Wade,* 410 United States Reports 113, decided Jan. 22, 1973.

11. See Ruth R. Faden and Tom L. Beauchamp, *A History and Theory of Informed Consent* (New York: Oxford University Press, 1986), 39.

12. Faden and Beauchamp, 40.

13. *Doe v. Bolton,* 410 United States Reports 211–13, cited by Faden and Beauchamp, 40.

14. Judith Jarvis Thomson, "The Right to Privacy," in *Philosophical Dimensions of Privacy,* ed. Ferdinand D. Schoeman (New York: Cambridge University Press, 1984), 284.

15. *Roe v. Wade,* see note 10.

16. At this point in time (April 1992), even the use of RU 486 requires medical supervision and approval. Moreover, because the drug is as yet unavailable in most countries (including the United States), the vast majority of pregnant women are unable to avail themselves of this option.

17. Jonathan B. Imber, *Abortion and the Private Practice of Medicine* (New Haven, Connecticut, Yale University Press, 1986), 57. See Gina Kolata, "Under Pressures and Stigma, More Doctors Shun Abortion," *New York Times* (Jan. 8, 1990), 1, 8.

18. Faden and Beauchamp, 40.

19. State funding for abortions is addressed in various versions of the Hyde amendment, which permits states to refuse medicaid coverage even for "medically necessary" abortions, that is, when continuation of pregnancy might result in severe, permanent damage to the woman's health. See Lynn T. Shepler, "The Law of Abortion and Contraception—Past and Present," in *Psychiatric Aspects of Abortion,* ed. Nada L. Stotland (Washington, D.C.: American Psychiatric Press, Inc., 1991), 60.

20. *Webster v. Reproductive Health Services,* excerpted in *New York Times* (July 4, 1989), 10.

21. *Roe v. Wade,* see note 10.

22. Sandra Day O'Connor, dissenting in No. 81-746, *Akron v. Akron Center for Reproductive Health,* and No. 81-1172, *Akron Center for Reproductive Health v. Akron.*

23. For example, *Stedman's Medical Dictionary,* 23d ed. (Baltimore: Williams and Wilkins, 1976), 1551.

24. National Commission for the Protection of Human Subjects of Biomedical and Behavioral Research, *Report and Recommendations: Research on the Fetus* (Washington, D.C.: United States Department of Health, Education and Welfare, 1975).

25. *Roe v. Wade,* see note 10.

26. Peter Singer, *Practical Ethics* (Cambridge, England: Cambridge University Press, 1979), 108.

27. Nelson and Milliken, 1061.

28. Rhoden, "The Judge in the Delivery Room," 1989–94.

29. Veronika E. B. Kolder, Janet Gallagher, and Michael T. Parsons, "Court Ordered Obstetrical Interventions," *New England Journal of Medicine* 316, no. 19 (May 7, 1987): 1193.

30. Kolder, Gallagher, and Parsons, 1193.

31. See, for example, John Kleinig, *Paternalism* (Totowa, New Jersey: Rowman and Allanheld, 1984).

32. "Selections from the Hippocratic Corpus," in *Ethics in Medicine,* ed. Stanley J. Reiser, Arthur J. Dyck, and William J. Curran (Cambridge, Massachusetts, MIT Press, 1977), 7.

33. See Martin S. Pernick, "The Patient's Role in Medical Decision Making: A Social History of Informed Consent in Medical Therapy," in President's Commission for the Study of Ethical Problems in Medicine and Biomedical and Behavioral Research, *Making Health Care Decisions,* vol. 3 (Washington, D.C.: United States Government Printing Office, October 1982), 1–35.

34. See Joel Feinberg, "The Mistreatment of Dead Bodies," *Hastings Center Report* 15, no. 1 (Feb. 1985): 31–37.

35. For example, as developed by John T. Noonan in "An Almost Absolute Value in History" and Baruch A. Brody in "Abortion and the Sanctity of Life," both in *The Problem of Abortion,* ed. Joel Feinberg (Belmont, California: Wadsworth Publishing Company, 1973), 10–17 and 104–120; Mary Anne Warren, "On the Moral and Legal Status of Abortion," *The Monist* 57 (Jan. 1973): 43–62; and Lawrence Becker, "Human Being: The Boundaries of the Concept," *Philosophy and Public Affairs* 4 (1975): 334–59.

36. George J. Annas, "Pregnant Women as Fetal Containers," *Hastings Center Report* 16, no. 6 (Dec. 1986): 13.

37. See Paul Starr, *The Social Transformation of American Medicine* (New York: Basic Books, Inc., 1982); and Mary B. Mahowald, "Moral Quandaries in Obstetrics and Gynecology," *Journal of Psychosomatic Obstetrics and Gynecology* 4 (1985): 59–66.

38. Annas, "Forced Cesareans: The Most Unkindest Cut," 16.

39. Kolder, Gallagher, and Parsons, 1194.

40. Guido Lucarelli, Maria Galimberti, Paola Polchi, et al., "Bone Marrow Transplantation in Patients with Thalassemia," *New England Journal of Medicine* 322, no. 7 (Feb. 15, 1990): 417.

9

Fetal Tissue Transplantation

Among ethical issues raised by recent advances in medical research, few if any are more complicated or controversial than that of fetal tissue transplantation. In great part the controversy is due to the relation between this issue and abortion, which remains a polarizing issue in the United States and elsewhere. Different positions about whether fetal tissue might be used for treatment of otherwise incurable neurological disorders have emerged from disagreement about whether its use would increase the incidence of elective abortions. Even among those who support the right of women to choose abortion, some are concerned that the availability of techniques for fetal tissue transplantation might encourage more abortions.

Early in 1986, I first learned of the exciting possibility that severely debilitating neurological disorders might be treatable through use of fetal tissue transplants. Jerry Silver, a colleague at Case Western Reserve University, was then studying regeneration of the central nervous system in mice. His comments on the pace at which his and others' work in transplantation was progressing, and his concern regarding human applications, suggested a significant project for our Ethics Center: to convene a forum of knowledgeable experts who might facilitate an informed public debate. Since the complexity and controversy surrounding the issue could be anticipated, I hoped that our forum might provoke prospective analysis of the issue, yielding useful input for researchers, caregivers, patients, and others who might be affected by the technique.

To that end, my colleagues and I convened a group of people whose expertise and background were relevant to fetal tissue transplantation. Our meeting on December 4–5, 1986, was supported by the School of Medicine at Case Western Reserve University. The presenters were neuroscientists, ethicists, and law professors. Other participants were drawn from different specialties of medicine, the National Institutes of Health (NIH), and the public-at-large. A statement summarizing the views of the presenters was published in *Science* on March 13, 1987.[1] It called for separation of timing and procedures for fetal tissue retrieval, anonymity between donors or their proxies and recipients, and adequate review of specific proposals to use fetal tissue for transplantation.

In the spring of 1988, the NIH ordered a moratorium on use of tissue from electively aborted fetuses by federally supported researchers.[2] The NIH also convened its own panel to address the issue. In December of that year, the NIH panel produced a report that mainly reiterated the recommendations of our 1986 forum.[3] Despite these recommendations, the issue has remained politically problematic, and the NIH has extended the moratorium indefinitely.[4] At this writing, efforts by legislators to reverse the moratorium have failed to override the president's veto. Measures designed to appease the concern of advocates of the moratorium have been described as obstructive of the therapeutic relationship and a woman's right to privacy.[5] One such proposal permits the federal government to inquire into a woman's reasons for abortion.[6]

Some may think it naïve to believe that the political stalemate can be resolved through the force of ethical arguments. If the arguments are valid and based on verifiable premises, however, their cogency is likely to influence public opinion, which in turn influences legislators who are dependent on public support for election and re-election. In this chapter, I provide an argument for conditions that should prevail if transplantation of fetal tissue is to be governed by egalitarian considerations. First, I develop a number of distinctions or features that are morally relevant to the issue. These include the empirical status of fetal "donors," different purposes and sites of tissue retrieval or implantation, therapeutic potential, abortion as means of access to tissue, and different motives for "donation."[7] I then examine opposing views on the morality of fetal tissue transplantation. As with other issues, careful and critical consideration of its complex dimensions leads to recognition that neither total restriction nor total permission is an appropriate moral stand. What is needed is a position that limits harm, promotes benefits, and respects autonomy in as equitable a manner as possible among those affected.

Empirical Status of the Source of Tissue

The human fetuses or abortuses used for tissue grafts may be distinguished as shown in Figure 9.1. In clinical texts the human fetus is defined as a postembryonic organism contained within the body of a pregnant woman.[8] Unlike the embryonic stage, when development of most organ systems is initiated, the fetal stage of development is marked mainly by elaboration of these systems. As already mentioned, the transition from embryonic to fetal stage occurs at about the eighth week of pregnancy, and the term *fetus* describes the developing organism from then until the end of pregnancy. Although many abortions involve embryos rather than fetuses, the term fetus is frequently used for either situation. The term *abortus* is used to describe the previable organism (previously an embryo or fetus) when it has been expelled or removed from the woman's body.

Fetuses and abortuses are human when the gametes from which they develop are human. This meaning of the term *human* is relatively uncontroversial because its validity is based on biology rather than philosophical agreement. In contrast, the term *person* remains philosophically controversial.[9] Whether persons or not, however, human fetuses or abortuses are genetically distinguishable both from the pregnant women in whom they develop and from other human organs or tissue that may be contained or developed within the bodies of women. Although the individuation of embryos is not definitively established until about fourteen days after fertilization, genetic distinctness is initiated by fertilization.

Besides the humanness and genetic distinctness of human fetuses and abortuses, ethically relevant aspects of their empirical status include questions of whether they are alive or dead, and if alive, whether they are viable or nonviable. If viability of the fetus is defined as "ability to survive ex utero, albeit with artificial aid,"[10] nonviable fetuses and abortuses are those unable to survive even with assistance. Nonviability thus implies that an entity is alive but dying. Thus far, social policy has excluded donation of vital organs or tissue from live donors, even when death is imminent.[11] The probability of success in transplantation increases through use of tissue from a living (rather than dead) fetus, but retrieval of vital organs or tissue before natural death is tantamount to killing the donor. Presumably, a departure from the long-standing precedent opposed to such retrieval (even if requested by a dying donor) can only be justified on grounds that the human fetus is not a person whose life should be supported until natural death occurs.

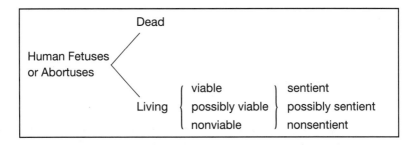

Figure 9.1. Possible status of human fetus or abortus.

Even if fetuses are not persons, another empirical factor related to the use of living fetuses for transplantation is their possible sentiency or capacity to experience pain. Recent research on infant pain suggests the capacity of mature fetuses in this regard.[12] Just when the nervous system is sufficiently developed for earlier fetuses to experience pain is not known. But even the possibility of experiencing pain is morally relevant. If retrieval of tissue from living fetuses were ever permissible (a doubtful possibility), it would still be morally incumbent on the "retrievers" to avoid pain on the part of the fetus, for example, through use of analgesics. Use of tissue from a dead fetus bypasses the problem of possible fetal sentiency as well as that of direct killing through transplantation. Accordingly, the consensus statement of the forum convened in 1986,[13] as well as that of the NIH panel that considered the issue in 1988,[14] assumes that only tissue from fetal remains will be used for transplantation.

Different Purposes, Sites of Tissue Retrieval, and Sites of Implantation

Neural grafting is a surgical procedure for transplanting tissues from various sources into specific areas of the nervous system that have been affected by a neurological disorder, disease, or injury. Note the following summary of differences in sites and purposes that may be morally relevant to decisions concerning grafts of fetal tissue:

From	*To*	*For*
Fetal brain	Recipient's brain	Research
Other parts of fetus	Other sites in recipient	Experimental treatment
		Therapy

The sites from which fetal tissue has been retrieved for neural grafting include the adrenal gland as well as the brain.[15] While most experimental work in neural grafting has involved transplantation into the recipient's brain, additional sites include the spinal cord and the peripheral nervous system.[16]

Adrenal fetal transplants into the recipient's brain, and transplants from or into parts of the body other than the brain (for example, fetal thymus, liver, pancreas), are ethically less controversial than transplants of fetal brain tissue into the recipient's brain. Unlike other organs, the brain is prevalently identified with an individual's distinct personality. To the extent that tissue removed from the fetal brain represents a distinctly different personality than that of the recipient, the problem of identity arises: Would the recipient thereby assume a different (or perhaps an additional) personality? According to the Office of Technology Assessment, "neural grafting may present a threat to personal identity and sense of self, aspects of the human mind that are, to some persons, the very essence of humanity."[17] Presumably, the import of the identity problem depends on the proportionate amount of tissue transferred and the degree of development achieved by the fetal brain. The size of a fetal brain is, of course, much smaller than the brain of an adult. Just how much fetal brain tissue should be transplanted for optimal results in the recipient has not yet been determined. Theoretically, it is at least possible for a minute portion of brain tissue to be removed stereotactically from a living donor, and implanted into the brain of a recipient without seriously jeopardizing the health or identity of the donor.Using so minute a portion of fetal brain tissue seems to trivialize the question of recipient identity. It may also be the case that tissue from several fetal brains are required for a successful transplant. In Sweden, Olle Lindvall and his colleagues have used tissue from four fetuses to treat one patient suffering from Parkinson's disease.[18]

The problem of identity is also undercut by the fact that the fetal brain is relatively undifferentiated in comparison with the recipient's brain. Obviously, the earlier the gestation, the more undifferentiated the fetal tissue. But neither the proportionate amount nor the degree of development of fetal brain tissue affects the recipient's identity unless the brain truly is the source of that identity. Because scientific evidence is apparently unable to explain this relationship solely on empirical grounds, it remains a matter of philosophical debate. Different concepts of a nonphysical "soul," "mind," "person," or "individual self," and its relation to the physical brain give rise to various theories, all of which are subject to critique and controversy.[19]

Even if it were established that the brain truly is the source of personal identity, the relevance of this view to fetal tissue transplantation must be weighed against the probability that the recipient's life would otherwise be cut short or lost. Thus the argument: Better to live even with another's identity (or partial identity, or identities) than not to live at all. Life, as affirmed in guideline 2, is the more fundamental value, on which identity itself depends.

In contrast with brain grafts, neural grafts from or into the spinal cord or peripheral nervous system are no more likely to evoke concerns about personal identity than allografts of other tissue or organs. With all grafts, however, as with medical procedures in general, the purpose may be therapeutic, nontherapeutic, or both. Each of these possibilities needs to be considered.

Although the nontherapeutic object of research or experimentation is knowledge, the purpose of the knowledge to be obtained through medical research is generally therapeutic for future patients. In other words, it is hoped that future patients will benefit by the knowledge that researchers provide. One may distinguish, then, between the immediate or direct goal of medical research and its indirect goal or purpose. While knowledge is surely a good worth pursuing, some view it as an instrumental rather than intrinsic good. In contrast, the goods of health, life enhancement, and life-preservation are generally seen as worth pursuing in their own right. The therapeutic goal of medical research is thus more compelling than its nontherapeutic focus on knowledge.

Medical procedures become therapeutic when their success is adequately demonstrated through cure or amelioration of human disease, sickness, or disability. At present, fetal tissue transplantation for treatment of severe neurological disorders in humans is experimental because its efficacy is not yet established, even with regard to the disease where research has been most promising, namely, Parkinson's disease.[20] Successful treatment of induced Parkinsonian symptoms in subhuman primates suggests the likelihood of success in applying the technique to humans.[21] The Parkinsonian symptoms induced in monkeys through methylphenyltetrahydropyridine (MPTP) may not be adequately analogous to the actual symptoms of humans with Parkinson's disease.[22] Nonetheless, the NIH panel concluded that animal studies "already provide sufficient evidence to justify clinical fetal transplantation experiments for treating Parkinson's disease and diabetes."[23] Although the NIH has not supported their research, U.S. scientists have reported some success in transplanting both fresh and frozen neural fetal cells into Parkinson's patients.[24] Scientists from Mexico, Sweden, China, Cuba, Poland, and Great Britain have also reported preliminary success.[25] More work needs to be done, however, before the procedure might properly be considered therapeutic rather than experimental.

Although surgical procedures in humans are generally undertaken for therapeutic rather than research purposes, the first and early uses of new procedures are considered experimental because it is not known whether they will be therapeutically effective or not. Fetal tissue transplantation for treatment of neurological disorders can only be extended to human beings on an experimental basis; but the main purpose of even the early human applications may still be therapeutic. If a disease is not curable by any known means, as is the situation with Parkinson's and most other neurological diseases, an experimental treatment may be primarily intended to restore health or alleviate symptoms rather than to gain knowledge. In a broad sense, both motives are present in all medical treatments.

Therapeutic Potential of Fetal Tissue Transplantation

It has been known for some time that unique properties of fetal tissue enhance the prospect of success in transplantation.[26] Fetal tissue is less immunologically reactive than mature tissue and is thus less likely to be rejected than adult tissue. Fetal tissue also has a greater developmental capacity than adult tissue. This capacity may be attributed to the relative lack of differentiation and rapid growth of fetal tissue, which develops both physically and functionally after transplantation.

Neural fetal tissue grafts are expected to be more effective in treatment of Parkinson's disease than grafts of adrenal tissue from fetuses. Autografts of adrenal tissue bypass ethical problems raised by use of fetuses, but the efficacy of these grafts has been broadly challenged.[27] It is reasonable to postulate that neural grafts will generally be more efficacious than grafts from other sites for treatment of neurological disorders. Use of neural rather than adrenal tissue for allografts may provoke the question of personal identity, as previously discussed.

Although research involving neural grafts from immature to mature animals dates back to the early part of the century, only recently have we seriously entertained the expectation that severe human neurological diseases may be cured through neural grafts from human fetuses.[28] The accelerated pace and success of research by neuroscientists have precipitated this hope on the part of millions affected by these disorders. To estimate the impact that successful human applications of the technique may engender, consider the following potentially treatable neurological diseases, injuries, or disorders: Alzheimer's disease, amyotrophic lateral sclerosis, epilepsy, Huntington's disease, multiple sclerosis, Parkinson's disease, spinal cord injury, brain injury, and stroke.[29]

Alzheimer's disease alone has a prevalence rate of 1.9 to 5.8 cases per 100 of the population 65 years of age or older,[30] and the incidence of Parkinson's disease is 20.5 per 100,000 of the population in the United States.[31] The very fact that such diseases may be listed as potentially curable marks a departure from the view that had been a given of neurological practice. Although brain damage had always been considered irreparable, this may no longer be the case.

How Human Fetal Tissue Becomes Available

Generally speaking, the means through which fetal tissue becomes available for transplantation is abortion. Consider, however, the distinctions shown in Figure 9.2 regarding that choice. Although procedures such as D&C (dilatation and curettage) and surgical treatment of ectopic pregnancies have not always been defined as abortions,[32] they are abortive if they result in the removal of an embryo or fetus from a woman's body. The intent of these procedures is to provide appropriate medical treatment of the pregnant woman rather than to terminate embryonic or fetal life. In treatment for ectopic pregnancies, the fetus could not survive even if surgery for the woman were withheld. In D&C, standard practice includes determination of pregnancy before the procedure is performed.

The terms *induced* and *therapeutic* are sometimes used interchangeably for abortions that are not spontaneous. Yet not all induced abortions are therapeutic; some are elective. An abortion is elective if it is undertaken for nonmedical reasons, and therapeutic if undertaken for medical reasons. *Medical reasons* given for therapeutic abortions may be associated with the pregnant woman's health or

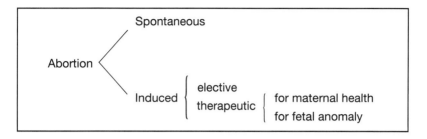

Figure 9.2.   Types of abortion.

with fetal anomaly.[33] Thus, the object of the therapy remains ambiguous. In some cases, the fetal anomaly that prompts an abortion is incompatible with neonatal survival (for example, anencephaly); in other cases, the fetal anomaly is compatible with a meaningful life (for example, trisomy 21).

The first treatment of Parkinson's disease through use of human fetal tissue relied on spontaneous abortion as the means through which the tissue was acquired.[34] There is concern, however, that this source is inappropriate because tissue obtained from spontaneous abortions is more likely to be anomalous than tissue obtained from elective abortions.[35] Moreover, because spontaneous abortions usually occur outside of hospitals, the likelihood of retrieving viable tissue by this means is not very high. Ectopic pregnancies have also been proposed as a source of fetal tissue for transplantation. Over 75,000 ectopic pregnancies occur each year, and the fetal tissue removed through surgery for this life-threatening condition is less likely than tissue from spontaneous abortions to have an increased incidence of chromosomal abnormality.[36] From an ethical point of view, surgical removal of an ectopic pregnancy is comparable to therapeutic abortion for the pregnant woman's health, with the added caveat that the circumstances are already fatal for the fetus.

As we have already seen, elective (nontherapeutic but induced) abortion remains morally problematic for many despite its legality. Even those for whom elective abortion is morally justified may consider it a tragic alternative or last resort. This view underlies concern that the option of fetal tissue transplants might trigger a greater incidence of abortions. A key issue, therefore, is whether elective abortion, as the main means through which fetal tissue is acquired, is morally separable from the therapeutic goal of neurological regeneration for recipients. The issue calls for reexamination of the traditional moral dilemma involving the relationship between means and ends. Does the end of treating neurological disorders justify the means of using fetal tissue?

A simplistic version of utilitarianism supports an affirmative answer to the question. In other words, the tremendous good that might be accomplished through the new technique outweighs the harm that might be done through elective abortion. However, if endorsement of the procedure leads to widespread increase in elective abortions, a reduced sense of the value of human life, and to exploitation of women, it is possible that such an array of undesirable consequences would outweigh the potential benefit of the technique.[37] So, even if the end justifies the means, it is not clear that it does so in this case. Whether or not the overall consequences of treating neurological disorders through fetal tissue transplantation will generally constitute a preponderance of harms over benefits is an empirical issue for which more data is needed to support a credible utilitarian position.

From a deontological point of view, the end does not justify the means, but this does not necessarily imply that fetal tissue transplantation is morally justified. The individual who knowingly and freely pursues a specific end, also (knowingly and freely) chooses the means to its fulfillment. In other words, intention is crucial to the moral relevance of the relationship. My use of the term *intention* implies deliberateness or choice. If one were to deliberately become pregnant, choose abortion, or persuade another to do so solely for the sake of fetal tissue transplantation, one would then be responsible for both means and end because one would be intending both. Along with that intention, the motive of the decision may be altruistic (as in treatment of a relative or anonymous patient), self-interested (as in treatment of oneself, or profit through sale of the tissue), or mixed (that is, self-interested *and* altruistic). Although worthy motives are morally relevant at least from a subjective point of view, they do not alter the fact that the intention in such cases applies to both ends and means.

In other situations involving fetal tissue transplantation, the individual who intends to use the tissue does not intend the abortion through which the tissue becomes available. Presumably, he does intend the retrieval procedure. But just as a transplant surgeon may retrieve essential organs from the brain-dead victim of a drunk-driving accident, without any implication that she thus endorses the behavior that led to the availability of the organs, so may a neurosurgeon who is opposed to abortion transplant neural tissue from a dead fetus electively aborted into a severely impaired Parkinson's patient, without thereby compromising his moral convictions. In fact, one may argue that a truly pro-life position favors the affirmation of life that the transplantation involves, while acknowledging the negation of life that abortion implies. When the abortion decision has already been made by others, a decision not to transplant seems less in keeping with a pro-life position than its opposite.

It is possible, of course, that potential beneficiaries of fetal tissue transplantation are, or become, morally complicitous in the elective abortions through which fetal tissue becomes available. Moral complicity occurs to the extent that individuals allow their association with an immoral act to imply or engender approval of the act. James Burtchaell claims that this type of situation "is detectable when the associate's ability to condemn the activity atrophies."[38] Whether or not complicity is detectable, intention remains the basis for distinguishing between actual complicity and mere association, and whether or not complicity is immoral depends on whether the abortion is undertaken for immoral reasons. As Dorothy Vawter and her colleagues have remarked, "only those who directly intend to use tissue from immoral abortions may be charged with being complicit because they knowingly benefit from such abortions."[39] If elective abortion is morally justified, the charge of complicity is misplaced.

Possible "Donors," Recipients, and Motives for "Donation"

The preceding section alluded to possible motives, donors, and recipients regarding fetal tissue transplantation, but each of these topics deserves fuller consideration. Consider first, then, the following possible motives for "donation" of fetal tissue: self-interest (personal gain), altruism (pursuit of the welfare of another or others), or a combination of self-interest and altruism. The motives of concern in discussions of the morality of the procedure are those of the individuals who provide the tissue or give consent for its use, rather than those of the recipients or others involved in the procedure. Such an individual may be called the "donor," but the term seems inappropriate if the motive for providing the tissue is self-interest. Lest we think that commercialization impugns the morality of organ or tissue donation, or renders it morally neutral, it should be recognized that society represents a diverse range of positions on comparable matters. For example, we reward beauty queens and sports stars for their physical attributes and performance, while we legislate against prostitution and commercial surrogate motherhood. Ordinarily, no payment is made for organ or tissue donation, but as we indicated in Chapter 7, compensation constitutes remuneration for some egg and sperm donors.[40]

Selling fetal tissue for economic profit is only one way of illustrating the probable motive of self-interest. Psychological needs undoubtedly motivate some of those whose outward gestures appear altruistic. Thus a woman who wants to curry her elderly father's favor may offer to become pregnant to develop a fetus as source of treatment for his Parkinson's disease. Or a woman who feels troubled about an abortion decision may offer to donate fetal tissue as a means of assuaging her own discomfort. Since motives are subjective, it is impossible to determine solely on the basis of observation that specific behaviors are altruistic or self-interested; most behaviors probably exemplify both motives. Moreover, it is not at all clear that self-interest is indicative of immoral or morally neutral behavior. Reputable philosophers such as W. D. Ross maintain that "self-improvement" is a prima facie duty,[41] and several significant contemporary theories of social justice are based on a view of human nature as essentially self-interested.[42] It thus seems wrong to negate the possibility of obtaining tissue solely on grounds that the individual who provides it is motivated by self-interest.

Second, consider the following possible donors of fetal tissue: the pregnant (or previously pregnant) woman, the partner of the pregnant (or previously pregnant) woman, and potential recipients. According to Kathleen Nolan, the "donor is the patron who bestows the [fetal] cadaver, the one who has given up something to another."[43] Obviously, the fetus or abortus cannot donate itself for transplantation. But the concept of donor (of fetal tissue) applied to the pregnant woman who has chosen to abort a fetus is also problematic. Presumably, by electing abortion, the woman has disassociated herself from the role she would have maintained by continuing the pregnancy, that is, one of concern about the interests of the fetus. Nonetheless, even the woman who elects abortion may be asked to give up something, for which she would then be the donor. She may be asked to continue the pregnancy to a later stage of gestation, and undergo an abortive procedure that she otherwise would not have undergone—in order to optimize the chance of success of fetal tissue transplantation for the recipient.[44]

If tissue from an abortus is requested after a woman has aborted, her body will no longer be affected by the decision. On the view that an abortus is the property of the woman who aborts it, she may donate or sell that property to whomever she likes. If the abortus is not property and the abortion was spontaneous or therapeutic, the woman might serve as donor or proxy for the fetus. If the abortus is not property and the abortion was elective, the woman may thereby forfeit donor or proxy status.

A concept of the fetus or abortus as property may also apply to the man responsible for the pregnancy. Ordinarily this person is the woman's husband or her sexual partner, but it could be an anonymous sperm donor. If the man responsible for the pregnancy is seen as "part owner" of the fetus or abortus, then he may donate or sell his "share." If the fetus or abortus is not property, the same parameters that apply to women regarding the distinction between spontaneous and elective abortion apply also to men. To the extent that men join women in electing abortions, their status as donors or proxies for the fetus may be impugned. However, so long as women and not men gestate fetuses and undergo abortions, it hardly seems defensible to claim that men are more appropriate donors of tissue from fetuses and abortuses.

Third, consider the following possible recipients of grafted fetal tissue: an anonymous recipient (one unknown to the donor or the donor's proxy), a relative, or oneself. Proposed guidelines for fetal tissue transplantation recommend an anonymous relationship between donor and recipient.[45] This is consistent with prevalent practice regarding cadaver donation, as well as blood, sperm, or ova donation. Anonymity is a way of avoiding conflicts that might arise because donor and recipient know each other, for example, social pressures or concerns about identity. Ironically, anonymous donation may not only be compatible with an altruistic motive but more conducive to it than donation to one's relative. A gift to an anonymous recipient is less likely to benefit a donor than a gift to a known or (*a fortiori*) designated recipient.

All three types of recipients may be provided with fetal tissue for self-interested reasons on the part of the provider. But donation to a relative, particularly a close family member such as one's son or daughter, parent, or spouse, may be construed as morally obligatory precisely because of the relationship. A care-based model of moral reasoning supports this position, but would extend the argument to relationships based on ties of affection and commitment, which may be more binding than those based on biological ties.[46] One is ordinarily obliged to do all that is reasonably possible to save the lives of those to whom one is closely related; it is virtuous or heroic to do that for others. In order to keep tissue or organ donation a matter of generosity rather than obligation, therefore, anonymous donors are preferable to relatives. Although family members may exert pressure, the legal system has persistently insisted that organs and tissue be provided on a voluntary rather than coercive basis, even when the prospective donor is a close relative. This is consistent with guideline 4: those who can think and choose for themselves should not have their thoughts or choices ignored or impeded.

A Spectrum of Positions

Despite the complexity that the foregoing considerations introduce, their moral relevance seems undeniable, particularly from an egalitarian perspective that emphasizes differences. Depending on how different features are weighed, different moral positions result. Together they represent a spectrum of possibilities.

At one end of the spectrum is the view that the use of fetal tissue is always morally wrong and we are, therefore, morally obliged never to participate in the procedure. At the other end is the position that fetal transplantation is always morally permissible and perhaps even morally obligatory. Between these extremes are a variety of views stipulating that fetal tissue transplantation is right in some circumstances and wrong in others.[47]

Consider first the view that human fetal tissue should never be used for transplantation. The following claims support this position:

• Deceased human fetuses have as much right to respect for their bodily integrity as do other deceased human beings. The mere fact of their immaturity is not adequate grounds for treating them less respectfully.

- Human fetuses cannot provide consent for donation of their tissue after death. Consent of donors, either immediately or through advance directives, transforms the situation from one in which the "donor" is objectified to one of respect for her autonomy.
- Obtaining tissue from human fetuses for research or therapeutic purposes is likely to lead to trafficking in fetal tissue and general disregard for the value of human lives. Neither of these consequences is socially desirable or morally justifiable.
- Retrieval of tissue from dead fetuses implies approval of the means through which the tissue becomes available, namely, abortion. If we assume that human fetuses are persons, induced abortion is tantamount to murder.

Next, consider the view that use of human fetal tissue is always morally acceptable, and perhaps even morally obligatory. The following claims support this position:

- Human fetuses are not persons, and they are therefore not entitled to the respect which is due to donors who are persons. American law as well as laws of other nations view birth as establishing the rights of persons, and some ethicists argue that personhood is not established until certain developmental milestones are achieved after birth.
- The fetus is the property of the pregnant woman, in the sense that any part of a person's body belongs to that person.
- Important therapeutic benefit to the recipient justifies taking the tissue from the fetal donor. So long as the fetus is already dead or destined to die, the burden/benefit ratio strongly supports the possibility of ameliorating or curing severe and otherwise incurable neurological disabilities through transplantation.
- Acquisition of fetal tissue is separate from its use for transplantation, comparable to the separation between the availability of organs from brain dead cadavers and transplantation of those organs into adult recipients. Accordingly, the use of fetal tissue for transplantation in no way implies encouragement or moral tolerance of abortion.

Each of the preceding positions — those supporting as well as those opposing fetal tissue transplantation — can be validly defended at greater length. As with many ethical issues, positions at either end of the spectrum can be articulated more clearly than those that fall between the extremes. It is simpler, therefore, to maintain that fetal tissue transplantation is always wrong or always right, than to maintain a position that allows or prohibits the procedure only under certain conditions. However, complicated questions are likely to require complicated responses, and significant issues deserve the care and effort necessary to determine responses that reflect their significance.

By taking account of complexity, an egalitarian position necessarily falls within the opposite ends of the spectrum. The distinction and features elaborated earlier suggest a number of limitations that may be introduced. First and perhaps most obvious of these is that living fetuses or abortuses, whether viable or nonviable, should be excluded as sources of vital tissue. Retrieving tissue from a living survivor of an abortion is not only morally but empirically equivalent to retrieval of tissue from a preterm, nonviable newborn. Either act obviously violates guideline 2 (individual lives should not be shortened). Depending on the developmental status of the sensory system, a living fetus may be able to experience pain, and that implies a moral responsibility to avoid the pain that tissue retrieval might involve (see guideline 3: those that can suffer should not be be caused to suffer). Restricting the source of tissue to fetal cadavers or (dead) abortuses bypasses these problems, and parallels the situation of tissue retrieval from nonfetal cadavers or brain-dead donors.

Regarding different purposes served through fetal tissue transplantation (guideline 4), therapeutic justifications are generally more compelling than research justifications. Nonetheless, some form of experimental treatment provides a necessary link between research and established therapy, and fetal research has generally been judged morally acceptable under certain conditions, such as consent from the parental proxy for the fetus and from the prospective recipient of the tissue. Guideline 5, which suggests that the natural development of fetuses should not be obstructed, does not apply to those whose development has already been terminated by abortion.

Despite possible problems raised by the use of brain tissue, therapeutic efficacy should be the key factor in determining whether tissue is retrieved from the brain or from other parts of the fetus. The speculative issue of personal identity is less compelling in light of the perennial nature of the mind/body debate, and the immediate needs of severely impaired persons whose lives may be saved, extended, or improved through the technique of fetal tissue transplantation. The same rationale applies to the site of implantation into the recipient. An obligation to support life (guideline 2), especially human life (guideline 7), seems to prevail.

Clearly, a procedure as ethically complex as fetal grafts ought not to be undertaken without reasonable expectation of significant positive results. If there were equally efficacious and less ethically problematic ways of treating neurological disorders, use of fetal tissue transplants would probably not be morally justified. As it is, however, there is little basis for opposing its possible applications to severe neurological disorders for which no alternative effective therapies are known, as long as others are not harmed and their autonomy is respected.

The means through which the tissue becomes available is the most controversial and complex aspect of fetal tissue transplantation. Different types of abortions represent a range of ethical controversiality. As means of acquisition, spontaneous abortions are not problematic if the tissue they provide is likely to benefit the recipient. Abortions undertaken for fetal defect raise questions (also raised by spontaneous abortions) regarding obligations to the recipient to use only healthy fetal tissue. Therapeutic abortions for maternal health and surgical removal of ectopic pregnancies are comparable to spontaneous abortions in that they do not involve direct intent to abort the fetus, but rather an intent to preserve the pregnant woman's life or health. Although we are morally responsible for acts of permission as well as commission, use of tissue from fetuses already aborted does not imply permission for their abortions.[48]

As presented in Chapter 4, elective abortions are morally problematic especially when they are undertaken during late gestation for reasons such as maternal inconvenience. The optimum time for fetal tissue transplantation is probably during the first trimester. While decisions regarding elective abortion and transplantation of fetal tissue may be morally and practically separable, it may be that the two are not separate in people's minds.

Setting aside medical criteria that are morally obligatory for determining whether neurografts should be performed, tissue for neurografts is ethically obtainable from spontaneous abortions and therapeutic abortions. Regarding elective abortions, however, the choice, timing and procedures regarding abortion should be distinguished from decisions regarding transplantation. If permissibility of fetal tissue transplantation leads to an increase in elective abortions, this would be a morally undesirable consequence calling for greater restrictiveness in use of the technique.

Because motives for donation of fetal tissue are probably not clearly determinable, it is futile to restrict fetal tissue transplantation on that basis. Because motives are often mixed, and self-interested as well as altruistic motives may be morally justified, it seems unjust to do so. But straightforward commerce or trafficking in fetal tissue introduces another dimension, namely, the possible exploitation of women who might be paid to provide fetal tissue for transplantation.

Commodification of human fetal tissue represents a contrast with society's general refusal to endorse commercialization of other human organs or tissue. To commodify something is to treat it as property. To construe human organs or tissue, even during fetal stage, as property suggests a disvaluing of what is human. So, while different motives for fetal tissue transplantation may be acknowledged, buying or selling the tissue should not be permitted. Just as economically less advantaged women are more likely to be surrogate mothers for economically advantaged infertile couples or individuals, so might economically disadvantaged women be the main source of fetal tissue available for sale to others. As stated in Chapter 6, socioeconomic differences extend the possibilities for exploitation: what counts as donation for an affluent person is mere reimbursement for a middle class person, and real income for a poor person. The observance of anonymity between providers of fetal tissue and transplant candidates is a way of avoiding social pressures to donate, reducing possibilities for exploitation, and preserving separation between abortion and transplantation decisions.

A claim that fetal tissue should be considered the pregnant woman's property is perhaps credible so long as the fetus is still contained within her body. A comparable claim on the part of the man responsible for the pregnancy is less credible because he has contributed less to the development of the fetus. The claim for either man or woman is weakened when the tissue is electively aborted because the choice of abortion implies a desire that one's relationship to the fetus be severed. If the fetus or abortus is not property, the same parameters that apply to women regarding the distinction between spontaneous and elective abortion apply to men. To the extent that men join women in electing abortions, their status as donors or proxies for the fetus may be impugned. However, so long as women and not men gestate fetuses and undergo abortions, it hardly seems defensible to claim that men are more appropriate donors of tissue from fetuses or abortuses.

Possibly the most potent argument for choosing pregnancy and abortion in order to obtain fetal tissue occurs when the potential recipient is also the potential donor, and the motive for the undertaking is treatment of one's own severe and otherwise incurable disease. Self-preservation, as a fundamentally justified form of self-interest, is then the rationale for the prospective donor-recipient. Despite the appeal of this argument, its validity depends on a morally troublesome concept of the fetus as means. A preferable alternative that also respects the legitimacy of the self-defense rationale is the use of tissue from other (already aborted) fetuses.

Summarily, the question of fetal tissue transplantation for severe disorders is ethically complex and controversial despite its therapeutic promise. To proceed with research and experimental treatment of human subjects is justified under certain conditions. These conditions are the moral equivalent of wedges along a slippery slope leading to denial of fundamental moral values. It would be morally wrong to forego the benefits of a promising treatment for fear of slipping on the slope. It would likewise be wrong not to proceed cautiously. Certain ethical precautions that are generally accepted requirements for experimental treatment and transplantation are applicable here also: (1) ascertainment that the proposed transplant is expected to provide significant benefit to the recipient, (2) ascertainment that comparable benefit may not be obtained through other means, and (3) free and informed consent of participants in different aspects of the procedure.[49]

In light of the heightened complexity and controversiality of fetal tissue transplantation, further cautions are appropriate. The preceding account suggests the following means of avoiding ethical pitfalls: (1) that tissue from living fetuses or living abortuses not be used for transplantation, (2) that decisions regarding abortion and transplantation be kept separate, (3) that anonymity between donor and recipient be observed, and (4) that buying and selling of fetal tissue not be permitted. Beyond the prospective observance of guidelines, retrospective scrutiny is also necessary. In time we may see that fetal tissue transplantation fails to fulfill its therapeutic promise, or fails to confirm our fears about possible exploitation of women, encouragement of elective abortion, and disregard for immature human life. An egalitarian ethic requires ongoing effort to adjust to inequalities as they arise.

NOTES

1. Mary B. Mahowald, Judith Areen, Barry J. Hoffer, Albert R. Jonsen, Patricia King, Jerry Silver, John R. Sladek, Jr., and LeRoy Walters, "Transplantation of Neural Tissue from Fetuses," *Science* 235 (March 13, 1987): 1307–08.

2. Gina Kolata, "Federal Agency Bars Implanting of Fetal Tissue,"*New York Times* (April 16, 1988), 1.

3. Consultants to the Advisory Committee to the Director, National Institutes of Health, *Report of the Human Fetal Tissue Transplantation Research Panel* (December 1988), Vol. I, II.

4. "U.S. Prodded Again to Lift Fetal Tissue Ban," *New York Times* (April 20, 1989), B15; George J. Annas and Sherman Elias, "The Politics of Transplantation of Human Fetal Tissue," *New England Journal of Medicine* 320, no. 16 (April 20, 1989): 1079–82; and Joseph Palca, "Fetal Tissue Transplants Remain Off Limits," *Science* 246 (Nov. 10, 1989): 752.

5. See Bette-Jane Crigger, "Update," *IRB (A Review of Human Subjects Research)* 13, no. 5 (Sept.-Oct. 1991): 10; and Gwen Ifill, "House Approves Fetal Tissue in Federally Financed Research," *New York Times* (July 26, 1991), A9.

6. Dorothy E. Vawter, Karen G. Gervais, and Warren Kearney, "Bill to Resume Federal Funding of Fetal Tissue Transplantation Is Damaging to Women," *IRB (A Review of Human Subjects Research)* 13, no. 5 (Sept.-Oct. 1991): 11.

7. This section is taken with some revisions from my "Neural Fetal Tissue Transplantation— Should We Do What We Can Do?" *Neurologic Clinics* 7, 4 (Nov. 1989): 745–53. Portions of this chapter also appear in my "Fetal Tissue Transplantation: An Update," in *Biomedical Ethics Reviews 1991,* ed. James Humber and Robert Almeder (Clifton, New Jersey: the Humana Press Inc., 1991), 103–21.

8. *Dorland's Illustrated Medical Dictionary,* 27th ed. (Philadelphia: W. B. Saunders Company, 1988), 5, 543, 619, 833; see also *Williams Obstetrics* (18th ed.), ed. F. Gary Cunningham, Paul C. MacDonald, and Norman F. Gant (Norwalk, Connecticut: Appleton and Lange, 1989), 40.

9. See Mary B. Mahowald, "Abortion and Equality," in *Abortion—Understanding Differences,* ed. Sidney Callahan and Daniel Callahan (New York: Plenum Press, 1984), 182.

10. 410 U.S. 113, 93 S. Ct. 705, January 22, 1973.

11. A possible exception to this precedent is the use of tissue or organs from anencephalic newborns. As yet, however, such use is legally impermissible until/unless the newborn is declared dead.

12. K. J. S. Anand and P. R. Hickey, "Pain and Its Effects in the Human Neonate and Fetus," *New England Journal of Medicine* 317, no. 21 (Nov. 19, 1987): 1321–46.

13. Mahowald et al., "Transplantation of Neural Tissue from Fetuses," 1307–08.

14. Consultants to the Advisory Committee to the Director, National Institutes of Health, *Report of the Human Fetal Tissue Transplantation Research Panel* (December 1988), vol. II, A1–A5.

15. In the first reported case of human fetal tissue transplanted for treatment of Parkinson's disease, tissue was retrieved from two sites in a single fetus that had been spontaneously aborted. Fetal adrenal tissue was implanted into the brain of one patient; fetal neural tissue was implanted into the

brain of another. See Ignacio Madrazo, V. León, César Torres, et al., "Transplantation of Fetal Substantia Nigra and Adrenal Medulla to the Caudate Nucleus in Two Patients with Parkinson's Disease," *New England Journal of Medicine* 318, no. 1 (Jan 7, 1988): 51.

16. Alan Fine, "Transplantation in the Central Nervous System," *Scientific American* 255 (Aug. 1986): 52–58B.

17. U.S. Congress Office of Technology Assessment, *Neural Grafting: Repairing the Brain and Spinal Cord* (Washington, D.C.: U.S. Government Printing Office, September 1990), 163.

18. Olle Lindvall, Patrick Brundin, Håkan Widmer, et al., "Grafts of Fetal Dopamine Neurons Survive and Improve Motor Function in Parkinson's Disease," *Science* 247 (Feb. 2, 1990): 574–77.

19. See, for example, Michael B. Green and Daniel Wikler, "Brain Death and Personal Identity," *Philosophy and Public Affairs* 9 (1980): 105–33; and Warren Quinn, "Abortion, Identity and Loss," *Philosophy and Public Affairs* 13 (1984): 24–54.

20. Madrazo et al., 51.

21. John R. Sladek, Jr., D. Eugene Redmond, and Robert H. Roth, "Transplantation of Fetal Neurons in Primates," *Clinical Research* 36, no. 3 (April 1988): 209–14.

22. Roger Lewin, "Cloud Over Parkinson's Therapy," *Science* (April 22, 1988): 390–92.

23. Consultants to the Advisory Committee to the Director, NIH, vol. II, A26.

24. See Curt R. Freed, Robert E. Breeze, Neil L. Rosenberg et al., "Transplantation of Human Fetal Dopamine Cells for Parkinson's Disease," *Archives of Neurology* 47 (May 1990): 505–12; also, "Doctors Implant Frozen Fetal Cells," *New York Times* (Dec. 13, 1988), C8 (no author), and Gina Kolata, "Transplants of Fetal Tissue Seen Easing a Brain Disease," *New York Times* (May 7, 1992), A13.

25. See Madrazo et al., 51; Lindvall et al., 574–77; "British Fetal Inplants," *New York Times* (April 19, 1988), C9 (no author); and Sandra Blakeslee, "Fetal Cell Transplants Show Early Promise in Parkinson Patients," *New York Times* (Nov. 12, 1991), B6.

26. Fine, 52–58; Barry J. Hoffer, Ann-Charlotte Granholm, James O. Stevens, et al., "Catecholamine-containing Grafts in Parkinsonism: Past and Present," *Clincial Research* 36 (1988): 189–95; and John T. Hansen and John R. Sladek, "Fetal Research," *Science* 246 (Nov. 10, 1989): 775–79.

27. C. G. Goetz, C. Warren Olanow, William C. Koller, et al., "Multicenter Study of Autologous Adrenal Medullary Transplantation to the Corpus Striatum in Patients with Advanced Parkinson's Disease," *New England Journal of Medicine* 320, no. 6 (Feb. 9, 1989): 337–41; and Gina Kolata, "Latest Surgery for Parkinson's Is Disappointing," *New York Times* (Aug. 30, 1988), C1, C3.

28. Fine, 52–58; and Hoffer et al., 189–95.

29. See U.S. Congress Office of Technology Assessment, 93.

30. Walter A. Rocca, Luigi A. Amaducci, and Bruce S. Schoenberg, "Epidemiology of Clinically Diagnosed Alzheimer's Disease," *Annals of Neurology* 19 (1986): 415–24.

31. A. H. Rajput, "Epidemiology of Parkinson's Disease," *Canadian Journal of Neurological Science* 11 (1984): 156–59.

32. Concerning dilation and curettage, see Benedict M. Ashley and Kevin D. O'Rourke, *Health Care Ethics* (St. Louis: Catholic Health Association of the United States), 292–93. Treatment of ectopic pregnancies may be viewed as treatment of the woman's life-threatening condition. In *Williams Obstetrics,* ectopic pregnancy and its treatment are an entirely separate topic from abortion. See *Williams Obstetrics* (18th ed.), 511–31.

33. See *Williams Obstetrics,* (18th ed.), 501.

34. Lewin, 390–92.

35. Kathleen Nolan, "Genug Ist Genug: A Fetus Is Not a Kidney," *Hastings Center Report* 18, no. 6 (Dec. 1988): 18.

36. Nolan, 13–19.

37. It is possible but highly unlikely that the availability of fetal tissue transplantation would lead to an increase in elective abortions. A situation that *might* encourage abortions for fetal transplant purposes would occur if the procedure is seen as specific treatment for an otherwise incurable and seriously ill relative or oneself. Excessive remuneration, or poverty in conjunction with meaningful remuneration, might also serve as inducements.

38. James Burtchaell, "University Policy on Experimental Use of Aborted Fetal Tissue," *IRB (A Review of Human Subjects Research)* 10, no. 4 (1988): 9.

39. Dorothy E. Vawter, Warren Kearney, Karen G. Gervais, et al., *The Use of Human Fetal Tissue* (Minneapolis: University of Minnesota, 1990), 257.

40. Nadine Brozan, "Rising Use of Donated Eggs for Pregnancy Stirs Concern," *New York Times* (Jan. 18, 1988), A1, A17; and R. P. S., "Sperm and Ova as Property," *Journal of Medical Ethics* (1985): 123–26.

41. William David Ross, *The Right and the Good* (Oxford: Clarendon Press, 1930), 21.

42. Robert Nozick, *Anarchy, State and Utopia* (New York: Basic Books, Inc., 1974), 18, 33; and John Rawls, *A Theory of Justice* (Cambridge, Massachusetts: Harvard University Press, 1971), 19, 135.

43. Nolan, 15.

44. See Mary B. Mahowald, Jerry Silver, and Robert A. Ratcheson, "The Ethical Options in Fetal Transplants," *Hastings Center Report* 17, no. 1 (Feb. 1987): 13.

45. Mahowald et al., "Transplantation of Neural Tissue from Fetuses," 1307.

46. Marilyn Friedman refers to such relationships as "communities of choice," contrasting these with "communities of origin." See her "Feminism and Modern Friendship," *Ethics* 99, no. 2 (Jan. 1989): 275–90.

47. See Mary B. Mahowald, "Ethical Perspective," *Health Matrix* 6, no. 2 (1988): 84–86.

48. Consider, for example, the case of Gerry and Terri Walden, who maintained their strong opposition to abortion while agreeing to fetus-to-fetus transplant for a fetus suffering from Hurler's syndrome, a severe genetic disease from which two of their children had already died. See Philip J. Hilts, "Anguish Over Medical First: Tissue From Fetus-to-Fetus," *New York Times* (April 16, 1991), A1, B8. For an argument on moral responsibility for what we permit (our "collusions"), see Joan Callahan, "Acts, Omissions and Euthanasia," *Public Affairs Quarterly* 2, no. 2 (1988): 21–36.

49. *Code of (U.S.) Federal Regulations* 45C FR46, Protection of Human Subjects, revised March 8, 1983.

10

Decisions Regarding Disabled Newborns

During the 1980s, both the media and the federal government focused their consideration of infants on a few controversial cases, possibly to the neglect of ethical issues involving other children, and larger social and ethical problems affecting all of us.[1] In this chapter, I broaden the perspective on neonatal dilemmas by providing a brief account of the historical, cultural, and medical contexts in which they arise, and a description of alternative approaches to their resolution. I also discuss cases in which nontreatment of extremely ill, impaired, and low birth weight infants may be morally justified. Although these conditions are often addressed separately, in practice they often occur in the same patient. From an egalitarian perspective, each infant should be treated as an individual, just as older patients are to be treated as individuals. In both situations, the patient represents a unique embodiment of limitations, abilities, possibilities, and relationships.[2]

Historical and Current Context

Facilities and technologies for neonatal intensive care are a relatively recent phenomenon, still comparatively unavailable or inaccessible to the populations of less developed nations. The first treatment center for newborn care in the United States was established early in this century, when infant deaths were primarily associated with infection or malnutrition.[3] At that time, rudimentary incubators provided necessary warmth for preterm infants, and oxygen supplementation was introduced to combat respiratory difficulties due to immature lung development. Progress in survival rates was not devoid of setbacks. For example, by 1954 it was recognized that the high oxygen concentrations that had saved some preterm newborns had also caused blindness. Subsequent curtailment of this treatment was accompanied by a corresponding rise in the rate of infant mortality.[4] Similarly, diethylstilbestrol (DES) was initially thought by some to be effective in bringing problematic pregnancies to term, but the drug was later implicated in carcinogenic and reproductive problems of the offspring.[5]

Further advances produced sophisticated techniques for prenatal diagnosis and treatment, monitoring neonatal heart rate, blood gases, and chemistries, and microsurgical procedures for newborn anomalies. Neonatology became a major pediatric specialty, spawning a huge and ongoing research effort with impressive clinical results.[6] Techniques introduced during the 1960s facilitated successful treatment of infants weighing less than 1,500 grams. These methods included constant positive airway pressure monitoring, by which oxygen requirements are constantly measured without interruption, and hyperalimentation, a means of providing nutrition to those who cannot tolerate other types of feeding. By 1970, the mortality rate from hyaline membrane disease, a common problem of preterm newborns, had dropped from 60 percent to 20 percent of those affected. By 1978, the survival rate for very early and very small babies had improved to the point where those weighing less than 1,000 grams warranted treatment.[7] Since then, fetal viability has advanced earlier into pregnancy, resulting in smaller and younger survivors of preterm birth and even late abortions. Two new techniques, extracorporeal membrane oxygenation (ECMO) and surfactant therapy have decreased the mortality rate of newborns even further. ECMO has improved the outcome for newborns suffering from four common or highly lethal conditions: meconium aspiration, persistent pulmonary hypertension, beta streptococcal sepsis, and congenital diaphragmatic hernia.[8] Surfactant therapy has been remarkably successful in treatment of lung immaturity in preterm infants.[9]

The majority of very low birth weight babies (less than 1.5 kilograms) who survive sustain no serious permanent compromise to their motor or mental functions. For example, 65 percent of the 781 infants weighing less than 1,500 grams who were admitted to the Neonatal Intensive Care Unit at Rainbow Babies and Childrens Hospital in Cleveland between 1975 and 1978 survived; 80 percent of these had normal neurodevelopmental outcomes.[10] As might be suspected, morbidity increases with decreasing birth weight. From 1982 to 1988 a study at the same institution showed an 18 to 20 percent survival rate for newborns weighing less than 750 grams. Among the survivors, 22 to 50 percent had moderate to severe neurodevelopmental impairment.[11] A multicenter study published in 1991 showed a 34 percent survival for infants weighing 750 grams or less and 66 percent for those from 751 through 1,000 grams.[12] Morbidity factors for survivors in this study include chronic lung disease, severe bowel infection, and brain hemorrhage, all of which may lead to long-term severe impairment. Despite the impressive technological developments, not everyone agrees that all critically ill newborns should be provided with lifesaving or life-prolonging treatment. Infanticide is a long-standing practice with which nontreatment decisions may be compared.

Anthropologists tell us that infanticide has been practiced throughout history in many cultures, including those of the western world.[13] At times, the practice was deemed acceptable because it was undertaken indirectly rather than directly. In other words, infants were not killed outright, but were left to die — often because they were defective, sometimes because they were twins or female or illegitimate. Since abandonment of an infant inevitably leads to death, there is little practical distinction between killing and letting a newborn die. Just as euthanasia is morally problematic, regardless of whether it is characterized as active or passive, so is infanticide, whether characterized as direct or indirect.[14] Refusing to institute respiratory support in a newborn whose lungs are not yet mature may be construed as indirect infanticide. The refusal to provide intravenous nutrition to an infant who is incapable of normal digestion may be construed similarly.[15] The difficulty of maintaining a sharp distinction between direct and indirect termination of lifesaving treatment has led Robert Weir to argue that it is sometimes morally justified to terminate an infant's life directly and actively.[16]

Several factors in contemporary American society conspire to exacerbate moral problems regarding infants. One is the emphasis on patient autonomy, which is generally assumed to be captured in the concept of "informed consent." While this concept is obviously inapplicable to newborns, it is sometimes applied to parents who make decisions on behalf of their children. In fact, the distinction between informed consent and proxy or substitute consent is often overlooked, and parents are falsely assumed to provide the former rather than the latter.[17] Legal and moral grounds for requiring informed consent of competent patients are stronger than those for substitute consent. Nonetheless, parental rights regarding their children have generally been perceived as primary, requiring practitioners to respect their decisions even when these involve the refusal of life-prolonging treatment.[18] Since the Baby Doe controversies of the 1980s, this emphasis has shifted to a situation where some physicians see themselves as advocates for infants even if this pits them against parents.[19]

In the past, a variety of treatment options were unavailable for many infants, regardless of whether they were disabled. Reversible life-threatening medical problems, which occur more frequently in permanently impaired newborns than in other infants, are now routinely repaired through surgery. The development of antibiotic therapy, feeding techniques, and fluid exchange procedures has greatly increased the actual number of disabled children who survive to adulthood. Moreover, while greater numbers of preterm infants now survive to live normal lives, some pay for their survival with iatrogenically induced permanent disabilities. There is thus an inevitable connection between very low birth weight babies and disabled infants.[20]

Two conflicting social phenomena make neonatal ethical dilemmas even more prevalent and complicated. One is the "premium baby" mentality that has resulted from the trend toward reduced family size, as well as the availability of contraceptive measures and abortion, discussed in earlier chapters. Allowing severely compromised infants to die is consistent with this mentality. In contrast, the "right to life" ideology and movement affirm the primacy of fetal interests over those of other individuals. Not surprisingly, "right to life" activists have joined the government and organizations representing the disabled in arguing that infants should not be denied treatment on the basis of their disabilities.[21] Either of these positions is supportable by an egalitarian perspective. Which position weighs more than the other depends on whether survival of a severely disabled newborn is of greater value than parental autonomy, or whether the obligation to respect parental autonomy overrides that of beneficence toward their infant. The cases considered next are well-known illustrations of this dilemma.

The Doe Babies

In the spring of 1982, an infant afflicted with Down syndrome and esophageal atresia was born at Bloomington Hospital in Indiana. Surgical repair is usually undertaken to correct the latter problem, but the former condition, with its concomitant mental retardation, is not correctable. For individuals with Down syndrome, the degree of mental retardation is not predictable at birth. The obstetrician informed the mother that she might choose between two "medical options" regarding her newborn: (1) consent to the surgery necessary for survival, or (2) decline that consent and request that the baby not be fed so as not to prolong his dying. The parents chose the latter course. Hospital personnel respected their choice, and local and state courts reviewed and approved their decision. Local attorneys attempted to reverse the decision through appeal to the U.S. Supreme Court. When the child died while the attorneys were en route to Washington for a special hearing of the court, the case became moot.[22]

Although his parents had him baptized, presumably giving him a name, the public came to know this infant as "Baby Doe." During his six days of life, he became uniquely but anonymously famous because of media coverage and public reaction to it. As a resident of Bloomington at the time, I knew several of the principals associated with the case (the obstetrician, the pediatrician, the lawyer for the parents, the priest who baptized the baby, and the pathologist who performed the autopsy), but never learned the identity of the infant or his parents. To their credit, the press respected the family's privacy.

Following the infant's death, the government notified all federally supported institutions caring for infants that funding would be denied if they discriminated against the disabled, as had allegedly occurred with Baby Doe. In March of 1983, the Department of Health and Human Services issued a ruling that required all such institutions to post signs citing both the government statute prohibiting discrimination against the disabled and a phone number to use in reporting suspected violations of the statute.[23] This ruling was overturned one month later by U.S. District Judge Gerhard A. Gesell, who described it as conceived in "haste and inexperience," and "based on inadequate consideration of the regulation's consequences."[24]

The second Baby Doe was born in Port Jefferson, New York, in fall of 1983, this one distinguished from the other by being called "Jane Doe." Like her predecessor, she too became a subject of public controversy within her first days of life. Unlike him, her name (Keri-Lynn) was eventually revealed by the media, and the child survived despite her parents' initial refusal of treatment.

Baby Jane Doe was born with spina bifida (an open spine), hydrocephalus (excess fluid on the brain), and microcephaly (reduced brain size), conditions predictive of paralysis in her lower extremities, incontinence, and retardation. According to reports published during her first weeks of life, surgical intervention might allow the child to survive for approximately twenty years; without the surgery she was likely to die within two years. A physician who counseled the father told him that his daughter was so neurologically compromised that she "would never experience joy, never experience sorrow."[25] When both parents declined consent for the surgery, their decision was reviewed and approved at local and state levels, and supported by their priest counselor. Lawrence Washburn, an attorney from New Jersey, brought the case to the attention of federal authorities, who attempted unsuccessfully to obtain the medical records. As with the Bloomington case, the government considered nontreatment of Baby Jane Doe to be a violation of the 1973 statute prohibiting discrimination against the handicapped.[26]

Litigation relevant to the second Baby Doe case led to a denial of the government's right to require the surrender of medical records in order to investigate treatment decisions regarding disabled infants. During the summer of 1984, the government's Baby Doe regulations were permanently enjoined by the U.S. District Court in New York. However, in fall of the same year Congress passed legislation requiring state child protection agencies to intervene in cases where severely disabled infants are refused "medically indicated treatment." Exceptions to this requirement are situations where "the infant is irreversibly comatose or the treatment would be futile and inhumane or would only prolong dying."[27] According to Betty W. Levin, pediatric professionals tend to overestimate the degree of interventions required by this legislation.[28]

Conflicting positions regarding the role of government in "Baby Doe" cases reflect different constituencies: medical organizations, associations for the disabled, the Department of Health and Human Services, and the President's Commission for the Study of Ethical Problems in Medicine and Biomedical and Behavioral Research (hereafter, President's Commission). However, the documents in which these positions are articulated all invoke the same criterion, namely, the best interests of infants.[29] While disagreement continues about interpretation of, and procedures for implementing, the "best interest" criterion,[30] broader agreement may be reached through an examination of its meaning. Before addressing this, however, I wish to deal with an equally controversial question relevant to guideline 4: who should decide the fate of severely disabled newborns?

Who Should Decide?

"Informed consent" is often seen as a sine qua non of justification for medical interventions.[31] Competent adults may legally decline even lifesaving therapy by removing themselves from hospital treatment programs against medical advice. Exceptions have been based on the patient's responsibilities to others, or the claim that hospital personnel are not obliged to violate their own professional standards or commitments.[32] Since newborns are incapable of providing informed consent, their parents usually act as proxy or surrogate decision makers in their behalf. The right of parents to act as proxies may be overruled, however, if their decision opposes their child's best interests. For example, if a Jehovah's Witness parent declines a blood transfusion essential to the life of his child, hospital authorities will obtain a court order allowing hospital personnel to intervene in the child's behalf. Thus, the parents' right to decide about their infant's treatment is legally less binding than their right to decide about their own treatments.

This distinction between informed and proxy consent is also significant from a moral point of view. It suggests that a priority of decision makers be observed, based on the degree to which each decision maker is related to the infant. Typically, the child's parents hold first place. However, the child's caretakers are also related to the child through their professional commitment, as well as the personal and contractual relationships they maintain with the infant and family.

Despite the legal and moral requirement of informed or proxy consent, long standing practice assigns the role of principal decision maker for medical dilemmas to the physician.[33] The justification for this priority is sometimes comparable to the argument presented by the cardinal in Dostoyevsky's story of "The Grand Inquisitor."[34] By assuming control of people's lives, the cardinal claimed that his church had gradually removed the burden of freedom that Christ brought to the world. Similarly, the physician or the health care team may accept sole responsibility for difficult decisions in order to spare families the anguish and unnecessary guilt that often occurs in such situations. Despite its plausibility and appeal, this paternalistic reasoning has several crucial flaws. One is the failure to acknowledge that a sense of guilt may be experienced regardless of how a decision is made. If this is true, it is more helpful to focus on the moral justification for a decision to prolong or discontinue treatment—that is, the intent to do what is best for the patient. Both families and practitioners may need explicit reassurance that relinquishing the hold on another's life is sometimes the most loving and caring alternative available.

Another flaw in arguments favoring decisions made solely by physicians (or parents) is the fact that responsibility for decision making is inevitably shared by all of the autonomous participants in a dilemma. Even if an attending physician writes an order or parents indicate their wishes, others choose to implement, ignore, or challenge those decisions. At times, a practitioner does not consider her actions to be a matter of choice; rather, she may simply be following the order of the attending physician or supervisor. At other times, a practitioner may subtly, perhaps even inadvertently, interpret an "order" in a manner that compromises its intent. For example, in a situation where a physician has instructed staff to resuscitate a critically ill patient if necessary, a nurse or resident who disagrees with that decision may respond with deliberate slowness to a signal that the patient has suffered cardiac arrest.

In many cases regarding neonates, there is neither ambiguity nor controversy about what constitutes morally appropriate behavior. For example, the vast majority of pediatricians and pediatric surgeons agree that an anencephalic newborn who is afflicted with intestinal atresia should not have corrective surgery for the latter condition.[35] The invasiveness of surgery cannot be justified on the basis of benefit to the patient because the infant is already dying. In cases where agreement has been reached about moral aspects of treatment or nontreatment, it is probably neither necessary nor helpful to extend the decision base beyond the delivery room or nursery. In fact, involving others in the decision process increases the possibility of violating confidentiality or family privacy.

Moreover, treatment deferral sometimes involves a real risk of harming the patient. Possibly the most common example of such a situation involves intubation of very small (e.g., less than 650 grams) or very early (e.g., less than 24 weeks gestation) preterm newborns. Without intubation, the infant cannot survive. At such times, whoever is competent to provide the treatment is justified in making the decision on the patient's behalf. Subsequently, however, and in most chronic cases, there is time for discussion and broader input, which ought to be obtained in cases where ambiguity or disagreement continues. Since most decisions to terminate lifesaving treatment are irrevocable, treatment should continue until the conflict is resolved.

Why should there be broader input? Mainly because neither health care practitioners nor parents have any special moral expertise, and the possibility of arriving at well-reasoned moral decisions is increased by the collaborative efforts of reasonable people. Those who maintain a distance from the situation can sometimes provide a more objective perspective, which may complement and supplement the views of those whose involvement in the situation may preclude a totally rational analysis. Extending the decision base in unclear or controversial cases may also be reassuring to those closest to the patient because it represents one more attempt at responsible resolution of a difficult dilemma.

A decision base may be extended beyond the physician or parents through consultation with other clinicians, the entire health care team, a hospital-based review committee, or recourse to the courts. In the interests of maintaining confidentiality and family privacy, it is preferable to use the least public forum in which ambiguity or disagreement may be resolved. The widespread endorsement of the health care team's effectiveness in providing basic health care suggests that it might also be effective in dealing with medically related moral problems.[36] Hospital-based review committees are a newer phenomenon whose efficacy deserves to be tested.[37] Recourse to the courts is a particularly troublesome means of extending the decision base for ethical dilemmas. The legal system introduces an adversarial dimension into a set of relationships that should ideally be based on trust, openness, and consensus. Litigation threatens, and sometimes severs, those relationships, thwarting the therapeutic purpose of the practitioner–patient alliance. There are times when legal recourse may be the only way of resolving ambiguity and disagreement—for example, in cases involving blood transfusions for children of Jehovah's Witnesses. But court decisions are not necessarily morally correct. In the case of Bloomington's Baby Doe, for example, there is widespread agreement that the court's concurrence with the parents' decision to decline treatment was morally unjustified.[38]

Recent government attempts to impose investigative procedures on federally funded facilities that care for newborns seem to be intrusions on the right of privacy and the confidentiality of the physician–patient relationship, and may even be harmful to the patients affected. After investigating many anonymous reports of suspected neglect of impaired newborns, the Department of Health and Human Services concluded that appropriate medical, legal, and moral decisions had already been made in the vast majority of cases. In cases at Vanderbilt University in Nashville, Tennessee, and Strong Memorial Hospital in Rochester, New York, however, it was reported that the government investigation obstructed care of the infants who had allegedly been neglected as well as other patients. The time required for personnel to respond to the queries of investigators could only be purchased at the price of time spent in caring for patients.[39]

In January, 1984, the Department of Health and Human Services strongly encouraged the formation of hospital ethics review committees to consider cases of suspected neglect of disabled infants through denial of treatment.[40] In addition to health care professionals from various disciplines, it was recommended that representatives of the disabled also serve on these committees. In general, the government's encouragement of the committee review mechanism supported the recommendation of the President's Commission.[41] The American Academy of Pediatrics also recommended the formation of local review committees and suggested appropriate procedures and principles.[42] However, the Commission had proposed the local review mechanism as an alternative to federal investigative procedures, arguing that the latter was unlikely to promote the best interests of infants and might actually impede the achievement of that purpose.

The continuing legal controversy surrounding "Baby Doe" cases evoked fairly widespread interest in the use of hospital review committees to address difficult cases. The extent of this interest and the influence of committees on practice remain to be seen.[43] Regardless of how decisions are made, however, we must also deal with the substantive issue of criteria for ethical decisions regarding disabled neonates. These reflect egalitarian considerations and traditional principles of biomedical ethics. An emphasis on the best interests of others also reflects an ethic of care for them.

Prolonging Life in Others' Interests

Since life is commonly perceived as a great gift, it may credibly be maintained that loss of life is always negative for the patient, and therefore the loss can only (possibly) be justified on the basis of others' interests. Indeed, in certain cases, the interests of others may be primarily served through the prolongation of an infant's life. Consider, for example, the fact that fees paid to neonatologists, hospitals, and hospital personnel are partly dependent on the patient population, which is incremented through preservation of lives, no matter what their quality. So long as infants survive, the possibility of obtaining new knowledge through experimental therapies and further clinical data also continues. Beyond these results, there are more subtle rewards that accrue to clinicians who succeed in prolonging infants' lives. First is the feeling of accomplishment borne of the experience of doing rather than just letting go. Most doctors, after all, are activists, more inclined to cure than sustained caring. As one neonatologist put it, "It is easier for me to live with the consequences of something I've done than it is to worry about something I have not done which might have given better results."[44] Second is a perceived consistency between the end of health care and the prolongation of life. Conversely, for some clinicians the death of any patient evokes a sense of professional failure.[45] And third, effective ties build up between clinicians and child (as well as between parents and child, and clinicians and parents), sometimes reaching a point where the emotional needs of the concerned adults obfuscate their recognition of the infant's interests. A nurse thus made a pertinent and poignant comment concerning an infant whose life had been prolonged for two years, despite a preponderance of anguish to him with no expectation of ultimate relief or survival: "We have been doing this for ourselves rather than for him."[46]

Legally it may well be in the interests of clinicians and parents to prolong the life of an infant. Although malpractice suits may be pressed for prolonging life, suits are more likely when treatment has been withdrawn.[47] Even in that case, however, the probability of a successful suit is very slim so long as "letting die" (passive euthanasia) is distinguished from active euthanasia. It is also possible that efforts to prolong the lives of severely disabled infants serve the interests of politicians or political parties. Support for the Reagan administration was surely enhanced in some quarters by the steps it took to prevent a "Baby Doe" situation from recurring.

While motives for prolonging life are sometimes mixed, it may still be maintained that the prolongation is always in the infant's interests. This position is justified if life is assumed to be an absolute value, separable from any quality of life consideration. The assumption has often been associated with an essentially religious perspective, such as that of the Roman Catholic Church in its teaching regarding abortion. Yet religious reasons may also be given, from that tradition as well as from others, for the contrary view, namely, that life is an important but relative value.[48] Christian Science argues against any kind of medical intervention as impeding the natural course of God's plan among human beings.[49] Jehovah's Witnesses argue more selectively against blood transfusions, allowing that deaths which occur through loss of blood fulfill God's will.[50] If faith in divine omniscience and omnipotence is assumed, it may in fact be blasphemous to maintain that human beings can either prolong or shorten life. If faith in an afterlife is affirmed, death may sometimes be construed as preferable to life on earth.

Several nonreligious factors also support an assertion of the absolute value of an infant's life. One may be described as "the uncertainty principle," which applies to infancy more than to other periods of (extrauterine) human existence. While neonatology has achieved wondrous things in recent years, and programs for facilitating maximum development of disabled children have yielded impressive results, it remains impossible to predict with certainty what the subjective or objective future experience of a particular newborn will be.[51] Most clinicians have in their reservoir of experience recollections of minor and major "miracles," that is, cases whose happy outcome was totally unexpected in light of the facts known at the time and the technology available. I think, for example, of an infant born with heart defects so grave that none similarly afflicted had ever been known to survive, whose recovery after surgery changed the mortality rate applicable to others. However, in this particular case, the issue was mainly a choice between probable death and an extremely slim chance of survival with neurological normalcy (or relative normalcy), rather than a choice between death and survival at a level of extreme neurological compromise.

Another relevant feature of newborn status or the status of children in general is the obvious contrast between them and adults with regard to the span of life already lived and that anticipated. The "right to life" is sometimes more compelling when asserted on behalf of those who have scarcely lived, which partly explains why children's deaths seem more tragic than those of the elderly. In some respects, new life signifies the fullness of hope, which may be dashed through death. On the other hand, in the case of a neonate, there has been little time and opportunity to build the affective ties that make death so painful for a loved one's survivors.

If the interests of the infant are primary, I do not believe that features peculiar to infants provide adequate justification for the preservation of any and every newborn's life. Nonetheless, these features do argue persuasively for a conservative approach to the irreversible decision to terminate or not initiate life-prolonging treatment. By "conservative" approach I mean one that seeks to prolong life if there is some real, although small, chance that the continuation is in the infant's own interests. Where there is high probability that this will not be the case, then the same criterion argues against prolonging life. To the extent that prolongation is likely to increase suffering for the child, a decision not to prolong life through technological support may be morally mandatory. It reflects our realization that those that can suffer should not be caused to suffer (guideline 3), and that individuals should not be treated as other than who or what they are (guideline 5). Just as the right to die may be construed as part of an adult's right to life, the same claim may be made with regard to infants. To deny this right to children is to practice what Richard McCormick has called a "racism of the adult world."[52]

An Infant's Right to Die

While the priority of the infant's interests suggests that decisions to prolong life will be made much more frequently than decisions to the contrary, that priority also suggests the relevance of "quality of life" considerations. Three types of cases are relevant in this regard. The first and simplest type is where therapy is futile because the underlying condition is irremediably fatal, and therapy would in no way reduce pain to the infant. In other words, survival beyond a few hours or days or weeks is not expected, no matter what is done or not done. Anencephaly, a condition where the infant's brain has failed to develop, is a commonly accepted example of this situation. Even those who claim "quality of life" factors are irrelevant agree that the life of an anencephalic infant need not be prolonged.

The second type of case is less simple: one where repeated, intrusive, painful interventions would prolong the infant's life, possibly for years, but continued life is of dubious benefit to the child. In such a situation, efforts to preserve life are likely to result in a preponderance of negative experiences for the infant without realistic expectation of improvement. The prolongation itself can only occur through multiple medical and surgical intrusions that are sometimes iatrogenic, and generally interfere with the natural course of the body's function. Not infrequently, despite use of analgesics and anesthesia, the interventions are also painful.

Consider, for example, an infant with the chromosomal abnormality trisomy 13, which involves profound mental retardation and frequent seizures for the 18 percent who survive beyond the first year of life.[53] Often, these infants face immediate life-threatening problems such as severe heart defects. Correction of these and concomitant problems requires multiple surgeries, medical and orthopedic interventions, and permanent hospitalization. To prolong life through invasive procedures might serve the interests of parents and clinicians, but can scarcely be judged to serve those of the infants themselves. The decision to let such an infant die is usually based on the fallible judgment that prolonging life will cause the child a preponderance of suffering.

The third type of case is more problematic than the preceding: one where the required therapy is not itself a source of pain, but neither is it curative, and the life thus prolonged is probably devoid of any qualitative satisfaction for the infant. Consider, for instance, a newborn who has had a Grade IV cranial hemorrhage (bleeding into the cerebral tissues) with uncontrollable seizures, whose intestines have necrotized. The child can only be fed through parenteral hyperalimentation, a process by which predigested food is infused into the body. The combination of the hemorrhage and seizures indicates high probability that the child might survive but only at a vegetative level of existence, that is, without any cognitive function or capacity for social interchange. Although such a child might survive for years, it is doubtful that his survival is in anyone's interest, including his own. Unless life is an end in itself, rather than a necessary condition for the actualization of human values and potential, maintaining life in these circumstances may be exploiting the child, that is, using him as a means of furthering others' ends.

A claim that infants' interests include the right to die may be based on a conception of life that is not merely quantitative.[54] Life is then perceived as a crucial but relative value, extremely significant as the basis of all other human values, but not absolute. This view necessarily involves the notion that quality of life factors are essential to any full affirmation of the value of human life. However, which factors are relevant, and how they are relevant, remains problematic.

Decisions made in behalf of incompetent or unconscious adults may enlighten us with regard to infants. Either of two approaches is generally followed with adults: (1) the decision is based on the patient's history, that is, an understanding of the patient's desires or values as applicable to such a situation (as expressed, for example, in a living will or other form of advance directive), or (2) utilization of the "reasonable person" standard, that is, determination of treatment and nontreatment on the basis of what any competent, conscious person would reasonably choose in similar circumstances. Obviously the first approach is inapplicable to newborns, but the second seems appropriate even though infants may not be described as "reasonable persons."[55] If, for example, a reasonable person would decline surgery that could in no way benefit her, and might in fact prolong a predominantly painful existence, why might we not invoke this criterion to justify a similar decision for an infant? In all three types of cases, that criterion would apply. Thus, in situations where (1) therapy is futile, (2) where it would prolong a life of predominant anguish, or (3) where the patient has suffered irreparable neurological devastation, the decision not to prolong life beyond its natural limits is reasonable, and should be respected as such. To reject the applicability of this standard to infants or children suggests complicity in what we have already described as adult racism. It thus stands opposed to an egalitarian perspective.

Giving Priority to Infants' Interests

Where individuals attempt to observe the priority of infants' interests, the nuances of particular cases may be interpreted in light of certain distinctions. For example, natural law theology has long invoked the distinction between "ordinary" and "extraordinary" treatment, claiming that the former is obligatory while the latter is not. Ordinary and extraordinary treatment is explained as relative to the unique circumstances of the case, including the accessibility of necessary technology and therapy.[56] Thus, what might count as extraordinary treatment of a cancer-ridden elderly patient who has indicated a desire to die may be ordinary in dealing with a newborn, whether seriously ill or not. Similarly, a distinction between optimal and maximal care is pertinent: maximal care means prolonging life no matter what the cost to the patient; optimal care means prolonging life only to the extent that the prolongation is in the patient's interests.[57] Maximal care may (inadvertently) serve the interests of others—for example, students who can gain more clinical experience by continuing care for the dying; it may simultaneously impede optimal care for the patient. An obligation to provide optimal care implies that others' interests do not constitute a sufficient criterion for refusal of treatment, while those interests may be relevant in applying the ordinary versus extraordinary distinction.

Clinical interpretations that have served as guides for individuals addressing problematic cases include a distinction between "coercing" and "helping" someone to live, and between "doing to " and "doing for" a patient.[58] Roughly, "coercing" and "doing to" constitute unjustifiable intrusions, while "helping" and "doing for" are justifiable because they are oriented toward the patient's own interests. Determination of where the distinctions apply remains difficult, and may never be made with absolute certainty, but some cases involve a very high expectation that survival will mean prolonged and unmitigated misery for a particular patient, child or adult. An example of coercing someone to live might be a situation where a patient experiencing the terminal stage of an incurable cancer has suffered kidney failure that is treatable by dialysis. Performing corrective cardiac surgery on an infant with an incurable and fatal genetic abnormality may be another. In such cases, the right to die, as part of the right to life, seems an undeniable component of patient rights.

Another relevant distinction is between "defensive" and patient-centered medicine. Increasingly and unfortunately, "defensive medicine" (that is, medicine practiced to avoid legal entanglements) has motivated clinicians to prolong lives in cases where there is persuasive evidence that this is not in the best interests of the patient. In 1983, James Strain, as president of the American Academy of Pediatrics, wrote that today's pediatricians have a different view from those interviewed for a 1977 national survey that disclosed that the majority would acquiesce to parental refusal of lifesaving surgery for seriously disabled infants.[59] At this point in time, he alleged, physicians would not accede to the refusal. A 1988 survey of pediatricians in Massachusetts by I. David Todres confirmed Strain's thesis.[60] Todres found that physicians were less inclined to give priority to parents' wishes and more inclined to treat disabled infants than they were ten years earlier. Unfortunately, some erroneously believe they are legally obliged to treat disabled infants more aggressively than others. In the interest of "defensive medicine," disabled newborns are then subjected to the discrimination of overtreatment.

Medical as well as moral decisions continually need to be reassessed in light of the changing condition of the patient. Thus a decision to prolong life may be reversed because a patient's condition has so gravely deteriorated that the prognosis is one of overwhelming misery for him. For example, extremely premature or very low birth weight babies who have been kept alive through intubation immediately after birth may fail to develop independent respiratory function, and suffer further internal malfunction such as renal failure and cerebral hemorrhage. As already suggested, such instances, which are increasing in our neonatal intensive care units, argue that a distinction between a very ill and impaired infant is not a clear one; in fact, illnesses that can be cured may induce permanent and profound disabilities through the very process by which the infant's life is prolonged.[61]

Similarly, decisions not to prolong life need to be continually reassessed in light of the infant's progress. Because clinical judgments are fallible, certain patients may outlive (both qualitatively and quantitatively) a decision not to provide lifesaving treatment in their behalf. Where that occurs, a decision to terminate or not to initiate treatment needs to be reviewed, and aggressive treatment instituted, continuing so long as there is reasonable expectation that the infant's interests will thus be served. The fact that mistakes in judgment occur in ethical as well as clinical dilemmas in no way argues that the judgments themselves are wrong or right in the context of what was known at the time. Only in the long run do such results justifiably exert an influence on subsequent decisions. They do so then because of the knowledge built up over time, providing the rational basis for a general way of acting.

Babies Doe Revisited

The decision to allow Bloomington's Baby Doe to die could not be justified on the basis of the priority criterion discussed here. In light of what we know about children with Down syndrome, it is more likely that this infant's interests would be best served by overriding the parents' refusal of lifesaving treatment for him. The therapy was not futile, the prognosis was not one of neurological devastation, and the predominant future experiences of the child might well have been positive for him. Moreover, while the parents preferred not to raise the child, adoption was a viable option. In fact, a number of couples, two of whom already had children with Down syndrome, offered to adopt Baby Doe before he died. This is a significant factor because actual promotion of infants' interests depends on the attitudes and resources of those who might (or might not) care for them, including government agencies. If a challenge to parental preference for withholding treatment does not address ways by which ongoing care will be provided, the challenge itself may pose a threat to the child's best interests.

In contrast to the Bloomington situation, the case of Baby Jane Doe was one where the priority of the infant's interests may have justified refusal of treatment. *If* the reported facts were correct (e.g., the prognosis that the child "would never experience joy, never experience sorrow"),[62] the option here was between intervention that would allow an extended period of life in a neurologically devastated state; and nonintervention that would permit gradual deterioration, with death occurring sometime during infancy. Obviously, the degree of uncertainty regarding these possibilities was a critical factor. Whether the child's medical problems and condition predicted a predominantly negative experience for her was crucial in determining the applicability of the criterion. However, if there were a high probability that this was so, the priority of the child's interests would be observed by foregoing medical interventions. In other words, the infant's right to life in such circumstances might best be respected by supporting her right to die. The parents' decision to respect that right was apparently motivated by a desire to place the interests of their child before their own: their ethic of care reasoned that love (especially parental love) occasionally means letting go of the one loved.

If the facts reported are correct, then the main difference between these cases is the probability that the future experience of one child would be predominantly positive for him, and the other child's future experience would probably be predominantly negative for her. Because of their differing prognoses, one infant's right to life had priority, and the other's right to die had priority. From an egalitarian perspective that respects differences among individual infants, each deserves to be treated differently.

NOTES

1. "Nondiscrimination on the Basis of Handicaps; Procedures and Guidelines Relating to Health Care for Handicapped Infants," *Federal Register* (Jan. 12, 1984) 49: 1622–54; "Big Brother Doe," *Wall Street Journal* (Oct. 31, 1983), 20; "Baby Jane's Big Brothers," *New York Times* (Nov. 4, 1983), 28; and Mary B. Mahowald and Jerome Paulson, "The Baby Does: Two Different Situations," *The Cleveland Plain Dealer* (Dec. 31, 1983), 9-A.

2. Much of the material in this chapter is adapted from two of my earlier articles: "Ethical Decisions in Neonatal Intensive Care," in *Human Values in Critical Care Medicine,* ed. Stuart Youngner (Philadelphia: Praeger Publishers, 1986), and "In the Interest of Infants," *Philosophy in Context* 14, no. 9 (1984): 9–18.

3. William H. Tooley and Roderick H. Phibbs, "Neonatal Intensive Care: The State of the Art," in *Ethics of Newborn Intensive Care,* ed. Albert R. Jonsen and Michael J. Garland (San Francisco: Health Policy Program, University of California, 1976), 11–15.

4. Tooley and Phibbs, 11–15.

5. Barbara C. Tilley, "Assessment of Risks from DES," in *The Custom-Made Child?* ed. Helen B. Holmes, Betty B. Hoskins, and Michael Gross (Clifton, New Jersey: Humana Press, 1981), 29–39. Concerning ineffectiveness of the therapy, see W. J. Dieckmann, M. D. Davis, L. M. Rynkiewiez, et al., "Does Administration of Diethylstilbestrol During Pregnancy Have Therapeutic Value?" *American Journal of Obstetrics and Gynecology* 66, no. 5 (Nov. 1953): 1062–81. Concerning cancer and reproductive complications in offspring, see Arthur L. Herbst and Diane Anderson, "Clear Cell Adenocarcinoma of the Vagina and Cervix Secondary to Intrauterine Exposure to Diethylstilbestrol," *Seminars in Surgical Oncology* 6 (1990): 343–46; and Raymond H. Kaufman, Kenneth Noller, Ervin Adam, et al., "Upper Genital Tract Abnormalities and Pregnancy Outcome in Diethylstilbestrol-Exposed Progeny," *American Journal of Obstetrics and Gynecology,* 148, no. 7 (April 1, 1984): 973–82.

6. Marshall H. Klaus and Avroy Fanaroff, *Care of the High-Risk Neonate* (Philadelphia: W. B. Saunders, 1973), xi.

7. Mildred T. Stahlman, "Newborn Intensive Care: Success or Failure," *Journal of Pediatrics* 105 (1984): 162–67.

8. Jay Goldsmith and Robert Arensman, "Predicting the Failure of Mechanical Ventilation: New Therapeutic Options," *Neonatal Intensive Care* 3 (1990): 40–47. Data supporting success rates in treatment of all four conditions are available through the Extracorporeal Life Support Registry, University of Michigan, 1991.

9. Richard Martin, "Neonatal Surfactant Therapy — Where Do We Go from Here?" *Journal of Pediatrics* 118, no. 4 (April 1991): 555–56; and T. Allen Merritt, Mikko Hallman, Charles Berry, et al., "Randomized, Placebo-Controlled Trial of Human Surfactant Given at Birth Versus Rescue Administration in Very Low Birth Weight Infants with Lung Immaturity," *Journal of Pediatrics* 118, no. 4 (April 1991): 581–94.

10. Maureen Hack, B. Caron, Ann Rivers, and Avroy Fanaroff, "The Very Low Birth Weight Infant: The Broader Spectrum of Morbidity during Infancy and Early Childhood," *Journal of Developmental and Behavioral Pediatrics* 4, no. 4 (Dec. 1983): 243–49.

11. Maureen Hack, Ann Rivers, and Avroy Fanaroff, "Outcomes of Extremely Low Birth Weight Infants between 1982 and 1988," *New England Journal of Medicine* 321, no. 24 (Dec. 14, 1989): 1642–47.

12. See Maureen Hack, Jeffrey D. Horbar, Michael H. Malloy, et al., "Very Low Birth Weight Outcomes of the National Institute of Child Health and Human Development Neonatal Network," *Pediatrics,* 87, no. 5 (May 1991): 587–97.

13. Laila Williamson, "Infanticide: An Anthropological Analysis," in *Infanticide and the Value of Life,* ed. Marvin Kohl (Buffalo, New York: Prometheus Books, 1978), 61–75.

14. James Rachels, "Active and Passive Euthanasia," *New England Journal of Medicine* 292, no. 2 (Jan. 9, 1975): 78–80.

15. John J. Paris and Anne B. Fletcher, "Infant Doe Regulations and the Absolute Requirement to Use Nourishment and Fluids for the Dying Infant," *Law, Medicine and Health Care* 11 (1983): 210–13.

16. Robert Weir, *Selective Nontreatment of Handicapped Infants: Moral Dilemmas in Neonatal Medicine* (New York: Oxford University Press, 1984), 215–21.

17. Anthony Shaw, "Dilemmas of 'Informed Consent' in Children," *New England Journal of Medicine* 289, no. 17 (Oct. 25, 1973): 885–90.

18. President's Commission for the Study of Ethical Problems in Medicine and Biomedical and Behavioral Research, *Making Health Care Decisions,* vol. 3, Appendices: Studies on the Foundations of Informed Consent (Washington, D.C.: U.S. Government Printing Office, 1982), 175–245.

19. Gina Kolata, "Parents of Tiny Infants Find Care Choices Are Not Theirs," *New York Times* (Sept. 30, 1991), 1 and A11.

20. Hack, Rivers, and Fanaroff, 243–49.

21. H. E. Ehrhardt, "Abortion and Euthanasia: Common Problems — the Termination of Developing and Expiring Life," *Human Life Review* 1 (1975): 12–31.

22. See articles in *The Herald-Telephone,* Bloomington, Indiana, April 23 and May 1–3, 1982, and *The Criterion,* Indianapolis, Indiana, April 23, 1982. Also see Weir, 128–29.

23. U.S. Department of Health and Human Services, "Interim Final Rule 45 CFR Part 84, Nondiscrimination on the Basis of a Handicap," *Federal Register* 48 (March 7, 1983), 9630–32.

24. Barbara J. Culliton, "Baby Doe Regs Thrown Out by Court," *Science* 220 (April 29, 1983): 479–80.

25. It should be noted that the reported facts and media coverage of this case have been disputed, and the infant fared better than had been anticipated. In time, the parents consented to surgery for treatment of the hydrocephalus, her spinal lesion closed naturally, and her parents took her home from the hospital the following spring. See Steven Baer, "The Half-told Story of Baby Jane Doe," *Columbia Journalism Review* (Nov./Dec., 1984), 35–38; and "Baby Doe at Age 1: A Joy and Burden," *New York Times* (Oct. 14, 1984), Sect. 1, 56.

26. See note 1.

27. U.S. Department of Health and Human Services, "Child Abuse and Neglect Prevention and Treatment Program; Final Rule," *Federal Register* 50 (Jan. 11, 1985), 1487–92.

28. Betty W. Levin, "Consensus and Controversy in the Treatment of Catastrophically Ill Newborns," in *Which Babies Shall Live?* (Clifton, New Jersey: Humana Press, 1985), 169–205.

29. See President's Commission for the Study of Ethical Problems in Medicine and Biomedical and Behavioral Research, *Deciding to Forego Life-Sustaining Treatment, A Report of the Ethical, Medical, and Legal Issues in Treatment Decisions* (Washington, D.C.: Government Printing Office, March 1983), 214–22; James Strain, "The American Academy of Pediatrics' Comments on the

'Baby Doe II' Regulations," *New England Journal of Medicine* 309, no. 7 (Aug. 18, 1983): 443–44; and *Handicapped Americans Report* (July 14, 1983), 6.

30. That neonatologists' interpretations of the best interest standard are widely divergent is evident in statements attributed to them by Elisabeth Rosenthal in "As More Tiny Infants Live, Choices and Burden Grow," *New York Times* (Sept. 29, 1991), 1. John Arras has addressed the limitations of this standard in his "Toward an Ethic of Ambiguity," *Hastings Center Report* 14, no. 2 (April 1984): 30–31.

31. See, for example, Paul Ramsey, *The Patient as Person* (New Haven, Connecticut: Yale University Press, 1970), 1–11; see also note 18.

32. Bernard M. Dickens, "Legally Informed Consent," in *Contemporary Issues in Biomedical Ethics,* ed. John W. Davis, Barry Hoffmaster, and Sarah Shorten (Clifton, New Jersey: Humana Press, 1978), 199–204.

33. David Thomasma, "Beyond Medical Paternalism and Patient Autonomy: A Model of Physician–Patient Relationship," *Annals of Internal Medicine* 98 (Feb. 1983): 243–48; and Thomas S. Szasz and Marc H. Hollender, "The Basic Models of the Doctor–Patient Relationship," *Archives of Internal Medicine* 97 (1956): 585–92.

34. Fyador Dostoyevsky, "The Grand Inquisitor," in *The Brothers Karamazov* (New York: Modern Library, 1950), 255–74.

35. Anthony Shaw, Judson G. Randolph, and Barbara B. Manard, "Ethical Issues in Pediatric Surgery: A National Survey of Pediatricians and Pediatric Surgeons," *Pediatrics* 60 (1977): 590.

36. Lawrence A. Rosini, Mary C. Howell, David Todres, and John J. Dorman, "Group Meetings in a Pediatric Intensive Care Unit," *Pediatrics* 53 (1974): 371–74.

37. Richard A. McCormick, "Ethics Committees: Promise or Peril?" *Law, Medicine and Health Care* 12 (1984): 150–55; and Mary B. Mahowald, "Hospital Ethics Committees: Diverse and Problematic," *Newsletter on Philosophy and Medicine (American Philosophical Association)* 88, no. 2 (March 1989): 88–94, reprinted in *HEC Forum* 1 (1989): 237–46 and in *Bioethics News* 2 (1990): 4–13.

38. Alan R. Fleischman and Thomas Murray, "Ethics Committees for Infants Doe?" *Hastings Center Report* 13, no. 6 (Dec. 1983): 5–9.

39. James Strain, "The American Academy of Pediatrics' Comments on the 'Baby Doe II' Regulations," *New England Journal of Medicine* 309, no. 7 (Aug. 18, 1983): 443–44.

40. U.S. Department of Health and Human Services, "Nondiscrimination on the Basis of Handicap: Procedures and Guidelines Relating to Health Care for Handicapped Infants," *Federal Register* 49 (Jan. 12, 1984), 1651.

41. President's Commission for the Study of Ethical Problems in Medicine and Biomedical and Behavioral Research, 227.

42. American Academy of Pediatrics, *Guidelines for Infant Bioethics Committees* (Evanston, Illinois: American Academy of Pediatrics, 1984).

43. See Mary B. Mahowald, "Hospital Ethics Committees: Diverse and Problematic," 88–94. I have attempted to evaluate infant ethics committees in "Baby Doe Committees: A Critical Evaluation," *Current Controversies in Perinatal Care* 15, no. 4 (Dec. 1988), 789–800.

44. Joan E. Hoggman, "Withholding Treatment from Seriously Ill Newborns: A Neonatologist's View," in *Legal and Ethical Aspects of Treating Critically and Terminally Ill Patients,* ed. A. Edward Doudera, J.D., and J. Douglas Peters, J.D. (Ann Arbor, Michigan: AUPHA Press, 1982), 243. Note, however, that living more *easily* with the consequences of something one has done is not equivalent to moral justification for doing it.

45. See August Kasper, "The Doctor and Death" in *Moral Problems in Medicine,* ed. Samuel Gorowitz, Andrew L. Jameton, Ruth Macklin, John M. O'Connor, Eugene V. Perrin, Beverly Page St. Clair, and Susan Sherwin, (Englewood Cliffs, New Jersey: Prentice Hall, Inc., 1976), 69–72.

46. Brenda Miller, at a health care team meeting, Pediatric Intensive Care Unit, Rainbow Babies and Childrens Hospital, Cleveland, Ohio, September 27, 1983.

47. See Susan Schmidt, "Wrongful Life," *Journal of the American Medical Association* 250, no. 16 (Oct. 28, 1983): 2209–10: "Of all the birth-related legal theories, wrongful life, and action filed on behalf of the infant born with a genetic or other congenital birth defect, has met with the most disapproval."

48. For example, Richard McCormick, "To Save or Let Die," *Journal of the American Medical Association* 229, no. 2 (July 8, 1974): 174–45.

49. *Academic American Encyclopedia,* vol. 4 (Danbury, Connecticut: Grolier Press, 1983), 412.

50. *Academic American Encyclopedia,* vol. 11, 394.

51. See Carson Strong, "The Tiniest Newborns," *Hastings Center Report* 13, no. 1 (Feb. 1983): 14–19.

52. Richard McCormick, "Experimental Subjects—Who Should They Be?" *Journal of the American Medical Association* 235, no. 20 (May 17, 1976): 2197.

53. Kenneth Lyons Jones, *Smith's Recognizable Patterns of Human Malformation,* 4th ed. (Philadelphia: W. B. Saunders, 1988), 20–21.

54. See Hans Jonas, "The Right to Die," *Hastings Center Report* 8, no. 4 (Aug. 1978): 36: "Fully understood, it [i.e., the right to life] also includes the right to death."

55. Norman Fost applies this standard to infants under the aegis of "ideal observer theory" in "Ethical Issues in the Treatment of Critically Ill Newborns," *Pediatric Annals* 10, no. 10 (Oct. 1981): 21. Jonathan Glover has a similar suggestion for dealing with infants. He claims that the best substitute for asking whether they wish to go on living (since they cannot register their own preferences) is "to ask whether we ourselves would find such a life preferable to death." See Jonathan Glover, *Causing Death and Saving Lives* (New York: Penguin, 1977), 161.

56. See Gerald Kelly, *Medico-Moral Problems* (St. Louis, Missouri: The Catholic Hospital Association, 1958), 129. For an excellent critique of this distinction, see James Rachels, *The End of Life* (New York: Oxford University Press, 1986), 96–100.

57. My formulation here is different from that of David Smith, who identifies "maximal" with "extraordinary," and "optimal" with "ordinary." See David H. Smith, "On Letting Some Babies Die," *Hastings Center Studies* 2, no. 2 (May 1974): 44. I return to the concept of "optimal care" in Chapter 15.

58. These are distinctions employed by pediatricians with whom I have worked: the first by Donald Schussler, M.D., the second by Jeffrey Blumer, M.D., Ph.D., both working in the Division of Critical Care, Rainbow Babies and Children's Hospital, Cleveland, Ohio, during the 1980's.

59. James Strain, "The Decision to Forego Life-Sustaining Treatment for Seriously Ill Newborns," *Pediatrics* 72, no. 4 (Oct. 1983): 572. Strain was comparing the pediatricians of 1983 with those interviewed for studies published in 1977 based on data obtained several years earlier. See, for example, Shaw, Randolph, and Manard, 588; and I. David Todres, Diane Krane, Mary C. Howell, et al., "Pediatricians' Attitudes Affecting Decision-Making in Defective Newborns," *Pediatrics* 6, no. 2 (Aug. 1977), 197–201.

60. Kolata, 1, A11; and I. David Todres, Jeanne Guillemin, Michael A. Grodin, and Dick Batten, "Life-Saving Therapy for Newborn: A Questionnaire Survey in the State of Massachusetts," *Pediatrics* 81, no. 5 (May 1988): 643.

61. See note 7.

62. My analysis here is crucially dependent on the accuracy of the reported facts and of the prognosis associated with them. As indicated in note 25, both were disputed in subsequent accounts of the case.

11

Children and Moral Agency

"Recent moral philosophy," according to Geoffrey Scarre, "has been guilty of child neglect."[1] In this chapter I attempt to redress that neglect by focusing on the possibilities for moral agency in children, particularly in the health care setting. I examine the meaning of moral agency and its applicability to children in light of specific examples, studies of child development, and variables that influence moral agency in adults as well as children. I also consider children's concepts of illness and death, and possibilities for promoting their moral agency in pediatric practice. What motivates my analysis is a concern that children, like women, are often treated stereotypically, without taking account of the differences in maturity, family circumstances, talents, and needs that exist among them as individuals. From an egalitarian standpoint, the differences are surely relevant.

What Is Meant by Moral Agency?

Many well-respected philosophers have developed careful responses to the above question. The results are mixed and remain controversial. As a starting point, therefore, I use one of those sources whose views are well-developed, well-respected, and persuasive.[2] In *Reason and Morality,* Alan Gewirth defines morality as "a set of categorically obligatory requirements for action that are addressed at least in part to every actual or prospective agent, and that are concerned with furthering the interests, especially the most important interests, of persons or recipients other than or in addition to the agent or the speaker."[3]

Despite the Kantian ring of this definition, it allows us to bypass the ongoing debate between utilitarians and deontologists because "obligatory requirements" may be determined either by rules or by ends, or by both rules and ends. To set the definition in the context of an ethic of virtue rather than one of obligation, we can substitute the phrase "requirements of virtue" for "categorically obligatory requirements." The reference to "actual or prospective agent" permits application to children as one or the other. I argue here for the stronger claim, that children (at least some of them, at least some of the time) are actual moral agents. According to Gewirth, "well-being" and "freedom" are the necessary goods to be promoted by moral agents. These may be viewed as "the most important interests" to which he refers.[4]

"Well-being" and "freedom" also represent the traditional ethical principles of beneficence and respect for autonomy.[5] For Gewirth, the interests to be promoted through morality are necessarily those of others, and not, or not only, those of the agent. In other words, morality is essentially a social enterprise. As with Kurt Baier's "moral point of view,"[6] this position may be supported on the basis of Kantian universalizability as well as utilitarian reasoning: we are thus looking at the world from the point of view of everyone, and for the good of everyone. It is also supported by a care model of moral reasoning, because attachments to others call for pursuit of their interests as well as our own.

Although Gewirth uses different terminology, his concept of agency is basically Aristotelian and consistent with the concepts of human beings elaborated in Chapter 1. According to Aristotle, the essential characteristics of humans are intellect and will (rational inclination).[7] Gewirth analyzes the generic structure of human action as dependent on purposiveness and voluntariness. Purposiveness embraces the cognitive element of agency, requiring that an individual act for some intended end. This is comparable to Aristotle's notion of reason or intellect. As intended, the end must be known by, and known to be related in some way to, the individual who selects it. Moreover, the end pursued, for Gewirth as for Aristotle, always appears good to the pursuer.

An action is voluntary, Gewirth claims, if an individual "unforcedly chooses to act as he does, knowing the relevant proximate circumstances of his action."[8] Voluntariness is thus tied to rationality, or Aristotle's notion of rational inclination, precluding the possibility that sheer impetuosity or spontaneity is truly voluntary. The "relevant proximate circumstances" of an action include the effects of the action, but not necessarily far-reaching or remote effects. However, knowing the effects does not imply action based on that knowledge; a utilitarian rationale is possible but not essential through the voluntariness that agency entails.

Gewirth's definition of voluntariness is open to the criticism that it lacks positive content beyond that which is already implied by purposiveness (cognitive content). The term *choice* may be understood to imply freedom or individual liberty (as in a pro-choice position on abortion), but these are notoriously controversial and often unclear concepts. Nevertheless, freedom is generally viewed as an essential condition of moral responsibility. The term *responsibility,* connoting blameworthiness or praiseworthiness, may be used synonymously with agency, but this term may be construed as begging the question of freedom. Praise or blame are sometimes attributed (correctly or incorrectly) to inanimate objects that apparently exhibit cause–effect relationships. For example, a computer may be "blamed" or "praised" for a "good" or "bad" job because it did (or didn't) do what it was programmed to do. Such terms may correctly describe nonmoral relationships, without implying choice or freedom on either side.

Since the concept of, and possibilities for, freedom have been dealt with extensively and critically by others, I wish here simply to acknowledge my concurrence with the position that morality is only possible if individuals are free to choose among recognized alternatives.[9] Like Kant, then, I postulate that freedom is, without claiming to know what it is.[10] In that context, it is more appropriate to define (free) choice negatively rather than positively, as "uncoerced" or "unforced." Free choice occurs to the extent that behavior is not determined by the multiple influences that affect individuals from without (environment, society) and from within (genes, bodily mechanisms, etc.).

What then is moral agency? To synthesize the crucial elements of the preceding discussion, it means the capacity for (a) voluntary, (b) purposeful actions recognized as (c) influencing the well-being or freedom of others. Points (a) and (b) refer to agency; (c) refers to the social context of agents' choices. Morality necessarily involves social relationships. An ethic of care is committed to maintaining those relationships, an ethic of justice to ensuring that the relationships are not exploitative. Both modes of reasoning are observable in children; nonetheless, children are sometimes thought to be incapable of moral agency because their reasoning powers are not sufficiently developed. In the next section I explain why this view is mistaken.

The Child's Capacity for Moral Agency

Consider the following cognitive components of the exercise of moral agency: (a) understanding of the cause–effect relationship, (b) understanding of oneself as a causal agent, and (c) understanding of moral values or principles. Two of the components (b and c) involve one side of a relationship, and the other (a) involves the tie between the two. Minimally, the exercise of moral agency displays this threefold understanding, even where that understanding is not, or cannot be, articulated.

In the mind of the agent, the understanding of moral values or principles must be linked with the sense of self as causative. Individuals thus identify moral values with effects that they can cause, or moral principles with the effect of adherence (to those principles) that they can cause. The recognition of, and link with, moral values permits utilitarian reasoning. The recognition of, and link with, moral principles permits deontological reasoning. The latter, as Kohlberg suggests, is the more complicated reasoning process because it entails an additional and subtle conception, that of adherence to principles.[11] Moreover, the utilitarian reasoning process is often (some would say always) empirically based, whereas deontological reasoning cannot be exclusively that, and may be totally nonempirical.

Understanding oneself as a moral agent necessarily involves recognition of special relationships with others, for example, through ties of kinship or friendship. Such relationships are values in their own right, and their preservation and nurturance are a priori moral obligations. To the extent that this dimension of morality supplements the traditional model of utilitarian or deontological reasoning, the agent reflects a care-based model of moral reasoning, illustrating the "different voice" that Carol Gilligan describes. Both voices are available to male or female children, as well as adults.[12]

The prevailing descriptive accounts of moral reasoning do not purport to be philosophical in an analytic or critical sense. They do not question the meaning of moral agency in its own right, or examine the child's capacity for moral agency on that basis. It may be helpful, therefore, to consider the actual behaviors of children as indicative of their capacity for moral agency. Situations such as the following are suggestive of that capacity.

A. A group of preschoolers are playing hide-and-seek. A "hider" complains that the "seeker" has cheated because she peeked while counting to ten.
B. A 10-year-old is a potential blood marrow donor for his sibling who has leukemia. On having the procedure with its harms and risks to him and possible benefits to his brother explained, he says he wants to do this for his brother.
C. A 14-year-old who is mentally retarded scolds her younger sisters for fighting with each other.
D. A 7-year-old secretively takes cash from his mother's purse, intending to buy candy at the neighborhood store.
E. A 9-year-old is aware that she is dying, but does not broach the subject with her parents because she does not want to upset them.
F. A 16-year-old mother is asked whether she would agree to lifesaving treatment of her seriously ill newborn.

In each of these examples the behavior of the child or adolescent[13] suggests a capacity to act as a moral agent. Whether the child is in fact expressing that capacity remains an open question, because ability does not necessarily imply actualization of the ability in a particular situation. In each case, however, the child apparently understands the cause–effect relationship as well as certain moral values or principles, and exhibits a sense of self as a causative being. In each case, the requirements of purposiveness, voluntariness, and recognition of influence of one's action on the well-being or freedom of another appear to be met.

Situations (A) and (C) present the most problematic scenarios for defending the child's competence to make intelligent, free decisions: for (A), the child's very young age seems to preclude the possibility; for (C), mental retardation seems to preclude it. However, leaving aside the question of a child's moral consciousness, the preschooler's perception that cheating is something to complain about coincides with the general perception of adults in that regard. The validity of the child's perspective seems evident here because an ethical principle called fairness has been betrayed. But fairness only makes sense in the context of social awareness: the child thus recognizes that the rule of not peeking applies to all of the players of the game. The preschooler's complaint is at least as voluntary as many adult complaints about cheating, and purposive in that she wishes to end the seeker's cheating behavior, or incite another (for example, the teacher) to end it.

Adolescents are generally credited with capacity for making more responsible decisions than younger children.[14] The factor that further complicates situation (C), however, is that the 14-year-old under consideration is retarded. Retardation, as discussed in Chapter 5, spans a broad spectrum, from profound to mild. Scolding a sibling for fighting suggests that the scolder is not severely retarded. Not only does the scolding indicate some degree of competence; it also suggests a recognition of the importance of relationships and moral cognition that fighting is wrong, at least in some circumstances. Mental retardation means a slower pace of learning or development but does not imply that understanding of relationships, values, or principles is impossible.

Situation (B) describes a borderline case of competence for a decision to undergo a low risk procedure for the sake of another. The moral content of the decision is obvious, but whether a particular 10-year-old child understands the risk and intent of the procedure is questionable. In the case of a 5-year-old potential donor, most practitioners would follow the parents' wishes whether the child "agreed" with them or not. They would not oppose the dissent of the 15-year-old, even with parental permission. But the case of the 10-year-old would need to be settled on an individual basis.[15] Whatever the age, assessment of competence should be distinguished from assessment of the moral content of the decision, because one does not imply the other. Moreover, if we are to respect the autonomy of competent potential donors on a consistent basis, this respect applies to refusal as well as consent to bone marrow donation. If a refusal is challenged for the sake of the prospective recipient, this should be acknowledged as a violation of autonomy whether the competent donor is a child or an adult. From an egalitarian perspective, the violation of autonomy may be morally defended in either case on grounds of obligatory beneficence toward the recipient.

Legally, no adult would be forced to undergo bone marrow donation for the sake of another.

The 7-year-old who secretively takes cash from his mother's purse indicates moral awareness through his secretiveness. Recognition of need for money, of where to find it, and of how to get it connotes basic understanding of the cause–effect relationship. Developmental studies clearly establish a sense of the self as independent or autonomous occurring even before age three.[16] Nonetheless, it is as true for children as for adults that consistency of behavior is not determinative that an individual's action or decision is autonomous. No correlation has been demonstrated between rational processes concerning morality and the actual behaviors selected by individuals. In the absence of empirical evidence to the contrary, then, we cannot legitimately disenfranchise children from the moral enterprise solely on grounds that their behaviors are inconsistent in themselves, or with the moral principles they claim as theirs.

The young girl described in situation (E) is aware that she is dying, and apparently has some meaningful concept of death. In fact, her decision not to broach the subject with her parents suggests that she well understands its seriousness and sadness.[17] Her parents are probably aware of the gravity of the situation, and she may realize that. Even if she does not, however, her decision not to discuss the matter with them illustrates one of the crucial requirements of moral agency, that is, recognition that one's own behavior influences the well-being of others. It also reflects the sensitivity to relationships on which a care model of reasoning is based.

Regarding the last situation, (F), most would not doubt that a normal 16-year-old is capable of making moral (and immoral) decisions. Indeed the law has prevalently acknowledged this probability through the "mature minor precedent," even while setting 18 years as the age of legal majority. Basically, this precedent of the U.S. judicial system is a long-standing practice by which parental consent for treatment of minors over 14 years of age has not been required so long as the adolescent is judged competent, and there is some impediment (which may simply be the adolescent's request that parents not be informed) to obtaining parental consent.[18] The probability of competence for a 16-year-old, the status of emancipated minor, and the assumption of a special relationship between (even a teenaged) mother and her child constitute a convincing case for claiming that the young woman described in situation (F) is legally and morally capable of responsible decision making for her newborn. As for the content of a decision to provide lifesaving treatment to a seriously ill infant, the moral dimensions of the decision are obvious and momentous.[19]

The range of ages, mental ability, and circumstances of situations (A) through (F) illustrate the differences that influence not only the capacity for moral decision making, but the content of these decisions as well. From an egalitarian standpoint, attention to such differences is important with regard to adults and children alike. All of the differences or variables discussed next are applicable to moral agency.

Variables Affecting Moral Agency in Specific Situations

Two types of variables are applicable to assessment of moral agency: the first type determines capacity; the second determines expression. Capacity is fundamentally a function of internal factors such as intellectual ability, knowledge, and emotional stability, but external factors such as the extent of an individual's experience and the range of options actually available also influence one's capacity for moral decisions. The expression of moral agency is a function of internal as well as external factors, but more importantly of the latter. The individual must know how to articulate choices, have the physical ability to do so, and have the physical, intellectual, and emotional ability to actualize whatever moral decision is made. Exercise of these abilities is subject to limitations by circumstances, some of which are changing (such as sleep or wakefulness, time to learn and think about alternatives and their implications), and some of which are permanent (such as permanent physical hindrances or obstacles).

One external variable is the content of particular decisions, which includes the magnitude and complexity of the options considered. Understanding the content of a decision may be totally beyond the ken of the agent, eliminating the possibility of a truly informed or competent decision. Actually, many decisions made by adults have this character—for example, decisions to have children and decisions to terminate life sustaining treatment for oneself or for another. These are irreversible decisions whose impact and broad implications cannot be known ahead of time. Yet we generally assume that adults who make such decisions are acting as moral agents. In other momentous but reversible decisions, such as choice of a profession or a marriage partner, we also make that assumption, even though some individuals claim subsequently that they did not choose freely or knowledgeably. If we allow that such decisions made by adults illustrate moral agency even though they do not express full knowledge of implications of the choice, we have thereby made a good case for the ability of some children to make meaningful decisions in situations where they do not or cannot fully understand the outcome.

Age is a variable that is often construed as determining capacity for, and expression of, moral agency. To Willard Gaylin, age is "the summation of intelligence, rationality, perception, and experience."[20] Laws, both civil and religious, define policies based on age—for example, the "age of reason," legal majority, voting age, marriage age. While some of the variables cited above (such as intellectual ability) are not subject to change simply on the basis of the passage of time and, therefore, age, most of them (such as knowledge and experience) undoubtedly require some time lapse. Accordingly, it is generally agreed that even a very bright (extremely intelligent) newborn, who is learning at a faster pace than she will ever learn subsequently, is not capable of decision making in a sense that is meaningful. It is not agreed that this is so when the child has had more time to learn and experience life. Categorical use of the age variable may lead to what Alfred North Whitehead calls the fallacy of misplaced concreteness,[21] by applying the age criterion to individuals who do not fit the generalization. The fallacy is most likely to be committed at either end of the aging spectrum, through exclusion of quite young and very old individuals from acknowledgment of their capacity to act as decision makers in their own regard.

Different variables overlap and affect each other. For example, the life span of an individual, as identified by age, typically provides opportunities for experiences crucial to knowledgeable decisions. Thus a teenager who has never been pregnant, no matter how bright she is, is likely to make a less informed decision about abortion than her mother might make. On the other hand, some teenagers who are already mothers may make more informed decisions about pregnancy than older women who have not had similar experience. A 13-year-old friend of mine is very bright, but less knowledgeable, by virtue of lack of experience, than many less bright peers who may legally choose abortion if they become pregnant. She may, however, be more knowledgeable with regard to other moral questions about which her judgments are rarely asked or respected. For example, this particular teenager has read and thought about political moral problems such as the discrepant distribution of food around the globe. When she took a chunk out of her savings to send to African famine relief, this surely suggested her capacity to act as a moral agent. Although she has never experienced severe hunger herself, she is probably as knowledgeable about the moral content of issues regarding world hunger as are many adults who have never experienced severe hunger.

The content variable has influenced the enactment of quite different laws regarding the capacity of children for decision making, whether moral, immoral, or amoral.[22] Sexuality and drug or alcohol rehabilitation are the main areas in which the law supports the right of most adolescents to make their own decisions. Admittedly, the purpose here is mainly utilitarian, and in a certain sense paternalistic: the requirement of parental consent might lead to more adolescent pregnancies and reduce enrollment in rehabilitation programs. But there is concomitantly an increased sense that many teenagers are capable of responsible moral decisions about these matters. Categorical refusal to recognize that strong possibility is untenable in light of our society's emphasis on respect for individual autonomy. It is ironic, therefore, that we persist in preventing teenagers from exercising moral agency in less significant contexts. For example, we insist on parental permission for an adolescent's participation in a research project that involves no risk,[23] while asking the same teenager for permission to treat her child. Because the latter decision is more momentous in its consequences, it provides the more compelling case for overruling the teenager's autonomy.

In light of the variables previously described, as well as the conditions of moral agency outlined at the start of this chapter, it seems clear that some children are moral agents some of the time. To the extent that they act as moral agents, they deserve blame or praise for their actions, but this does not imply that they should receive the same punishments or rewards as adults.[24] Moreover, whether individuals exercise moral agency depends on other variables in addition to their capacity for exercising it. In the health care setting, the pertinent variables include concepts of illness and death.

Children's Concepts of Illness and Death

Although the variables involved in children's concepts of illness and death may lead to different concepts for each child, there are generalizable features that correspond with developmental stages. Both kinds of knowledge, general and specific, are useful for those who wish to respect the moral agency of children in the health care setting.

Children's concepts of illness have been correlated with Jean Piaget's stages of cognitive development: prelogical thinking, which is characteristic of children between 2 and 6 years old; concrete logical thinking, which is manifest in those between 7 and 10; and formal logical thinking, which is typical in children who are 11 years and older.[25] Prelogical thinking involves a rudimentary sense of the cause–effect relationship and is immediately triggered by familiar spatial or temporal cues. In their studies of children's concepts of illness, Roger Bibace and Mary E. Walsh identify two types of prelogical explanations: phenomenism and contagion.[26] The phenomenist explanation interprets the cause of illness as an external concrete phenomenon that is spatially or temporarily remote. Asked what causes colds, the child at this stage might answer: "The sun," or "God." When the cause of illness is seen as proximate but not touching the child, the concept is one of contagion. Still at a prelogical stage, the child may think that colds are caught "from the outside," or "from someone who gets near you."

Children who have reached the stage of concrete logical reasoning are generally able to distinguish between what is internal and external to themselves. According to Bibace and Walsh, their explanations of illness then include concepts of contamination or internalization.[27] The child may think people "catch" colds from other specific persons (contamination), or from harmful bacteria that have entered the body by swallowing or inhaling (internalization). At the stage of formal logical thinking, children have a better sense of the distinction between themselves and the world, allowing them to point to the source of illness within the body even while describing an external agent as its ultimate cause. Physiological explanations show an understanding of the nonfunctioning or malfunctioning of an internal organ or process. Psychophysiological explanations extend the recognition of internal physiological causes of illness to recognition of psychological causes as well.[28] A child who is 11, for example, is likely to understand that a heart attack may be caused by underlying tension.

Bibace and Walsh's account of children's developmental concepts of illness support my claim that children recognize the cause–effect relationships essential to capacity for moral agency at an early age. It also suggests possibilities for fostering moral agency by attending to the ways in which children understand illness. For example, physiological explanations of the virus that causes chicken pox are probably futile for a 7-year-old, whereas a concrete description of how more pox will appear, are likely to itch, and so on, will be quite meaningful. Because young children tend to focus on immediacy, preparation of a 5-year-old for surgery might concentrate on observable events such as the bright lights in the operating room and the doctors' gowns and masks, whereas preparation of a 12-year-old might include a description of anatomical details involving the surgery itself.

Preparing a child for death may be more difficult than preparing one for surgery, but the preparation may again be facilitated by recognition of developmental stages of children's concepts. According to Susan Carey, experts agree on three periods in that development.[29] The first and earliest (which Carey attributes to youngsters 5 years old and under), is a concept of death as sleep or separation. While the emotion this concept evokes is a sad one, death is not recognized as final or inevitable because people ordinarily wake from sleep and may return after their separations.

In the second stage of the child's emerging concept (early elementary years), the child views death as a permanent cessation of existence. The cause of death, however, is generally thought to be external. There is as yet no sense that death may be caused solely from within the body, or as a result of internal processes initiated by external events. In the third stage (about 10 years and older), death is seen as an inevitable biological process. The notion of death's finality is sometimes tempered by belief in an afterlife (as it is in some adults), but there is still the recognition that the dead person will not return "in this life."

Despite Carey's claim of universal agreement about the stages of children's concepts of death, some authors maintain that most children understand the essential components of a concept of death (irreversibility, nonfunctionality, and universality) as early as 7 years of age. Regardless of whether the experts agree, however, some children are obviously more precocious than others in their conceptual development. Unfortunately, for some it is the experience of their own terminal illness, or being closely associated with dying persons and deaths, that explains their precocity. It is important, therefore, to pay attention to the individual as well as generic variation in children's concepts of death.

With regard to both serious illness and impending death, Gareth Matthews argues that disclosure to children of their prognoses is morally more defensible than withholding such information. He claims that the usual arguments for nondisclosure are based on two worries: first, that the bad news will be successfully communicated to the child; and second, that the child will be unable to cope with the news. Regarding the first worry, Matthews maintains that "one will simply not succeed in getting across the mature message, that is, the full meaning of impending death."[30]

Concerning the second worry, children may be facilitated to cope with the news of death by their inability to fully comprehend it. Even for young children who conceive of death as sleep from which one will eventually awaken, the news of death is probably not so awful as it is for adults. According to Matthews, adults' own difficulties in dealing with the deaths of children probably underlie their inclination to withhold information from children.

Marilyn Bluebond-Langer has amply demonstrated the capacity of older children to understand and deal with their impending deaths.[31] She describes five stages through which children dying of cancer progress. First, they think: "I am seriously ill"; second, "I am seriously ill but will get better"; third, "I am always ill but will get better"; fourth, "I am always ill and will not get better"; and fifth, "I am dying." While categorizing the stages, Bluebond-Langer found that chronological age was irrelevant to their sequence. The experience of the disease itself, with all the unpredictable elements of that experience, prompted the different stages. This is a strong argument for recognition of the limitation of generalizations based solely on chronological age. While drawing on the advice of experts in cognitive development, it is thus important to relate their finding not only to individual variations in cognitive ability but also to children's experiences as individuals. To the extent that each child's capacity for moral agency is recognized and supported, the pitfalls of paternalism and maternalism are avoided (see Chapter 2). Beyond parenting, the field of pediatrics presents an important challenge in this regard.

Implications for Pediatric Practice

Children often meet the requirements of informed consent, despite lack of acknowledgment to that effect by the legal system. The usual requirements for informed consent are competence for responsible decision making, understanding of pertinent information, and voluntariness. All three factors are found in some children at least some of the time, and not found in all adults all of the time.

Consistent with guideline 4, moral agency should be respected in whomever it occurs, and to the extent that it occurs. Depending on the child and the circumstances, respect for moral agency may mean their participation in decision making, assent or dissent to decisions made by others, or their full informed consent. Truly informed and free consent may thus be viewed as an ideal to be approximated by children as well as adults.

As I have already suggested, participation in decision making is possible even at a very young age. The participation may be as minimal as the child's developmental awareness allows. The following case is an example of failure to respect a young child's developmental awareness by allowing her to participate in a tragic situation and its implications. As the case unfolded, a seriously ill mother's right to see her child was also denied.

A family of four experienced a devastating auto accident in which the 7-year-old boy and his father were killed. The 5-year-old girl was transported to the nearest tertiary care center, where she was placed in the pediatric intensive care unit after undergoing emergency surgery. Her mother, also in critical condition, was transported to the same hospital, treated, and placed in the surgical intensive care unit. Although the child was expected to recover, the woman was thought near death.

When told of the death of her husband and son, and of her daughter's and her own condition, the woman asked to see her daughter. Concurrently, the little girl asked several times to see her parents. She was told that they were in another part of the hospital and not able to visit her yet. The pediatric surgeon insisted that a visit between mother and child not be arranged, maintaining that the child would be too upset by the sight of her mother's injuries and illness. "As the child's physician," she claimed, "I am obligated to prevent anything that might compromise my patient's welfare."

Whether the surgeon had a legal right to refuse the parent's request is doubtful, but that is not my point here. An a priori moral obligation not to ignore or impede the thoughts or choices of others (guideline 4) seems to have been violated with regard to both patients: the woman's autonomy was overridden, and the child's desire was also overruled. Nurses challenged the physician's refusal, but by the time a meeting was convened to resolve the disagreement, the woman had lapsed into unconsciousness, making her request unfulfillable. Although the child's desire to see her mother might still have been granted, the surgeon's refusal was honored.

If seeing her mother would have compromised the child's chance of recovery, the physician had a valid but inadequate reason for opposing both patients' wishes. Obviously, the degree of compromise and its risk needed to be assessed. If these were minimal, respect for both patients' wishes should have overridden the principle of beneficence; if the compromise and risk were grave, beneficence might have overridden the reduced capacity for autonomy of the child, and possibly also that of her mother.

Apparently, the pediatric surgeon in this case was unaware of the literature regarding children's concepts of illness and disease and of their capacity for dealing with such tragic events.[32] Had she recognized her limitation, she might have consulted someone with expertise in these matters. In fact, a child psychologist who heads a program for counseling children and their parents on how to deal with the deaths and dying of loved ones was available at the pediatric surgeon's own institution. Had he been asked, he would have visited both child and mother and offered advice based on generalized knowledge as well as his specific knowledge of this child's capability for dealing with the situation. Unfortunately, he was not invited to do so.

Participation of young children in health care decisions is mainly a matter of sensitive disclosure. There are times of course when the child has *no* capacity for understanding. Even at an early age, however, most children have at least rudimentary understanding of death, life, health, and disease, concepts basic to communication with them about their care and that of their family members.[33] Such communication is an essential component of respect for their developing autonomy. If the child cannot understand these concepts, then mere disclosure is unlikely to be harmful; if the child *may* understand them, disclosure is obligatory in proportion to the probability of understanding and in proportion to the risk of harm from the disclosure. Generally, the thought of what might happen is worse than the reality. But if disclosure would precipitate resistance to a procedure necessary to a child's health, foregoing it is justified. The assumption in such circumstances is that the child's limited understanding has reduced her capacity for autonomous decision making.

While disclosure is a component of respect for autonomy, it is not equivalent to the obligation of obtaining assent, or respecting dissent, once disclosure is provided. The pediatrician who explains to a 7-year-old child what a lumbar puncture involves does not require the child's assent to the procedure so long as it is crucial to proper diagnosis and treatment of a serious infection such as bacterial meningitis. But the same child may decide whether to present his left or right arm for a blood sample. He may also decline to participate in a nonrisk research project that his treating physician proposes. Assent and dissent are modes of participation, but not equivalent to fully informed and free consent.

Fully informed consent of children is best approximated in cases such as the following:

A 12-year-old boy had congenital strabismus (eye muscle imbalance) for which he had been operated on at 2 years of age. Although the surgery achieved partial correction, the imbalance became more obvious as the boy

got older, and he was self-conscious about it. At a routine visit the ophthalmologist told the boy that new surgical techniques could be utilized to permanently improve the appearance of his eyes. It would not, however, improve
his vision. The boy said he did not want surgery.

Several weeks later, the boy changed his mind and told his parents he might
like to have the operation, but would first like to ask the doctor some questions about it. Responding to a message left at his office, the physician called
the boy and said: "Fire away. Ask me anything you want because you should
know what it would be like before you decide to go through with this kind of
operation. You don't have to have it unless you want it."

Pressures from parents might, of course, reduce the autonomy of children capable of informed consent to elective procedures such as the preceding. Such
pressures may come from caregivers as well, and adult patients endure pressures
that compromise autonomy also. Economic constraints are sometimes more limiting than social constraints. Thus the voluntariness essential to consent is present in various degrees in different situations. The caregiver or parent who wants
to maximize respect for a child's autonomy attempts as far as possible to reduce
those pressures.

Admittedly, had the 12-year-old declined lifesaving rather than elective
surgery, the argument for respecting his autonomy would not have been compelling. But what about situations in which treatment is medically indicated, yet
not immediately necessary in order to preserve life? Consider, for example, the
following case.

A 15-year-old girl had severe scoliosis, requiring surgery in order to prevent
respiratory compromise. The girl's mother had consented to the surgery, but
the girl insisted she did not want it. Despite her objections, the girl was taken
to the operating room on the appointed day. As the orthopedic surgeon arrived, he heard her scream "Don't do this to me. I told them I didn't want it."
The surgeon declined to perform the surgery, and sent the patient back to the
floor, where he later apologized to her for not having learned of her wishes
beforehand. At that point he initiated the first of several discussions in which
he "made a deal" with the patient. He explained that she might die within the
next few years if she didn't have the surgery, and that she could choose the
time for it so long as this occurred within the next half year. The physician's
effort to give the adolescent more control over her medical course was effective. Two months later, she voluntarily underwent the surgery.

In neither of the two preceding cases were the physicians legally obligated to
obtain consent for surgery from their pediatric patients. If they were to follow
the egalitarian guideline of respecting autonomy of children as well as adults,

however (guideline 5), both were morally obligated to elicit the consent of their pediatric patients. One of the more subtle challenges of pediatric practice is to determine the extent to which children are capable of moral decision making, to respect and foster that capacity while weighing it against the risks that this involves.

To return to the notion of parentalism discussed in Chapter 2, the challenge of pediatric practice is thus to be parentalistic rather than paternalistic, which means that clinicians as well as parents should maximally respect and encourage the developing autonomy of children. Grounds for justifiable intervention are mainly utilitarian rather than paternalistic. Although laws systematically deny that most children are capable of "informed consent," it is surely not true that "children are not moral agents." [34] To their credit, both the ophthalmologist and the orthopedist in the above cases distinguished between legal and moral aspects of decision making, recognizing that the two may be at odds in children as well as adults. They contributed to the moral development of their patients by maintaining a more discerning and critical attitude toward differences in individuals, regardless of their age.

NOTES

1. Geoffrey Scarre, "Children and Paternalism," *Philosophy* 55 (1980): 117. Much of this chapter is taken from my "Possibilities for Moral Agency in Children," in *Freedom, Equality, and Social Change,* ed. Creighton Peden and James Sterba (Lewiston, New York: Mellen Press, 1989), 275–85.

2. The most recent scholarly analysis and defense of Alan Gewirth's views is Deryck Beyleveld, *The Dialectical Necessity of Morality* (Chicago: University of Chicago Press, 1991).

3. Alan Gewirth, *Reason and Morality* (Chicago: University of Chicago Press, 1978), 1.

4. Gewirth, 61, 48.

5. See Tom L. Beauchamp and Laurence B. McCullough, *Medical Ethics* (Englewood Cliffs, New Jersey: Prentice Hall, Inc., 1984), 14–16.

6. Kurt Baier, *The Moral Point of View* (Ithaca, New York: Cornell University Press, 1985).

7. See Aristotle, *On the Soul* II, 3, and *Nichomachean Ethics* I, 1. For Aristotle, human beings have a rational soul rather than merely a sensitive or vegetative soul as have other animals or plants. What distinguishes humans from other animals is *nous* or "mind," which is expressed through scientific thought and through deliberation. Scientific thought has truth for its own sake as its object, while deliberation has practical or prudential purposes. See Frederick Copleston, *A History of Philosophy,* vol. I, part II (Garden City, New York: Image Books, 1962), 70.

8. Gewirth, 27.

9. I thus reject the thesis of "hard determinism" that denies human freedom. Neither, however, would I subscribe to the existentialist view that freedom (absolute freedom) defines human nature. Freedom, I believe, is partial, as is moral responsibility. See my "Beyond Skinner: A Chance to be Moral," *Journal of Social Philosophy* 4 (1973): 1–4.

10. For example, in Immanuel Kant, *Critique of Practical Reason,* trans. Lewis White Beck (Indianapolis: Bobbs-Merrill Company, 1956), 107–9.

11. See Lawrence Kohlberg, "The Child as Moral Philosopher," *Moral Education,* ed. Barry I. Chazan and Jonas Soltis (New York: Teachers College Press, 1973), 131–43.

12. Carol Gilligan, *In a Different Voice* (Cambridge, Massachusetts: Harvard University Press, 1982).

13. When I use the term *child* throughout this chapter, I intend the designation to refer to anyone who is legally a minor. In other words, I generally mean both children and adolescents.

14. See Willard Gaylin, "The Competence of Children: No Longer All or None," *Hastings Center Report* 12, no. 2 (April 1982): 33–38.

15. Personal communication, Peter Coccia, Director, Division of Pediatric Hematology and Oncology, Rainbow Babies and Childrens Hospital, Cleveland, Ohio, June 1985. Dr. Coccia has since moved to the University of Nebraska School of Medicine in Omaha.

16. See Robert Schell and Elizabeth Hall, *Developmental Psychology Today,* 4th ed. (New York: Random House, 1983), 392–93.

17. See Arlene B. Brewster, "Chronically Ill Hospitalized Children's Concepts of Their Illness," *Pediatrics* 69, no. 3 (March 1982): 355–62.

18. Sanford L. Leikin, "Minors' Assent or Dissent to Medical Treatment," *Journal of Pediatrics* 102, no. 2 (Feb. 1983): 169–76.

19. "Momentous" in the sense that William James attaches to "genuine options," that is, as opposite of "trivial." See his "The Will to Believe," in *Essays on Faith and Morals* (Cleveland: World Publishing Company, Meridian Books, 1962), 34.

20. Gaylin, 34.

21. Alfred North Whitehead, *Process and Reality* (New York: Harper and Row, 1957), 11.

22. See Alexander Morgan Capron, "The Competence of Children as Self-Deciders in Biomedical Interventions," in *Who Speaks for The Child? The Problems of Proxy Consent,* ed. Willard Gaylin and Ruth Mackin (New York: Plenum Press, 1982), 57–114.

23. See Capron, note 22, and Leikin, note 18. See also Patricia Keith-Spiegel, "Children and Consent to Participate in Research," in *Children's Competence to Consent,* ed. Gary B. Melton, Gerald P. Koocher, and Michael J. Saks (New York: Plenum Press, 1983), 179–207.

24. It is not inconsistent, therefore, for laws to permit execution of adults but not children for comparable crimes. In part because variables such as knowledge and experience influence the exercise of moral agency, rehabilitative goals are more compelling in the case of children than adults.

25. Jean Piaget, *The Moral Development of the Child* (New York: Free Press, 1969).

26. Roger Bibace and Mary E. Walsh, "Children's Conceptions of Illness," in *Children's Conceptions of Health, Illness, and Bodily Functions,* ed. Roger Bibace and Mary E. Walsh (San Francisco: Jossey-Bass, Inc., Publishers, 1981), 35–36.

27. Bibace and Walsh, 36–37.

28. Bibace and Walsh, 37–38.

29. Susan Carey, *Conceptual Change in Childhood* (Cambridge, Massachusetts: MIT Press, 1988).

30. Gareth B. Matthews, "Children's Conceptions of Illness and Death," in *Children and Health Care: Moral and Social Issues,* ed. Loretta M. Kopelman and John C. Moskop (Boston: Kluwer Academic Publishers, 1989), 140.

31. Marilyn Bluebond-Langer, *The Private World of Dying Children* (Princeton, New Jersey: Princeton University Press, 1978).

32. Bluebond-Langer, note 31, and Bibace and Walsh, 31–48.

33. See, for example, Charles A. Corr and Donna M. Corr, "Pediatric Hospice Care," *Pediatrics* 76, no. 5 (Nov. 1985): 774–80; Ellen Perrin and P. Susan Gerrity, "There Is a Demon in Your Belly: Children's Understanding of Illness," *Pediatrics* 67, no. 6 (June 1981): 841–49; and Barbara Brodie, "Views of Healthy Children Toward Illness," *American Journal of Public Health* 64, no. 12 (1974): 1156–59.

34. See Paul Ramsey, "The Enforcement of Morals: Nontherapeutic Research on Children," *Hastings Center Report* 6, no. 4 (Aug. 1976): 21–30.

12

Gender Socialization
And Adolescents

In Chapter 2, I examined the relationship between sex-role stereotypes and health care, and in Chapter 11, the capacity of children for exercising moral agency in the health care setting. In this chapter, I draw these themes together through a focus on two issues that particularly affect adolescents. One is the possibility of providing growth hormone therapy to those who have no medical need for it; the other is the growing incidence of anorexia nervosa among young adults and teenagers. Preliminarily, however, I provide a brief account of sex-role or gender socialization among adolescents as influences on their need or desire for health care.

Sex-Role Socialization Among Teenagers

While there is general agreement about the content of gender stereotypes, they have been differently formulated by different authors. Presumably, one's gender *identity,* or sense of oneself as a man or woman, is distinguishable from sex-*role* or *gender* stereotype. Nonetheless, some have maintained a necessary connection between sex-role conformity and psychological health. Alfred Heilbrun, for example, begins a discussion of "Sex-Role Identity in Adolescent Females" by defining the sex-role identity of the child as "the degree to which his or her behavior and attitudes coincide with cultural stereotypes of masculinity and femininity."[1] He then assigns traits as follows: "Typical of the behavior subsumed under the adult masculine sex-role are: achievement, autonomy, dominance, and endurance; feminine adult sex-role behavior would include deference, abasement, succorance, nurturance, and affiliation."[2] After linking sex-role identity with gender stereotypes, Heilbrun goes on to define psychological "adjustment" or "health" as "the degree to which the individual is capable of maintaining herself interpersonally without seeking professional help for personal problems." He describes the "best adjusted girls" as those who are in "better psychological health" in comparison with others; these need not seek professional help.[3]

What constitutes pyschological health? It is sometimes thought that identification with their mothers facilitates the psychological adjustment of daughters, but research has not borne this out. In fact, when Heilbrun and Donald Fromme studied girls whose parents were atypical with regard to sex-roles (feminine father, masculine mother), they found no relationship between the daughters' parental identification and their level of adjustment. When their parents represented sex-typical models (masculine father, feminine mother), the best-adjusted girls identified more strongly with their fathers than with their mothers.[4] They thus diverged from conformity to sex-role stereotypes.

Nevertheless, some clinical studies reflect the widespread practice of measuring adolescent health and psychological maturation by gender specific behavior.[5] From an egalitarian perspective, this view is troublesome, to say the least. It suggests that the feminine traits that have led to and maintained the oppression of women are crucial to their healthy maturation, and that the masculine traits that have permitted men to pursue personal power unselfconsciously are reserved to them. To be healthy, girls must then be compliant, while boys should be in charge. The developing capacity for autonomous decision making is thus more likely to be frustrated in girls than in boys. In either case, the stereotype of conformity to sex-role belies the fact that sexuality does not entail a simple division of humanity into two disparate parts, but a continuum of human traits, various assortments of which are necessary to the health of unique individuals of either sex. On one end of the spectrum are those that epitomize the stereotype of femininity, at the other the stereotype of masculinity. Between the two are androgynous blends whose proportion of masculinity or femininity determines one's gender type. On such an account, sex-role conformity is surely not necessary to adolescent health.

Jeanne Humphrey Block suggests a more egalitarian approach to adolescent socialization when she writes that

> the ultimate goal in development of sexual identity is not the achievement of masculinity or femininity as popularly conceived. Rather, sexual identity means, or will mean, the earning of a sense of self in which there is a recognition of gender secure enough to permit the individual to manifest human qualities our society, until now, has labeled as unmanly or unwomanly.[6]

By such an account, a healthy female adolescent is one who does not need professional help because she tends to be dominant rather than deferential in her interactions with others, or because affiliation is less important to her than auton-omy. Nor would a male adolescent be construed as needing professional help solely because he is more oriented towards nurturance than self-interest, or more compliant than autocratic in his behavior. From an egalitarian point of view, it is important to resist imposition of stereotypic roles on individuals whose natural propensities lie in different directions.

The challenge of such resistance is particularly strong in dealing with teenagers because socialization based on gender distinctions reaches a high point during that time. Without settling the question of whether observed behavioral differences between adolescent boys and girls are triggered more by socializa-tion than by the onset of puberty, many writers support a gender-intensification hypothesis as the explanation for such changes. According to John Hill and Mary Ellen Lynch, adolescence is a period when new domains may become "the object of gender-differential socialization pressure," and when increased de-mands for conformity to such pressure arise.[7] The pressures intensify because of various developmental tasks faced by teenagers: establishment of a stable sense of self and self-worth, reflecting acceptance of the bodily changes that accom-pany puberty; achievement of strong friendships and comfortableness in relating to the opposite sex; and transition from physical and emotional dependence on parents to relative independence.[8] Definitions of gender identity relate to each of these tasks, with different implications for boys and girls. The differences, un-fortunately, have inegalitarian consequences.

Consider, for example, the fact that early pubertal change tends to be an ad-vantage for boys and a disadvantage for girls. Roberta Simmons and her col-leagues suggest that the key differences are:

> whether the changes lead one to approximate the cultural ideal or not. For the boy, the physical changes of puberty render him more muscular and athletic and thus more in line with the American physical ideal for males. For the girl, the changes at first lead her to be bigger than all her male and female peers and then on average to be shorter and heavier than the later-developing girls. The result is that she is less likely to approximate the female ideal of beauty than the late-developing tall and slim girl.[9]

When the "cultural ideal" is rigidly imposed on male and female adolescents, it affects them not only differently but unequally. Even if the ideal perfectly re-flected the natural propensities of boys and girls as they move towards adulthood, it would entail prejudicial consequences for those of either sex who do not or cannot conform to the stereotype. Moreover, since the stereotype generally rele-gates women to secondary social status, its overall influence on female adoles-cents would impede rather than facilitate the development of their full potential.

As already suggested, the reinforcement of gender roles extends beyond phys-ical differences. According to James Coleman, girls generally have higher grades than boys and have aspirations equal to theirs, but receive fewer rewards for their achievement. He claims that a pattern of early or frequent dating inhibits the display of competence on the part of girls.[10] In contrast, a spirit of braggadocio is typical for boys. Although socialization practices have changed in recent years, parents still sometimes encourage behaviors that fit boys and girls for traditional sex roles. As they become increasingly watchful and protective of their daughters, they are more tolerant of independence for their sons; many expect their daughters to be more concerned about their appearance, and their sons to be the initiators of opposite-sex interactions. The gender-based double standard thus evoked tends to become a self-fulfilling promise that has clinical as well as social repercussions. One of the ways in which this standard is reflected is in parental requests for growth hormone treatment for their sons.

Growth Hormone Therapy for Boys

Synthetically produced human growth hormone (HGH) has introduced the possibility of using a medical technique to solve a nonmedical problem.[11] Until recently, HGH could only be obtained from the pituitary glands of human cadavers. Because this source yielded only a minute amount of the hormone, its use was restricted to those suffering from growth deficiency syndrome, or pituitary dwarfism. This condition is usually seen as a medical problem with disabling consequences. Since the advent of recombinant DNA technology, a synthesized form of HGH has been available. Its availability has led to broader interest in, and willingness to provide, synthesized HGH to other patients.[12]

Who are those other patients? For the most part they are boys who are short of stature in comparison with their peers, and whose parents seek HGH even though medical grounds for the treatment are lacking. Among those whose height is two standard deviations or more below the mean for their age, only 10 percent suffer from a deficiency of growth hormone.[13] Most of the others owe their shortness either to genetic determination or constitutional delay. "Genetic determination" refers to the predictable influence of parental height on the height of their offspring. "Constitutional delay" refers to the condition of children who grow more slowly but for a longer stretch of time than others. Their final height is often within normal range.

Why do parents seek HGH for their sons when it is not medically indicated? Typically, they do so because shortness is considered a disadvantage. So long as tallness is not extreme, it tends to bestow a social and economic advantage on people, particularly men and boys. Tall boys are less likely to be teased or bullied by their peers, and they are often more successful in sports. As for adults, consider the following items as indicative of the advantage of tallness: (1) the average difference in starting salary between taller and shorter male library science graduates is more than three times greater than the difference between the more and less academically qualified; (2) of all U.S. presidents, only two have been shorter than average for an American male of their era; and (3) when job recruiters were asked to choose between equally qualified candidates, the taller candidate had nearly a three-to-one advantage over the shorter.[14]

If tallness is a social and economic advantage, it is understandable that parents would seek HGH as a means of promoting that advantage for their offspring even in cases where there is no medical need, and even where their children are expected to reach average stature. However, two further considerations are relevant to their decisions: the cost of the treatment and the burden of the treatment to the child. The cost of HGH is about $20,000 per year for a 30 kilogram (about 66 pound) child, and a treatment program may last five years.[15] Third-party payment will only cover the cost for patients for whom the treatment is recommended on medical grounds. The injections are painful and may produce adverse side effects such as diabetes mellitus, hypertension, and atherosclerotic coronary artery or cerebrovascular disease. Although the therapy is usually effective in accelerating adolescent growth, it may not result in a significant increase in the patient's final height.[16]

Extreme shortness of stature, no matter what its etiology, is more than a disadvantage; it is a disability. For example, to someone whose shortness makes it impossible to perform ordinary tasks of daily living such as driving a car or reaching for items easily accessible to others, shortness is disabling. Although disabilities are not diseases, we ordinarily use medical means to correct disabilities when this is feasible. We do so out of a sense of justice or fairness as entailing an obligation to provide all individuals with equal opportunities for fulfilling basic human needs. However, our sense of justice does not extend to obligations to promote advantages for others. It does not, therefore, imply a moral obligation to respond to the requests of parents who seek to enhance their children's life prospects by treating them with HGH to make them taller. To pursue the overcoming of shortness as a disadvantage necessarily involves the pursuit of advantages over those who remain short. The fact that tallness is a greater advantage for men than for women suggests general acceptance of men's dominance over women, not only physically but in other respects as well. It thus reinforces gender stereotypes.

An argument defending the use of HGH for those who are short of stature but not disabled by it may be based on the principle of respect for autonomy, applied to children and their parents. The familiar requirement of "informed consent" supports this view by attributing to patients or their families the final word in decisions that most affect them. Usually, however, patients assert their decision-making authority as "informed *dissent*," or refusal of medical treatment. In the case of HGH, we may consider the possibility of "informed *request*," that is, a situation in which the patient asks for a specific mode of treatment that has not been offered as a medical option. Patients are entitled to ask for specific treatments, but whether practitioners ought to respond affirmatively in such instances is another matter.

On egalitarian grounds, practitioners should decline to provide HGH for patients who are not disabled or have no medical reason for their shortness of stature. The crucial premise of an argument supporting this view is that justice overrides autonomy in such instances. In a society that officially opposes racism and sexism as unjust, we need to recognize that "heightism" is another type of prejudice to be deplored.[17] In other words, egalitarianism rather than individualism should prevail. Physical characteristics are at times relevant criteria for exclusion or inclusion, but tallness is clearly irrelevant in many situations in which it represents an advantage. Where reasons for seeking HGH are tied to the expectation of advantages based on irrelevant reasons, provision of the treatment reinforces and perpetuates prejudice, including sexism.[18] Even where tallness is a relevant advantage, its expected benefit may not outweigh or be proportionate to the burden of treatment and its impact on others. If individuals are to be treated as who or what they are (guideline 5), decisions regarding them should not be based on irrelevant criteria.

Beyond egalitarian considerations, the provision of HGH to adolescent boys who are neither disabled nor diseased may not constitute an ethic of care because it may not be in the interests of the boys themselves. While describing extreme shortness as a "psychologically disabling condition" for some children and adults, a report in the *New England Journal of Medicine* (1984) maintains that unrealistic expectations of patients and families involved in HGH therapy sometimes lead to depression.[19] Accordingly, psychological problems induced by gender socialization are not likely to be resolved through HGH treatment. The "psychologically disabling" effect of shortness may thus be more effectively treated by promoting realistic expectations on the part of candidates for the therapy and by addressing the social causes of prejudice regarding height.

Neither is HGH treatment adequately justified on grounds of parental autonomy. As already discussed, proxy decisions are less binding than decisions made by patients for themselves. The burdensomeness of the treatment, its experimental status, and the capacity of adolescents to make free and informed decisions for themselves argue in favor of respecting the autonomy of the teenager if he is opposed to his parents' request for HGH. Obviously, if the patient's and proxy's decisions are congruent, this strengthens the case for respecting their autonomy.

Should physicians then provide HGH for nonmedical reasons? Note the difference between this question and another: Do physicians have a right to provide such treatment? The preceding argument supports a "no" answer to the first question, but not the second. Based on a distinction between an ethic of obligation and an ethic of virtue, I would say "yes" to the second question. An ethic of obligation draws a line between what is ethically right and wrong, and moral behaviors fall on either side of the line. An ethic of virtue defines a moral ideal, such as equality, that individuals can approximate through personal decisions.

In light of this distinction, justice represents an egalitarian ideal that the virtuous physician pursues by refusing to reinforce the social prejudice that confers advantages on tall persons for irrelevant reasons. That same physician has a right not to be as virtuous as she might be, so long as the treatment is not in itself morally wrong. Thus the physician is not *obliged* to promote social equality by refusing to provide the advantage of medical technology to specific patients. Her right, as well as the right of patients to request such treatment, is consistent with the individualistic ethic that a capitalist society presupposes. So long as society refrains from critiquing the individualism of professionals and patients in other respects, it seems unfair, and possibly hypocritical, to impose an ethic of social virtue on the individual practitioner.

On the other hand, an egalitarian perspective invites professionals and patients alike to critique a practice that reinforces discrimination based on shortness. Such a critique would be provided through support for efforts to alter the stereotypic expectations that prompt some individuals to undertake health risks to improve their social advantage. As with the issue that I discuss next, the challenge is to address the causes rather than merely the symptoms of the disease or need for treatment. While tallness is especially desired for boys, thinness is particularly desired for girls. Both socially reinforced desiderata involve psychological burdens for individuals, but the latter is more devastating in its clinical implications. Treatment for shortness of stature remains optional; treatment for anorexia nervosa, a life-threatening disorder of epidemic proportions, is not.[20]

Anorexia Nervosa and a Feminine Ideal of Thinness

Despite increased public awareness of its prevalence, anorexia nervosa (AN) is neither new nor rare.[21] The cardinal feature of this disorder is "the individual's marked pursuit of thinness with the associated conviction that her body is too large."[22] Paradoxically, the anorexic joins her refusal of food with a preoccupation with food. The latter characteristic sometimes shows itself in a detailed study of nutrition and gourmet cooking, a tendency to eat one's own food secretly and to spend an excessive amount of time at meals.[23] While weighing well below the average for her frame, height, and age, the anorexic persists in viewing herself as fat. This reinforces the low self-esteem that accompanies the disorder. As Paul Garfinkel and David Garner observe,

> The feeling of self-worth in anorexic patients is closely bound to external standards for appearance and performance....Pressures on women to be thin and to achieve, and also conflicting role expectations which force women to be paradoxi-

cally competitive, yet passive, may partially explain why anorexia nervosa has increased so dramatically. Patients with anorexia nervosa respond to these pressures by equating weight control with self-control and this in turn is equated with beauty and "success."[24]

If the gender intensification hypothesis is true, it is easy to see why teenage girls, in an age of media saturation with the cult of feminine thinness, are particularly susceptible to this disorder. Only 5 percent of anorexics are men or boys.[25]

AN is difficult to cure, and in some cases impossible. It is both psychologically and physically debilitating, and sometimes fatal. Ironically, although anorexics deliberately starve away 20 percent or more of their body weight, the disease prevalently occurs in cultures where food is abundant—in white, affluent, well-educated, intense, hard-working families who are conscientiously protective of their daughters.[26]

Cultural expectations have long exerted unhealthy influences on women, presenting them with a forced option between conformity to an ideal of the feminine and physical normality. In prerevolutionary China, for example, the footbinding of a woman was a status symbol for her husband, indicating that his wealth was sufficient for her not to work. This practice not only limited the wife's mobility and impaired her capacity to work, but restricted her social involvement as well.[27] In the nineteenth century, the wearing of tight corsets was a way of promoting the thin look expected of women. These not only induced discomfort and interference with digestion, but sometimes caused serious injury.

As suggested in the previous section, men are also affected by expectations that impose health risks. Tattooing and scarification are historical examples of this. Susan Sontag describes "the tubercular appearance" among upper classes of the last century as an "index of being genteel, delicate and sensitive."[28] Such attributes were mainly esteemed in artists and in women. Note, however, that the characteristics often attributed to artists conform with feminine rather than masculine stereotypes. Pallor was a fashionable attribute which women used whitening powders to promote. According to Garfinkel and Garner, consumption was the precursor of AN. The consumptive appearance pursued by women gave rise to the anorexic look currently glamorized by the media as an ideal of feminine beauty.[29]

Many authors report the growing incidence of AN, and its occurrence in older women as well as men.[30] For example, a study of dieting behavior among adolescents in Sweden found that the "feeling of fatness" increased with age among females from a 50 percent incidence at 14 years of age to a 70 percent incidence among 18-year-olds. A 10 percent prevalence of "mild cases" of AN was observed among girls, with one "serious case" out of every 155 of those considered.[31] Boys of comparable ages seldom reported feeling fat or dieting. Arthur Crisp and his colleagues studied a relatively large school population in London and found a prevalence rate of 1 in 100 among girls aged 16 and over.[32]

Despite the severity and increased incidence of AN, few authors have argued that gender stereotypes should be targeted in pursuit of its cure. For the most part, they simply document the epidemiology of the disorder, and some note the concurrence of gender socialization and anorexia.[33] From an egalitarian standpoint, however, preventive treatment is called for, and such treatment requires not only rejection of gender stereotypes but positive efforts to thwart pervasive social tendencies in this regard. In treating those who already have the disorder, a comparable critique is crucial. So long as girls "buy into" a feminine ideal of thinness that threatens their psychological and physical well-being, their low self-esteem is bound to be reinforced, and with that a prolongation or exacerbation of their illness. Insofar as the disorder is provoked by gender socialization, they will continue to fall prey to AN until the ideal is changed or relinquished.

In *Fasting Girls,* Joan Brumberg describes three types of factors that contribute to the development of AN: biological, psychological, and cultural. No single theoretical model, she maintains, is adequately explicative of "the current rash of eating disorders and the place of anorexia nervosa in the long history of female food refusal."[34] While Brumberg gives most weight to cultural influences, she clearly views biological vulnerability and psychological predisposition as the villains also. Her account of multicausality seems more plausible than explanations that attribute AN solely to one cause or the other. Among biomedical influences, a number of more specific causal candidates emerge: hormonal imbalance, dysfunction of the hypothalamus, lesions in the brain's limbic system, irregular output of vasopressin and gonadotropin, and excess cortisol production.[35] Neither singly nor together, however, do these features adequately explain the distinctive socioeconomic and gender status of anorexic patients, or why its incidence is so great at this time in history.

Psychosocial explanations interpret AN as "a pathological response to the developmental crisis of adolescence."[36] The teenager's refusal of food is seen as an expression of a struggle for autonomy, individuation, and sexuality. Following Freud, psychoanalysts have equated an unwillingness to eat with the desire to suppress one's libido or sexual drive, and resistance to the inexorable progress towards adulthood. Ironically, while repressing that drive, the anorexic nonetheless seeks to control the only thing she feels she can control, her body. Much of her behavior is obsessively compulsive. Family systems theory imputes the behavior traits of anorexics to other family members as well. The "psychosomatic family" is then viewed as "controlling, perfectionistic, and nonconfrontational," descriptives that apply equally to their anorexic daughter.[37]

As with many psychiatric disorders, mothers are often viewed as responsible for the condition. Allegedly, preoedipal mother conflict arises from identification with a kind, passive father and hostility towards an aggressive castrating mother who is ineffective and unhappy.[38] The anorexic's intense, unconscious hatred of her mother leads to rejection of femininity in general. Keeping thin makes her more boylike in appearance, and the amenorrhea (interruption of menstrual cycle) that extreme thinness effects is a further sign of success in her avoidance of femininity.

The same limitations found in the biological model of AN are present in the psychological model as well; incidence, gender, and socioeconomic features of the disorder are not thereby explained. Which returns us to the socialization issue and the cultural model, providing a limited response to those problems. In general, the incidence of AN has increased because society has intensified and exaggerated a perception of thinness as a sign of fitness and attractiveness.[39] It has done this to a greater degree with women than with men, and its message, like most new styles, has been "bought" most prevalently by those who have the time, energy, and interest to buy it, that is, affluent or moderately affluent white women.

Susan Bordo describes AN as a crystallization of the psychopathology of contemporary culture. It illustrates "the social manipulation of the female body" that has emerged "as an absolutely central strategy in the maintenance of power relations between the sexes over the last hundred years."[40] The strategy has its roots in the metaphysical dualism that characterizes the writings of Plato, Augustine, and Descartes. It is a mind-body dualism that does not simply separate the two but prioritizes mind over body. To the extent that women are viewed (by themselves as well as men) as sex objects valued solely or primarily for their bodies, sex inequality will inevitably prevail. To Bordo, the patriarchal Graeco-Christian tradition provides a "particularly fertile soil for the development of anorexia."[41] Popular American culture contributes to this development through its emphasis on control of the body, and the overcoming of its vulnerability through physical fitness. Paradoxically and tragically, although the will to conquer and subdue the body sometimes expresses an aesthetic or moral rebellion, "powerlessness is its most outstanding feature."[42] A sexual double standard is clearly to blame for the fact that anorexia predominantly afflicts women rather than men. Many anorexics are self-consciously aware of two selves that are in constant conflict: one a dominating male self that represents greater spirituality, intellectuality, and will power; the other a female self that represents uncontrollable appetites and flaws. It is the male self that the anorexic seeks to develop, precisely because it has been so ingrained in her that male is better. [43]

Feminists are particularly prone to blame sex-role socialization for anorexia, and then to propose socially corrective measures as preventive therapy and feminist consciousness-raising as individual therapy. For some feminists, however, AN is not only caused by a sexist culture, but constitutes a behavior that is antithetical to that culture. Construed in this way, anorexia does not involve enslavement to an unhealthy idea of the feminine, nor does it simply mean refusal to accept femininity. It also means rebellion against patriarchy through rejection of one's own sexuality.

Liberal feminists are likely to view sexism as a cause of anorexia, and to argue for elimination of the sex-role stereotyping that imposes an unequal burden on women. In order to liberate women to develop their full potential as individuals, their options need to be expanded rather than restricted by some ideal of feminine behavior or appearance. Liberal feminism thus supports the right of individual women to pursue this or any ideal, even if it leads to unhealthy consequences for themselves, as long as the pursuit does not impede the liberty or welfare of others. In a capitalist society, the means to fulfill the ideal of thinness are evident in the huge commercial success of diet, weight-loss, and fitness programs that generally attract more women than men. Selling thinness has clearly become a profitable industry. The goal of liberal feminism is to preserve this social structure while rectifying its sexist flaws. Given the imbeddedness of stereotypic sex-roles within the structure, it is doubtful that so limited a critique could much reduce their input as a cause of AN.

Radical feminism would extend its criticism of sexism as a cause of AN to a critique of the patriarchal structure that pervades society as a whole. This critique applies to women as a class rather than as individuals. Women become an exploited class in a system wherein men profit by the feminine ideal of thinness. As Susie Orbach put it, "Fat is a feminist issue" because it exemplifies resistance to that ideal. A cult that has enshrined "slimness" as "the new god" is blasphemously antifeminist and antiwoman.[44] The goal, then, is to recognize the patriarchal nature of the new god, to topple it by substituting a goddess that embodies an ideal of femininity that takes account of the fact that healthy women come in all shapes and sizes, even as men do. In Orbach's words,

> Some of us are short, some of us are tall. We can have short legs, medium-length legs, long legs, big breasts, medium-size breasts, small breasts,... large, medium or small hips; we can be pear-shaped, broad or rounded, have flat stomachs, full stomachs, even teeth, crooked teeth, large eyes, dimpled cheeks.[45]

Men, of course, come in all shapes and sizes also. While tallness remains an advantage to them as well as to some women, society generally accepts a broader range of physical shapes and sizes for men than for women.

A socialist version of feminism would concur with the radical version, focusing on the inequality of a social system that creates needs out of desires provoked by a profit motive. AN is a tragic example of the dehumanizing effects of such a system. If the feminine ideal were stripped of its debilitating impact on the development of women as individuals, and a masculine ideal were similarly cleansed of its impeding influence on men, then a socialist ideal might be approximated. It would not eschew concern about obesity and health for women, but it would have equal concern about such matters for men. In other words, equality, as respect for differences as well as potential among individuals of either gender, would be the criterion governing any ideal regarding body size or fitness.

Tallness, Thinness, and Their Disparate Consequences

An egalitarian critique of gender stereotypes applies to members of both sexes because the health, achievement, and happiness of men as well as women are impeded by such stereotyping. On the whole, however, the impact on women is more disastrous, as suggested by the contrast between the two issues used here as examples: tallness in fact confers an advantage, while thinness may be a sign of weakness or ill health. The two ideals have separate effects not only on physical health, but also on emotional well being. The self-image of taller boys is stronger than their peers, while that of thin girls remains relatively weak. Psychologically, short women probably suffer less from their shortness than short men do from theirs, but obese women apparently suffer more than obese men.

Being heavy is more socially acceptable in men than in women, and generally involves physical power that is lacking in women who pursue and achieve the feminine ideal of slimness. In an age in which the most horrendous powers of evil can be unleashed by so slight a movement as the flick of a switch, it is amazing that we still tend to think that bigger is better and judge those who are puny or small as weak in significant ways—even when their intelligence, creativity, and endurance outstrip those of their larger counterparts. Human beings appear to be simplistic in this tendency to confuse quality with quantity. While women themselves are not often viewed as better when bigger, they nonetheless tend to join men in assessing bigger as better in other matters.

Physiologically, it is better to be slender at any height than fat at any ideal height. Psychologically, too, unless one's profession or life-style demands weight as well (as in football), it is better for both men and women to be slender. Because tallness and slimness are sometimes advantages for adults, they are also advantages for those approaching adulthood, that is, adolescents. Would that such advantages might be equally distributed among individual teenagers, but then they would not be advantages because they would not empower some over others. (I have more to say on empowerment in the last chapter.)

Adolescents are more likely than adults to feel insecure about their gender identity and thus be influenced by sex-role sterotyping. For young women, however, the ambiguity described by Simone de Beauvoir as characteristic of the female sex[46] is particularly troubling because of its apparent association with such a severe disorder as anorexia nervosa. Mature women struggle to maintain a healthy balance between home and work responsibilities, while their male counterparts are rarely expected or inclined to devote comparable energies to both sets of tasks. Their daughters, observing their mothers, confront the discrepancy between their own and their brothers' future prospects, and experience similar ambiguity.

According to Carol Gilligan, ambiguity is greater in girls because of their desire to preserve attachments while establishing their own identity in their progress towards adulthood. In *Mapping the Moral Domain,* she writes:

> In resisting detachment and critiquing exclusion adolescent girls hold to the view that change can be negotiated through voice, and that voice is the way to sustain attachment across the leavings of adolescence.[47]

The adolescent's "exit" from childhood leaves a problem of loyalty that she tries to negotiate through use of her own voice. The anorexic's voice is loud at first, but destined to be silenced by weakness, sickness, or even death from the disorder. Her desire to be loyal to herself as well as to others inevitably leads to ambiguity. Hilde Bruch uses the image of a golden cage to describe this ambiguity.[48] Different accounts of anorexic behavior suggest conflicting voices, one involving conformity to a socially induced unhealthy stereotype, the other involving refusal to accept the sexist connotations of being female. The anorexic may in fact affirm both messages, thus defining the content of her ambiguity: to be or not be a woman.

Perhaps there is something symbolic in the fact that the male voice changes naturally at adolescence. For some young men, however, the pressures of conformity to gender stereotypes are both painful and unhealthy. Their voices need to be heard as well. From an egalitarian perspective, respect for differences among adolescents requires silencing the competing noise of gender stereotypes to allow them to speak for themselves. Listening to the unique voices of youth is a significant challenge for all of us. But no less a challenge awaits us as we turn our attention in the next chapter to a group of people who are often rendered voiceless by their poverty. While anorexia nervosa and growth hormone treatment are rare in this group, other important health issues are abundant.

NOTES

1. Alfred B. Heilbrun, Jr., "Sex-Role Identity in Adolescent Females: A Theoretical Paradox," *Adolescence* no. 3, 9 (Spring 1968): 80.

2. Heilbrun, 80.

3. Heilbrun, 80–81.

4. Alfred B. Heilbrun, Jr., and Donald K. Fromme, "Parental Identification of Late Adolescents and Level of Adjustment: The Importance of Parent-Model Attribute, Ordinal Position and Sex of the Child," *Journal of Genetic Psychology* 107 (1965): 49–59.

5. See, for example, studies by W. Emmerich, S. W. Gray, M. M. Helper, Miriam M. Johnson, P. Mussen, and L. Diestler, cited by Heilbrun in "Sex-Role Identity," *Adolescence* 3, no. 9 (1968). In contrast to clear and consistent findings that father-identification in boys is consonant with good subsequent adjustment, Heilbrun considers the findings of studies regarding mother-identification in girls to be equivocal.

6. Jeanne Humphrey Block, "Conceptions of Sex-Role: Some Cross-Cultural and Longitudinal Perspectives," *American Psychologist* 28, no. 6 (1973): 512.

7. John P. Hill and Mary Ellen Lynch, "The Intensification of Gender-Related Role Expectations during Early Adolescence," in *Girls at Puberty: Biological and Psychosocial Perspectives,* ed. Jeanne Brooks-Gunn and Anne C. Peterson (New York: Plenum Press, 1983), 201.

8. Roberta G. Simmons, Dale A. Blyth, and Karen L. McKinney, "The Social and Psychological Effects of Puberty on White Females," in Brooks-Gunn and Peterson, 235–36.

9. Simmons, Blyth, and McKinney, 264.

10. James S. Coleman, *The Adolescent Society* (New York: Free Press, 1961).

11. Much of the material in this section was initially developed in my article entitled "When Patients Request Treatment Not Offered," *Medical Ethics for the Physician* 3, no. 1 (Jan. 1988): 3–4.

12. Melvin M. Grumbach, "Growth Hormone Therapy and the Short End of the Stick," *New England Journal of Medicine* 319, no. 4 (July 28, 1988): 238. A yet more recent use of HGH is to halt the aging process. Although this application is preliminary and highly experimental, it raises even more complicated ethical and philosophical questions.

13. Martin Benjamin, James Muyskens, and Paul Saenger, "Short Children, Anxious Parents: Is Growth Hormone the Answer?" *Hastings Center Report* 14, no. 2 (April 1984): 5.

14. John Gillis, *Too Tall, Too Small* (Champaign, Illinois: Institute for Personality and Ability Testing, 1982), quoted in Benjamin, Muyskens, and Saenger, *Hastings Center Report* 14, no. 2 (April 1984): 8.

15. John Lantos, Mark Siegler, and Leona Cuttler, "Ethical Issues in Growth Hormone Therapy," *Journal of the American Medical Association* 261, no. 7 (Feb. 17, 1989): 1020. By the time this book appears, the cost is likely to be higher.

16. Grumbach, 239.

17. Grumbach, 239.

18. Because men are generally taller than women, heightism probably reinforces sexism, but I will not develop this point here. Nor will I develop a parallel argument against cosmetic surgery, which also involves risks that may be undertaken because of social expectations and gender stereotypes. But see Kathryn Pauly Morgan, "Women and the Knife: Cosmetic Surgery and the Colonization of Women's Bodies," *Hypatia* 6, no. 3 (Fall 1991): 25–53.

19. Louis E. Underwood, "Report of the Conference on Uses and Possible Abuses of Biosynthetic Human Growth Hormone," *New England Journal of Medicine* 311, no. 9 (Aug. 30, 1984): 608.

20. Joan Jacobs Brumberg, *Fasting Girls* (Cambridge, Massachusetts: Harvard University Press, 1988), 31; and Richard A. Gordon, "A Sociocultural Interpretation of the Current Epidemic of Eating Disorders," in *The Eating Disorders* ed. Barton Blinder, Barry Chaitin and Renée Goldstein (New York: PMA Publishing Co., 1988), 151.

21. Portions of this section are included in my article entitled "To Be or Not Be a Woman: Anorexia Nervosa, Normative Gender Roles, and Feminism," *Journal of Medicine and Philosophy* 17 (1992): 233–51. As I remark in that article, much of this discussion of AN is also applicable to bulimia nervosa. Because differences between the two disorders would need to be further explored, I limit the discussion here to AN.

22. Paul E. Garfinkel and David M. Garner, *Anorexia Nervosa* (New York: Brunner-Mazel, Publishers, 1982), 2. See *Diagnostic and Statistical Manual of Mental Disorders,* 3d ed., revised (DSM-III-R) (Washington, D.C.: American Psychiatric Association, 1987), 67.

23. Garfinkel and Garner, 11.

24. Garfinkel and Garner, 10.

25. See DSM-III-R, 66.

26. Carol Lawson, "Anorexia: It's Not a New Disease," *New York Times* (Dec. 8, 1985), 93.

27. Garfinkel and Garner, 105.

28. Susan Sontag, *Illness as Metaphor* (New York: Farrar, Straus & Giroux, 1978), 28.

29. Garfinkel and Garner, 106, citing a seventeenth century treatise by Richard Morton, *Pthisiologica: On a Treatise of Consumptions* (London: Sam Smith and Benj. Walford 1694). See also Brumberg, 43–44.

30. Garfinkel and Garner, 100–101.

31. Garfinkel and Garner, 101–2.

32. Arthur Hamilton Crisp, R. L. Palmer, and R. S. Kabucy, "How Common Is Anorexia Nervosa? A Prevalence Study," *British Journal of Psychiatry* 128 (1976): 549–54.

33. In addition to Garfinkel and Garner, and Brumberg, see Gloria R. Leon and Stephen Finn, "Sex-Role Stereotypes and the Development of Eating Disorders," in *Sex Roles and Psychopathology,* ed. Cathy Spatz Widom (New York: Plenum Press, 1984), 317–37. Exceptions to this trend include Marlene Boskind-Lodahl, Susan Bordo, and Susie Orbach, as mentioned below.

34. Brumberg, 24.

35. Brumberg, 25.

36. Brumberg, 28.

37. Brumberg, 29.

38. Marlene Boskind-Lodahl, "Cinderella's Stepsisters: A Feminist Perspective on Anorexia Nervosa and Bulimia," *Signs: Journal of Women in Culture and Society* 2, no. 2 (1976): 4, 34.

39. Obviously, overweight is not healthy, and slimness is preferable to obesity. My point here is that the ideal of thinness promoted by mores and media is excessive, suggesting that individuals are fat when they are not.

40. Susan Bordo, "Anorexia Nervosa: Psychopathology as the Crystallization of Culture," *Philosophical Forum* XVII, no. 2 (Winter 1985–86): 76–77.

41. Bordo, 97, n. 18.

42. Bordo, 85.

43. Bordo, 86.

44. Susie Orbach, *Fat Is a Feminist Issue II* (New York: Berkeley Publishing Corporation, 1982), 27–31.

45. Orbach, 27.

46. Simone de Beauvoir, *The Second Sex,* trans. H. M. Parshley (New York: Alfred A. Knopf, 1972), 133.

47. Carol Gilligan, "Exit-Voice Dilemmas in Adolescent Development," in *Mapping the Moral Domain,* ed. Carol Gilligan, Jane Victoria Ward, and Jill McLean Taylor (Cambridge, Massachusetts: Harvard University Press, 1988), 148.

48. Hilde Bruch, *The Golden Cage* (Cambridge, Massachusetts: Harvard University Press, 1978), 150.

13

The Feminization of Poverty:
Its Impact on Women's and
Children's Health

While poverty clearly leads to poorer health and poorer quality health care for men as well as women, its impact on women and children is more significant both qualitatively and quantitatively. Women seek treatment more often than men and are generally more needful of health services for themselves and their children. The "feminization of poverty" has led to poorer health care for women and children alike.[1] In this chapter, I examine how and why this is so, using crack use and AIDS as examples in this regard. To redress the imbalance of poverty's effects on women and children, as well as the inequality that confronts both men and women because of racism, classism, and other forms of prejudice, a systematic change is necessary. Another form of social prejudice, ageism, also leads to poverty and poor health. In addressing the apparent conflict between health needs of the elderly and children, I argue for more equitable distribution with respect to life span.

Poverty and Its "Feminization"

In my younger years I taught eighth grade students in a poor area of Brooklyn, New York. The parents of most of these students had recently arrived from Puerto Rico. When I describe the area as poor, I mean that most people required financial assistance in order to meet the expenses of daily life. Most were supported by welfare payments, and even those who were not lacked many of the conveniences that most Americans have (such as cars and television sets). Nonetheless, the families in this Brooklyn neighborhood were better off in some ways than the relatives they had left behind. In New York, they lived in tenements rather than shacks and had more clothes, better food, and more job and education possibilities. Although their move to the north brought new needs (such as heat in the wintertime), they could travel more readily, and most had telephones in their apartments. In Puerto Rico, the options would have been more limited. They were thus poorer than other Americans, but wealthier than their counterparts who remained on "la isla encantada."

When the United States government establishes a poverty line, it assumes that poverty is absolute rather than relative.[2] The preceding description of my Puerto Rican students and their families suggests otherwise. It suggests that poverty is relative to some standard of what constitutes survival in a given place and time, *or* a standard of what constitutes the antithesis of poverty, namely, affluence. Poverty occurs when the former standard is not met, or when the latter standard is difficult or impossible for the poor to achieve. By the standard of affluence, most of us are poor but some are poorer than others. Our being poorer than others does not constitute poverty unless the disparity between us and them means that they can survive and function normally without assistance, while we cannot.

When poverty is defined in terms of survival, its meaning remains unclear or incomplete. A patient like Karen Ann Quinlan or Nancy Cruzan may be said to "survive" despite lack of cognitive or affective function.[3] But if survival requires ability (or potential ability) to interact with others, such patients have not survived through the mere feat of breathing, with or without technological assistance. Persistent vegetative state would be an indicator of poverty if survival meant mere respiration, with total dependence on others for continuation of physiological function.[4] Infancy and any life-threatening illness or disability would be additional indicators of poverty. We do not generally regard the latter situations as examples of poverty because the dependence they connote is partial or temporary, and others are present who ensure survival of some ability to interact. Those others may do so because of ties of blood or affection, or because they may be paid to do so by the dependent person or that person's loved ones. Poverty occurs where no such others are reliably present, making it necessary that impersonal others (such as the state or charitable organizations) step in so that individuals can continue to survive as persons.

To some experts, "absolute poverty" is definable in economic terms. As Ruth Sidel puts it, "absolute poverty means living below the official poverty line; absolute poverty is not having money for adequate food, clothing, and shelter."[5] The problem with this definition is that its two components may be at odds with each other. Although "the official poverty line" may be intended to coincide with the amount of money necessary for adequate food, clothing and shelter, it is unrealistic to think that such a standard set by the government is equally applicable to all individuals in all circumstances. Moreover, what counts as "adequate" can only be determined through comparison with a reference point that is empirically delineated. Does adequate food include appetizing as well as nutritious items? Does adequate clothing include sports shoes or bathing suits? Does adequate shelter include possibilities for family privacy? And what about money for health expenses, education, or recreation? Does inability to pay for these items have no relevance to "the official poverty line?"

The official poverty threshold in the United States is based on the Department of Agriculture's estimates of the costs of its Economy Food Plan for a given family composite. In 1955, when the Food Plan was developed, the average family spent one-third of its income on food. An income equal or greater than three times the amount needed for food was then, and is now, construed as above the poverty line.[6] Classification of individuals or families as "above or below the poverty level" does not reflect the fact that some receive government subsidies while others do not. In 1990, the Census Bureau poverty threshold for a family of four was $13,359.[7] Although the threshold is adjusted yearly according to the rise in the Consumer Price Index, it does not take into account the proportionately higher cost of housing and the influence of material hardships on individuals and families. Mayer and Jencks have found that official poverty levels overestimate hardship among the elderly and underestimate it among families with children.[8] The inaccurate estimates lead to a situation in which the funding of health care for the elderly may reduce the funding available for children, an issue I discuss later.

In 1978, Diana Pearce coined the term "feminization of poverty" to highlight the fact that a majority of the poor in the United States were women. She calculated that "nearly two out of three of the 15 million poor persons over 16 were women," suggesting that poverty would continue to spread among women.[9] A decade later Pearce speculated that if the trend continued, all of the poor would be women by the year 2000.[10]

Although women's absolute poverty did not escalate through the 1980s, their relative poverty persisted into the 1990s. Most of the poor continue to be women, especially during the childrearing and elderly phases of their lives. Most poor people are white, but black people are more likely to be poor than white people.[11] Black women are more likely to be poor than white women, but both are more likely to be poor or poorer than their male counterparts. While absolute poverty rates for men and women declined between 1950 and 1980, the poverty rate of women relative to men rose. Moreover, the sex/poverty ratios of blacks and whites were similar.[12] Although single persons tend to be poorer than those who are married, the income of women usually plummets after divorce, and that of men increases.[13] This income disparity prevails even when men retain custody of their children.[14]

Sara McLanahan and her colleagues explicitly link the poverty of women to that of children, noting that the term "feminization of poverty" is used "to describe changes in the proportion of the poor who live in female-headed families, or changes in the proportion of the poor who are female or who live in female-headed families."[15] In fact, most of those who live in female-headed households are children. Thus the feminization of poverty is accompanied by the pauperization of children, and because most women live longer than most men, pauperization extends disproportionately to elderly women as well.[16] At both ends of life's spectrum, women are reduced to poverty through living costs that escalate in the absence or loss of male support. Regardless of whether their income is supplemented by transfer payments (government benefits), the poverty persists.

Women who are single parents may be widowed, divorced, separated, or never married. Although each of these situations has different economic repercussions, all are considerably poorer than two-parent families. According to Irwin Garfinkle and McLanahan, three factors contribute to the disparity: the low earning capacity of single women, the lack of child support from noncustodial fathers, and the meager benefits provided by public assistance programs.[17] It is often impossible for single working mothers to support themselves and their children at a level above the poverty line. Health care costs are frequently beyond their reach.

Effects of Poverty on Health and Health Care

In addition to the difficulty of paying for health care, poor people, in fact, have more health care needs than others. As Sidel observes in *Women and Children Last:*

> If you're poor, you're more likely to be sick, less likely to receive adequate medical care, and more likely to die at an early age. The effects of poverty on health and general well-being are clearcut and profound.... A few of the key issues for poor women and children [are]: reproduction, mental illness, hunger, lead poisoning, and finally, how the child welfare system "takes away the future" of so many poor young people.[18]

Social problems such as unemployment, underemployment, poor nutrition, teenage pregnancy, drug use, and prostitution all contribute mightily to the health deficit of the poor. Ironically and tragically, however, the poor have the lowest use of health services in comparison with their need. Even when different health status is considered, income still determines quantity and quality of health care: "those better off receive more care than poorer persons in comparable states of health."[19]

Barbara Starfield cites epidemiological studies of children to illustrate the health hazard of poverty for infants with regard to specific diseases. For example, in comparing low income families with higher income families, the frequency of births to teenagers is triple, the frequency of infants with low birth weight is double, and the frequency of delayed immunizations is triple for the low income families. Asthma and bacterial meningitis are twice as common, rheumatic fever more than twice as common, and lead poisoning three times as common for low income families.[20] Studies also show that the severity of these and other diseases is exacerbated among the poor.[21]

Examples of the influence of poverty on women's reproductive health are obvious. Poor women who are infertile simply cannot pay for technologies such as in vitro fertilization and are more likely to "serve" as surrogates or egg "donors" in order to supplement their meager income.[22] Poor women who are fertile, particularly teenagers, are likely to be prevented from obtaining abortions solely because of their inability to pay. Contraceptive testing, involuntary sterilization, and coercive treatment during pregnancy all occur more prevalently among poor women than among other women.[23] Any one of these issues deserves fuller treatment than I can give here. In what follows, however, I briefly consider two problems with profound implications regarding poor women and children that are increasingly evident and troublesome: crack use during pregnancy, and pregnancy in women who are HIV positive.

Cocaine (Crack) Use, AIDS, and Pregnancy

Although there are important differences between the issues of crack use and AIDS, they are alike in their prevalence among poor women. They are also related to each other in that HIV positive status may be initiated by drug (including crack) use. Prostitution and use of unclean needles are the means through which poor women are likely to contract or communicate AIDS.

Illegal drug use is an international problem that precipitates crime and violence as well as loss of health among men and women of all ages. The problem becomes particularly poignant when one sees its effects on young children born disabled and sickly apparently due to their mothers' addiction.[24] Efforts to solve the problem have focused on law enforcement and education. Despite huge expenditures, these efforts have been relatively unsuccessful. Leaders of developing countries that produce the illegal drugs argue that their economies are to blame. They further claim that if the government of the United States wishes to stem the tide of illegal drug trafficking, it should provide economic assistance to enable poor people to resist the lure of money that can be obtained from participation in the drug trade.

The same argument is applicable to poor people in our own country, but the argument has not been effective enough here or elsewhere to obtain support for programs that would make a difference. Possibly the cost of transforming the economy to such an extent is beyond our reach. However, even if that is so (which is doubtful), an egalitarian perspective calls for reducing the gap between rich and poor, and this would reduce the power of the drug lords.

Crack use is particularly problematic because it is more accessible to potential users and apparently more addictive than other illegal drugs. Indeed, crack-using mothers have acknowledged that their drive to have the drug was so strong that it quashed their maternal instinct.[25] Some have abandoned their children, or sold them in pursuit of the drug. Obsession with crack overrides ordinary human emotion, leaving individuals without affect. So compelling is the need for it that addicts may no longer be capable of autonomous judgment. If autonomous judgment *is* impossible, then the principle of respect for autonomy is no longer relevant except as a goal to promote or restore. Beneficence and justice become the main ethical principles to consider in addressing conflicts.

As mentioned in Chapter 7, pregnant women often undergo considerable sacrifices to maximize the health and welfare of their future children. Some even risk their own health and welfare to that end. Crack use during pregnancy has been shown to cause preterm birth, which in turn is the main cause of infant mortality and morbidity. Accordingly, when crack-using pregnant women engage in behavior that threatens their newborns, their actions are contrary to the usual behavior of pregnant women, and ordinarily contrary to their own customary behavior, as judged by those who know them well. If autonomy is defined as authenticity,[26] the inconsistency between their crack-using behavior and their non-crack-using behavior suggests that they are not acting autonomously when they threaten the welfare of their fetuses through continued use of crack. If addiction is a condition over which a person lacks control,[27] and poverty reduces options as it surely does, we have further evidence for reduced autonomy.

An egalitarian approach calls for respecting autonomy to the extent that it is present (guideline 4), for respecting life in its different stages and forms (guideline 2), for avoiding the infliction of pain (guideline 3), and for respecting differences among individuals (guideline 5). Each of these guidelines is applicable to the pregnant crack-user. Beyond respect for her compromised autonomy is the obligation to support her health and her capacity for recovering full autonomy, as well as an obligation to support the fetus or future child. In accordance with the meaning of paternalism outlined in Chapter 2, it would not be paternalistic to treat a pregnant drug addict without consent in order to benefit her or her fetus *if* her addiction had in fact deprived her of autonomy.[28] In most cases, however, autonomy is limited but not totally absent, and the practical ethical question is the extent to which caregivers are then permitted or obliged to intervene. Although the question is not answerable in a definitive manner, its answer can be approximated through a parentalist perspective. This approach would attempt to discern the degree of autonomy expressed, and to respect that autonomy so long as it does not disproportionately impede the welfare of others. Intervention in order to protect others is not paternalistic but beneficent towards them. Distributing harms and benefits in an egalitarian fashion means weighing the different impacts of intervention on those involved.

A parentalist perspective is also applicable to women who are HIV positive, that is, those who are infected with the AIDS virus but may not yet be symptomatic. Homosexual activity places men at risk of HIV infection to a much greater degree than does sexual activity on the part of lesbian women. Unlike men, for whom the risk of AIDS through homosexual contact is as great or greater among those who are affluent,[29] the risk of AIDS is greater for women who are poor.[30] They are more likely than men to become prostitutes, and to prostitute themselves in order to satisfy their dependence on drugs. Tragically, prostitution is sometimes a means through which adolescents and young women who apparently know no other way to "earn" a living survive. Drug use (including crack) is perceived as a means of escaping temporarily from the emotional and physical oppression of poverty. Once a person is addicted, however, concerns about clean needles or protected intercourse are subordinated to the insatiable drive for more drugs. Just as the maternal instincts of crack-using women are suppressed by their addiction, so are their instincts to observe precautions against the fatal infection of AIDS.

One of the more problematic features of HIV positivity in women who are pregnant or potentially pregnant is the risk of AIDS to their offspring. American scientists and the U.S. Centers for Disease Control in Atlanta estimate that about one-third of the newborns of women who are HIV positive will be infected by the virus.[31] A recent study by a European team reports a risk of only 13 percent.[32] To many people this level of risk for so deadly a disease is too high to warrant continuation of a pregnancy. Arguments for abortion of pregnant women known to be HIV positive are comparable to those for abortion of fetuses with fatal genetic defects such as Tay-Sachs disease, trisomy 13, or trisomy 18. An important difference, however, is that the latter diseases can be detected prenatally with relative certainty, whereas AIDS infection is not detectable before birth. Since abortion in such cases would be based on the statistical odds, roughly 67 percent of the fetuses aborted in order to avoid AIDS would be free of the disease.

Those who regard both abortion and AIDS infection of infants as tragic events might claim that HIV positive women are morally obliged to avoid pregnancy.[33] Arguments supporting this view are the same as those applicable to performing abortions on pregnant women known to be HIV positive, with the caveat that in the former case, the risk of aborting healthy fetuses is eliminated. Generally, such arguments are based on the interests of the fetus or future child (fetal euthanasia) and on the costs to others of treatment of AIDS-stricken children. However, when women with AIDS, who are likely to be poor prostitutes or drug addicts, weigh their options, the possibility of bearing a healthy child may still seem attractive in comparison with other circumstances of their lives. While recognizing her own limited life span, a woman with AIDS may find comfort in bringing a new life into the world, loving her child regardless of whether he is also afflicted with AIDS, and trusting others to care for the child when she is no longer able to do so.[34] An egalitarian perspective supports her choice in this regard.

Feminist Views of Poverty

Although feminism stands opposed to inegalitarian aspects of the impact of poverty on women, some versions of feminism tacitly support poverty within society because they affirm the economic structure that makes it inevitable, namely, capitalism. The key factor in this endorsement is an emphasis on individual choice that precludes a class-based critique. Liberal, existentialist, and postmodern versions of feminism, while arguing for the right of individual women to pursue their own values, reject the limitations to individual liberty that are essential to genuine equality. In this section, therefore, I wish to explain how a version of feminism that gives higher priority to equal liberty than to equality (construed more broadly than equal liberty) finally fails both practically and theoretically to overcome the subordination of women. From an egalitarian standpoint, the feminization of poverty is intolerable because of its classist as well as sexist elements. Because of the nurturant tie between women and children, the two elements are usually related, suggesting that an effective version of feminism must take account of economic inequality as well as social inequality. Marxist and socialist versions of feminism exhibit both types of critique because they link one to the other. Radical feminism concentrates mainly on overcoming male domination in any social structure.[35]

Liberal feminism purports to uphold the rights of individuals to develop their full potential within society as it is. "As it is" in the United States means a market economy that relies on the spending habits of people whose incomes range from rich to poor. Since more women than men are poor, the disparity inevitably leads either to acceptance of the feminization of poverty or to a departure from the capitalist component of liberalism. Fortunately, capitalism has not been "purely" practiced; a full market economy has been compromised through government programs that reduce the inequality inevitably resulting from maximization of individual liberty. The result is a mixed economy, one that attempts to restrain the choices of individuals only to the extent that those of others are not thereby impeded. John Stuart Mill is a classic proponent of this version of liberal feminism.[36] Poverty, with its concomitant health risks to women, men, and children, is tolerable in such a system so long as it does not lead to a reduction in the greatest happiness of the greatest number. It may, in fact, be argued that poverty is sometimes freely chosen, as preferable to its alternatives. Drug addiction and prostitution, both associated with poverty, are viewed in this light by some liberal feminists.[37] As indicated earlier (Chapter 6), Lori Andrews defends surrogate motherhood on this basis, arguing that a ban on surrogacy would be inconsistent with the feminist principle of reproductive choice.[38]

Existentialist feminism is epitomized in the writings of Simone de Beauvoir, for whom the liberty that defines women as individuals requires that they refuse to accept traditional role definitions as wives and mothers.[39] Yet it is precisely because women frequently fulfill traditional feminine roles of caring for others, particularly parents and children, that they tend to be economically poorer than men. The solution to the feminization of poverty that existentialist feminism proposes is not acceptable to most women because they view their caring roles as better than the alternatives. In particular, they wish to exercise their capacity for motherhood. While they choose to care, however, they do not choose to be poor or to risk their own or their loved ones' health because of poverty. The flaw of existentialist feminist reasoning is its pretense that autonomy is not ultimately impeded by life's circumstances, including other choices.

Postmodern feminism is also individualistic, insisting that differences among women have not been adequately considered by those purporting to champion the rights of women as a class. Insistence on recognition of differences is surely supportable on egalitarian grounds. Although they develop their critiques of other feminists in different ways, postmodern feminists join in rejecting traditional assumptions about truth and reality as "phallologocentric," that is, ordered around an absolute word (*logos*) that is "male" in style (related to the *phallus*).[40] Even terms such as *feminist* and *lesbian* are considered parasitic on phallologocentric thought. Presumably, the term "feminization of poverty" would be unacceptable for similar reasons. Postmodern feminists refuse to construct or affirm any explanatory theory because any such theory is ultimately male-defined. Within a postmodern feminist context, therefore, it is impossible to discuss, let alone propose, solutions to problems regarding the poverty and health of women and children as a group.

While there are diverse expressions of radical feminism, each rejects in its entirety the patriarchal structure that has historically oppressed women. Some call for an androgynous solution to the patriarchal imposition of "femininity" and "masculinity"; others advocate feminist or lesbian separatism. Neither approach directly addresses the fact of poverty's feminization, and neither deals with disability or class or race as further sources of economic inequality. Rosemarie Tong illustrates the last point vividly:

> To suggest, for example, that a poor, male hired hand on a rich white woman's ranch is oppressing her because he has the physical power to rape her and is not

oppressed by her even though she is paying him at best a subsistence wage is to suggest something at most partially true. Likewise, to say that a white woman who loses out in a job competition to a slightly less qualified black man is discriminated against, all the while ignoring the ways in which she participates in his oppression, is to close our eyes to the complexity of the situation. [41]

In other words, radical feminism fails to recognize poverty as a generalized form of oppression that affects men as well as women. Its concern with women's health mainly involves rejection of male attempts to control women's sexuality and reproductive capacity. From an egalitarian perspective, this critique is justified but inadequate.

Because equality is the central concern of Marxist and socialist theory, these versions of feminism oppose poverty as well as its feminization. The unequal distribution of health care is an apt target of criticism. Health and life, after all, are such fundamental human needs that the ability to satisfy them represents a correspondingly basic responsibility. Capitalism is the culprit that Marxists and socialists alike identify as exploitative of women and their work, whether productive or reproductive. For traditional Marxist feminism, economic inequality is primary, and gender inequality follows from that; for socialist feminism, sex and gender inequality are primary, giving rise to various forms of economic inequality.[42] Egalitarian strategies for overcoming the oppression of women differ accordingly. Marxist feminist strategies would eliminate poverty by eliminating economic classes, considering the feminization of poverty only (if at all) as a secondary result. Socialist feminist strategies attempt to raise the economic status of women to that of men (or lower that of men to that of women) so that poverty is no longer a predominantly feminine phenomenon. Men and women might be equally poor in such a scenario, but their relative poverty is a social injustice to be overcome.

Marx is well-known for his suggestion that we can gauge progress in humanization by examining the man–woman relationship. The extent to which neither person is "owned" or treated as a means rather than an end "reveals the extent to which man's [sic] *natural* behavior has become *human*...."[43] But neither his nor Engel's nor Lenin's writings reflect appreciation for the informal work of caring. Like Plato, they tend to disparage biological reproduction and the work associated with childrearing, while emphasizing work outside the home. As a result, women in so-called Marxist societies have generally worked outside the home in greater numbers but with little reduction in the demands of their home life. For many women, to be "liberated" has meant two full-time jobs instead of one; for men it has meant tired wives still acting as housekeepers and principal caregivers of their children. Because Marx's ideal of a totally classless society promotes economic equality without social equality, a sexual double standard has prevailed in official Marxist regimes.[44]

Tong describes socialist feminism as a "confluence of Marxist, radical, and, more arguably, psychoanalytic streams of thought."[45] Its dissatisfaction with Marxism stems from the "tendency of Marxist patriarchs to dismiss women's oppression as not nearly as important as workers' oppression."[46] Socialist feminists join radical feminists in targeting patriarchy as well as capitalism as responsible for pervasive sexism; they join psychoanalytic feminists in their emphasis on relationships as an essential component of their egalitarian critique. Consistent with this approach, reduction of poverty's impact on women and children is crucial to the socialist feminist agenda. Equal access to fulfillment of basic human needs (nutrition, shelter, etc.) as well as health care for basic problems is indispensable to those who exercise the principal caregiving relationship towards others.[47] Government programs that provide free prenatal care for poor women, parent education, and support for home care of the ill and disabled illustrate this recognition that human health is dependent on relationships as well as medical condition.

Because socialist feminism attaches central importance to women's oppression as women, and not just as workers, the reproductive health of women assumes a higher priority than in traditional Marxism. The nuances of a complex concept of equality that distinguishes among different needs and abilities are thus evident. Socialist feminism respects this complexity more than a Marxist version that belittles reproductive tasks. Critiques of technological advances that allow more poor women to be exploited are a particular concern of socialist feminists; egg donation and surrogacy programs are foremost examples in this regard.

Ageism and Socialist Feminism

Socialist feminism also provides a framework for addressing the perennial problem of ageism in society. Ageism, which I define as the irrelevant use of age as a criterion for excluding members of certain groups from social benefits, leads to poverty and poor health among members of those groups. Within the two groups that poverty most affects, women and children, an apparent dilemma arises because of the increased resources necessary to provide health care for both the young and the old. At both ends of life's spectrum, the remarkable advances of medical technology may be denied to some because of their expense. However, the overall health of children may be compromised by the costs of caring for the elderly.

In 1986 Samuel Preston observed that conditions among the elderly in the United States had improved markedly during the previous two decades, but that conditions among children had deteriorated during that time. Citing the U.S. Department of Agriculture standard by which family income is the the measure of poverty, he wrote: "In 1970, 16% of those less than 14 years of age lived in poverty compared with 24% of those older than 65 years of age. By 1982 the situation had been reversed: 23% of children lived in poverty compared with 15% of the elderly."[48] When noncash transfer payments such as foodstamps and Medicare are counted as ameliorative or exacerbative factors, the fraction of the elderly living in poverty in 1982 falls from 15 percent to 4 percent, whereas the corresponding reduction for children is only from 24 percent to 17 percent.

Since 1979, public programs for the elderly have expanded, while those for children have been reduced. Medicare funding rose from $3.4 billion in 1967 to $57.4 billion in 1983, while in the period between 1979 and 1982, enrollment in the AFDC (Aid to Families with Dependent Children) program was reduced from 72 to 52 out of every 100 children living in poverty. Preston notes that the expenditure per child of federal programs was less than a tenth of their expenditure per older person. He cites reduced suicide rates among the elderly and an upward trend in suicides of children as an indicator of differences between the psychological well-being of the two groups. Using demographic models, he also points out that declines in mortality rates among the elderly during the past few decades have been roughly three times as great as the decline among children.[49]

Family instability, according to Preston, is one of the major risk factors for children as opposed to the elderly. While "the nuclear family remains the principal mode of support for children," its capacity for nurturance has declined because of the erosion of the family structure, particularly through noninvolvement or disengagement of fathers. In contrast, "the elderly are protected from changes in family structure by the fact that the state has assumed primary responsibility for their welfare."[50] Their political support comes from three sources: the elderly themselves, those who act on behalf of the elderly, and younger persons who act in self-interest to promote programs that will serve them in their own old age. Children have only the second type of support: parents, mostly mothers, acting in behalf of their progeny. Unfortunately, the poor mothers of poor children seldom have any political clout.

Two additional factors may contribute to the discrepancy between health care for the old and the young. One is the fact that people do not choose their parents, but they may choose to be parents. The advent of reliable contraception and the availability of abortion have led to a widespread view that other people's children are solely *their* responsibility, no matter how poor the parent or child. A second factor is that poor children are increasingly members of minority groups, with whom the majority find it difficult to identify. Where self-interest dominates a national ethic, as appears to be the case in the United States, these factors surely influence voters against support for programs that do not contribute to their own or their progeny's welfare.

A socialist critique directly confronts the self-interest motif of capitalist society. A socialist feminist critique does this with particular concern for the just treatment of women and their children, but not with exclusive concern for any group. How then does the criterion of age figure in a socialist feminist analysis of the allocation dilemma regarding the old and the young? Basically, it argues in favor of changing the current practice so that children and others receive at least as much support as the elderly. "As much" obviously demands clearer delineation. Minimally, it suggests that per capita expenditures be proportionate to the needs of each group to maintain a health level appropriate to their age. But what about cases where both groups, or individuals from either age group, cannot be supported? Suppose, for example, either of the following scenarios:

CASE I: A CHILD VERSUS AN ELDERLY ADULT.

Simultaneously, two calls are received by an emergency medical technician dispatcher: one from a frantic parent of a young child who has suffered respiratory arrest, the other from a frantic (adult) son whose parent has suffered respiratory arrest. If no further data are ascertainable regarding history or circumstances affecting prognosis, and only one emergency vehicle is currently available, to which person should help be directed first?

CASE II: PRETERM INFANTS VERSUS GERIATRIC PATIENTS.

Two bills are before the state legislature, which, like all state legislatures, has limited resources to utilize in support of public programs. One bill calls for funding of treatment for preterm infants in neonatal intensive care units; the other for funding of treatment for very old patients in intensive care units. If only one bill can be supported, which should it be?

Note that each of the cases is a hypothetical rather than actual case. My use of hypotheticals illustrates the point that many feminists, casuists, and others make regarding the impracticality of traditional ethics.[51] To prescind entirely from the contextual differences that are bound to arise in such cases is to deal with irrelevant abstraction. From a practical standpoint, what these cases point to is the need and obligation to avoid such dilemmas by preventive measures, including political action.[52] "Preventive medicine" may be broadly defined as encompassing such measures, but, unfortunately, this is an area of health practice that is generally underfunded despite its potential for reducing social inequality. Even when dilemmas are not preempted by ethical foresight, clinicians usually find relevant reasons for choosing among alternatives.

Although Cases I and II would probably never occur in as stripped a fashion as here recounted, consideration of them can serve to clarify priorities in distribution of health care resources. As a pragmatist, I am committed to the notion that such clarifications have practical relevance (I expand on this in Chapter 15). One of the factors to consider in establishing priorities is the fact that the age of patients is often medically relevant to determination of treatment and prospects for success.

As described in Chapter 11, age is also relevant, although not an adequate criterion, in the determination of competence for consent to treatment. In Case I, however, both of the patients in need are incapable of consent, and a suitable surrogate has apparently assumed the role of advocate or decision maker in behalf of the patient. In Case II, consent is not relevant because the bill is simply making treatment available to either age group.[53] The question that remains is whether age per se is also a *morally* relevant basis for inclusion or exclusion from treatment.

Guideline 6 proposes that "individuals be given an equal share of the resources available, insofar as these are pertinent to their needs, desires, capabilities and interests." Guideline 7 suggests that human beings have a more compelling claim to limited resources than other species. But neither or both of these guidelines is adequate to determine whether age is morally relevant when "needs, desires, capabilities and interests" are comparable for different persons or groups. Norman Daniels provides a cue in this regard in his explanation of why the age criterion should be distinguished from criteria of race or sex in addressing problems of discrimination. Citing R. J. Zeckhauser and W. K. Viscusi's observation that "the elderly comprise a minority group that we all hope to join," Daniels maintains:

> This basic fact points to the naturalness of the suggestion that we think about cooperative social schemes that bear on aging in prudential terms.... From the perspective of institutions operating over time, the age criterion operates *within* a life and not *between* lives.... Thus it is rational and prudent that I take from one stage of my life to give to another, in order to make my life as a whole better.[54]

Daniel's argument supports the legitimacy of voting for the bill that funds treatment of neonates rather than the elderly in Case II. A reasonable life span, which is a reasonable desire on the part of reasonable people, would be preempted by the unavailability of neonatal intensive care for preterm infants. Very elderly patients in intensive care units have already lived beyond their expected life span. So also have some very ill and incurable infants. If other features of their cases are comparable, however, an egalitarian account suggests that in the face of limited resources newborns get a chance to live at least part of the time allotment that their elders have had. Moreover, since the incidence of preterm birth is greater among the poor and among blacks, the health disparity between them and others would be reduced through such a measure.

Although this account may be construed as socialist in its emphasis on equal rights to a reasonable life span, it seems opposed to feminism in that it reduces benefits to the elderly, most of whom are women. At least two factors suggest otherwise. One is that extension of the life span is not an unmixed good. More than for infants, extension of the life span of the very old is likely to entail prolonged technological support in a hospital environment, compromised by pain, discomfort, medication, and other intrusive therapies, without realistic expectation of cure or return to one's previous life-style. Because of their socialization and because of double standard thinking, women are more likely than men to suffer the impositions of medical technology, even when they would prefer to be allowed to die.[55]

A second reason why favoring treatment programs for infants over those for the elderly is consistent with feminist advocacy for women is that mothers generally are more closely connected to their infants than are fathers. In an ethic that stresses the moral value of relationships, it is usually impossible to do good to infants without simultaneously doing good to their mothers. The majority of infants requiring intensive care are born before term, and the incidence of low birth weight among the poor is twice that of the nonpoor.[56] While many of the elderly are poor, the proportion of those receiving intensive care is less than that of poor children receiving intensive care. Accordingly, in voting for the bill that funds intensive care for infants, despite the fact that intensive care for the very elderly may not then be funded, one votes for the poor and for women.

Daniels suggests that preference for either individual in Case I is morally problematic "only if we take a time-slice perspective" on the issue.[57] If the dispatcher regards the question of whether to send the ambulance to the elderly person or the child as a policy question, then the decision would be similar to the one for Case II, favoring the child on egalitarian grounds relevant to life span. Because age is a feature of everyone's life, the allotment of limited pertinent resources across the expected life span of all individuals is just even though this sometimes results in some individuals having fewer resources at the end of their lives. In other words, *age* is a relevant criterion for distribution of scarce resources. *Ageism* occurs when the criterion is used irrelevantly. Determination of relevance necessarily involves consideration of other features that characterize individuals—such as their stipulated wishes and the network of relationships they have established.

Ageism also occurs when the costs of health care for the elderly are viewed as the cause of reduced health care for children. In the preceding case, I have attempted to depict situations where there are absolutely no relevant factors except the age difference for determining whether one individual or group should get assistance rather than another. As I have already acknowledged, however, this type of situation rarely if ever happens. When apparently confronted with an allocation dilemma, health professionals usually look for an "ethical bypass," that is, an alternative that avoids the dilemma. In Case I, for example, multiple factors related to the prospect of success would and should be considered by the dispatcher, such as distance to be travelled, whether someone already at the scene could provide treatment, previous health history, other current health problems, and so on. If a way can be found to save both individuals by tapping some other emergency resource, this would surely be preferable to saving only one. In most cases, such an ethical bypass is available. Inevitably, some risk is involved in attempts to save both simultaneously, such as the possibility that both may be lost or impaired. If the risk is small, equal concern for both needy persons justifies the attempt; if the risk is great, it does not. If all of the alternatives are exhausted and there is no other way of distinguishing between the cases, the differences in their life spans argue on egalitarian grounds for a decision in the child's favor.

Legislators in Case II are particularly likely to look for an ethical bypass. While they are aware that children do not vote, they also recognize that their parents do. Moreover, some would probably be persuaded by the ethical argument for funding the health needs of children rather than adults if both cannot be supported. For such legislators, the likelihood of finding a way of funding both types of program, at least in part, is high. If all else fails, Congress might even resort to a tax hike. Taxes, of course, are the usual means of obtaining support for the poor, whether they be aged or young or in between. Taxes, after all, are an essentially socialist mechanism because they aim to reduce the disparities between rich and poor in order to approximate a more egalitarian social ideal. The crucial limitation of laissez-faire capitalism is thus recognized: maximization of individual liberty leads to maximization of inequality. As already remarked, we have never had a thoroughly capitalistic system in the United States. Thankfully.

NOTES

1. See Diana Pearce, "The Feminization of Poverty: Women, Work and Welfare," *Urban and Social Change Review* 11 (Feb. 1978): 28–36; and Ruth Sidel, *Women and Children Last* (New York: Penguin Viking Books, 1986), 133–56.

2. The poverty index used throughout the United States today is based solely on cash income and family composition. See U.S. Bureau of the Census, *Statistical Abstract of the United States: 1991*, 111th ed. (Washington, D.C.: U.S. Government Printing Office, 1991) 429.

3. Karen Quinlan became comatose at the age of 21. Her family asked for and obtained court permission to have her removed from the respirator. Although it was expected that she would then die, Quinlan breathed on her own and "survived" for the next ten years without recovering consciousness. See Ronald Munson, *Intervention and Reflection* (Belmont, California: Wadsworth Publishing Company, 1988), 157–59. Nancy Cruzan also became comatose, lapsed into a persistent vegetative state, and was weaned from a respirator. After years of providing nutrition through mechanical means and of monitoring and treating infections, the Cruzan family sought legal sanction to have artificial means of nutrition removed. The U.S. Supreme Court denied the request on grounds that there was not clear and convincing evidence that Cruzan would have wanted such support dis-

continued. See *Cruzan v. Director, Missouri Department of Health,* U.S. Supreme Court No. 88-1503, argued Dec. 6, 1989, decided June 25, 1990.

4. Regarding medical aspects of a persistent vegetative state, see Ronald E. Cranford, "The Persistent Vegetative State: The Medical Reality (Getting the Facts Straight)," *Hastings Center Report* 18, no. 1 (Feb./March 1988): 27–32.

5. Sidel, 88.

6. U.S. Bureau of the Census, *Statistical Abstract of the United States: 1991,* 111th ed., 429.

7. U.S. Bureau of the Census, *Current Population Reports* Series P-60, 175, *Poverty in the United States: 1990* (Washington, D.C.: U.S. Government Printing Office, 1991), 195.

8. Mayer and Jencks, 1988, as cited in Claudia Coulton, Julian Chow, and Shanta Pandey, *An Analysis of Poverty and Related Conditions in Cleveland Area Neighborhoods* (Cleveland, Ohio: Center for Urban Poverty and Social Change, Mandel School of Applied Social Sciences, Case Western Reserve University, January 1990), 9.

9. Pearce, 28–36.

10. Diana Pearce, "The Feminization of Poverty," *Journal for Peace and Justice Studies* 2, no. 1 (Spring 1990): 1.

11. Barbara Starfield, "Child Health Care and Social Factors: Poverty, Class, Race," *Bulletin of the New York Academy of Medicine* 5, no. 3 (March 1989): 300.

12. Sara S. McLanahan, Annemette Sørensen, and Dorothy Watson, "Sex Differences in Poverty," 1950–1980," *Signs: Journal of Women in Culture and Society* 15, no. 11 (1989): 107, 108, 110.

13. See Susan Moller Okin, *Justice, Gender, and the Family* (New York: Basic Books, Inc., 1989), 4. Okin also notes that only about 10% of single parent households are headed by men, and 65% of these are a result of marital separation or divorce.

14. McLanahan, Sørensen, and Watson, 115.

15. McLanahan, Sørensen, and Watson, 102.

16. Sidel, 158.

17. Irwin Garfinkle and Sara McLanahan, *Single Mothers and Their Children* (Washington, D.C.: Urban Institute Press, 1986), 11.

18. Sidel, 36.

19. Sidel, 137.

20. Starfield, 300.

21. Starfield, 300.

22. "Poor" is used in a relative sense here. For various reasons, women who rank below the official poverty line are less likely than lower middle class women to be surrogates or egg donors. In either case, however, the women tend to be poorer than the infertile couples who pay them. See Alta Charo, "Legislative Approaches to Surrogate Motherhood," in *Surrogate Motherhood,* ed. Larry Gostin (Bloomington: Indiana University Press, 1990), 89.

23. Veronika Kolder, Janet Gallagher, and Michael Parsons, "Court-Ordered Obstetrical Interventions," *New England Journal of Medicine* 316, no. 19 (May 7, 1987): 1192–96.

24. Strokes in utero, prematurity, and neurological disorders are among the common problems affecting children born of crack-using mothers. See Deborah A. Frank, Barry S. Zuckerman, and Hortensia Amaro, "Cocaine Use during Pregnancy: Prevalence and Correlates," *Pediatrics* 82, no. 6 (Dec. 1988): 888–95; and Jan Hoffman, "Pregnancy, Addicted—and Guilty?" *New York Times Magazine* (Aug. 19, 1990), 34.

25. See Harmeet K. D. Singh, "To Lower Infant Mortality Rate, Get Mothers Off Drugs," *Wall Street Journal* (May 1, 1990), A10.

26. See Bruce L. Miller, "Autonomy and the Refusal of Lifesaving Treatment," *Hastings Center Report* 11, no. 4 (Aug. 1981): 23–24.

27. Although I have used the neutral term *condition,* many people consider addiction to drugs, including alcohol, "diseases." While I do not develop this point here, the classification of drug addiction as a disease can also be used in the argument that it reduces autonomy. See J. M. Fort, "Should the Morphine Habit Be Classed as a Disease?" in *Concepts of Health and Disease,* ed. Arthur

Caplan, H. Tristram Engelhardt, and James McCartney (Reading, Massachusetts: Addison-Wesley Publishing Co., 1971), 327–32.

28. This does not imply that *forced* treatment is justified.

29. Hemophilia, a sex-linked recessive condition that involves a defect in the coagulating function of the blood, is another risk factor affecting both affluent and poor men, but not women. Transfusions required for treatment of hemorrhages expose hemophiliacs more frequently to the HIV virus.

30. See Nora Kizer Bell, "Women and AIDS: Too Little, Too Late?" *Hypatia* 4, no. 1 (Fall 1989): 9–10; and Julien S. Murphy, "Women with AIDS: Sexual Ethics in an Epidemic," in *AIDS: Principles, Practices, and Politics,* ed. Inge B. Corless and Mary Pittman-Lindeman (Washington, D.C.: Hemisphere, 1988), 65–79.

31. Sheldon Landesman, Howard Minkoff, Susan Holman, Sandra McCalla, and Odalis Sijin, "Serosurvey of Human Immunodeficiency Virus Infection in Parturients," *Journal of the American Medical Association,* 258, no. 19 (Nov. 20, 1987): 2703; and Howard Minkoff, Deepak Nanda, Rachel Mendez, and Senih Fikrig, "Pregnancies Resulting in Infants with Acquired Immunodeficiency Syndrome or AID-Related Complex: Follow-Up of Mothers, Children, and Subsequently Born Siblings," *Obstetrics and Gynecology* 69 (1987): 288–91.

32. A. E. Ades, M. L. Newell, and C. S. Peckham, "Children Born to Women with HIV-1 Infection: Natural History and Risk of Transmission," *The Lancet* 337, no. 8736 (Feb. 2, 1991): 253–60. Also, see "European Study Finds Few Babies Get AIDS," *New York Times* (March 5, 1991), 36.

33. John Arras suggests that directive counseling may be in order for HIV infected women making reproductive decisions. See his "AIDS and Reproductive Decisions: Having Children in Fear and Trembling," *Millbank Quarterly* 68, no. 3 (1990): 353–82.

34. For example, see Mireya Navarro, "Women with AIDS Virus: Hard Choices on Motherhood," *New York Times* (July 22, 1991), A1, B7.

35. See Rosemarie Tong, *Feminist Thought* (Boulder, Colorado: Westview Press, Inc., 1989).

36. For example, in his *The Subjection of Women* as excerpted in my *Philosophy of Woman,* 2d ed. (Indianapolis: Hackett Publishing Company, 1983), 46–64.

37. Christine Overall, *Ethics and Human Reproduction* (Boston: Allen and Unwin, 1987), 116–19.

38. Lori B. Andrews, "Feminism Revisited: Fallacies and Policies in the Surrogacy Debate," *Logos* 9 (1988): 81–96; see also Chapter 6.

39. See Mahowald, "De Beauvoir's Concept of Human Nature" and excerpts from "The Second Sex," in *Philosophy of Woman,* 79–99.

40. Tong, 217–33. Also see Linda J. Nicholson, ed., *Feminism/Postmodernism* (New York: Routledge, Chapman and Hall, Inc., 1990).

41. Tong, 28–29.

42. Like Alison Jaggar, I believe that socialist feminism is more consistent with the central insights of Marx than traditional Marxism. See her *Feminist Politics and Human Nature* (Totowa, New Jersey: Rowman and Allanheld Publishers, 1983), 52.

43. Karl Marx, "Economic and Philosophic Manuscripts of 1844," in *The Marx-Engels Reader,* ed. Robert C. Tucker (New York: W. W. Norton and Company, Inc., 1972), 69.

44. This is one of the reasons why Marx's ideal of a truly egalitarian society has never been realized even in states that profess to follow his ideology. See, for example, Hilda Scott, *Does Socialism Liberate Woman?* (Boston: Beacon Press, 1974).

45. Tong, 173.

46. Tong, 173.

47. Liberals and liberal feminists support the concept of equal access to fulfillment of basic needs as well, but they do not argue for full social equality. For a critique of liberal feminism, see Joan C. Callahan and Patricia G. Smith, "Liberalism, Communitarianism, and Feminism," in *Liberalism and Community,* ed. Noel Reynolds, Cornelius Murphy, and Robert Moffat (Lewiston, New York: Edwin Mellen Press), forthcoming.

48. Samuel Preston, "Divergent Paths for Children and the Elderly," *Pediatrics* 77, no. 1 (Jan. 1986): 117.

49. Preston, 117–19.

50. Preston, 118.

51. For example, see Susan Sherwin, "Feminist and Medical Ethics: Two Different Approaches to Contextual Ethics," *Hypatia* 4, no. 2 (Summer 1989): 57–72; and Albert R. Jonsen and Stephen Toulmin, *The Abuse of Casuistry* (Berkeley: University of California Press, 1988), 279–332. Sherwin has further developed her critique of traditional ethics in her *No Longer Patient* (Philadelphia: Temple University Press, 1992), 43–57.

52. Sherwin laments the lack of a political dimension in contemporary medical ethics, arguing that a feminist medical ethics would necessarily reflect that dimension. See Sherwin, "Feminist and Medical Ethics: Two Different Approaches to Contextual Ethics," 62, and *No Longer Patient,* 84–88.

53. One could argue that a democratic process, that is, a representative voting system, is tantamount to consent ("consent of the governed"). But since children cannot vote, this hardly represents their consent.

54. Norman Daniels, *Just Health Care* (Cambridge, England: Cambridge University Press, 1985), 96. R. J. Zeckhauser and W. K. Viscusi, "The Role of Social Security in Income Maintenance," in *The Crisis in Social Security: Problems and Prospects* ed. Michael J. Boskin (San Francisco: Institute for Contemporary Studies, 1978), 54.

55. Steven H. Miles and Allison August, "Courts, Gender and 'The Right to Die'," *Law, Medicine and Health Care,* 8, no. 1–2 (Spring/Summer 1990): 85–93.

56. Starfield, 300.

57. Daniels, 97.

14

The "Family" and Health Care

Clinicians of various stripes tend to describe their practices as imbedded in a family context. Social workers, for example, often define their role as family advocates; their efforts to assist individuals necessarily take account of family influences and supports. Similarly, pediatric nurses and pediatricians sometimes pay more attention to parents than to their patients, because they recognize that children's health is more affected by their home life than by clinical interventions. As one pediatric oncologist put it, "I'd be irresponsible if I did not regard each child's family as my patient."[1] While specializations within modern medicine focus more on individuals than on families, a renewed appreciation of the family is evident in the development of "family medicine" as a specialty. Not only is family recognized as an important influence on health, but the health of each family member and the family as a whole is viewed as the responsibility of "the family doctor."[2]

This emphasis on the family among practitioners and within health care institutions may well be seen as consistent with a care-based model of moral reasoning. It may also be seen, however, as problematic for an egalitarian perspective. Both theoretical and practical problems arise because the meaning of family is unclear, and because women's role within that context is generally more burdensome than men's. Children's interests are not well served by the unclarity or by the inegalitarian consequences. Consider, for example, the following possible family arrangements:

- A nuclear family, that is, a married couple with genetically related or adopted children
- A blended family, that is, a married couple with children from previous marriages, with or without children related to both
- An extended family, that is, at least three generations in the same household, or other relatives in addition to parents and their children
- An unmarried couple living together, with or without children
- A divorced couple with joint custody of their minor children
- A couple of the same sex living together with or without genetically related or adopted children[3]
- A married couple, a surrogate, and child who is genetically related to the husband and surrogate
- A couple whose genetic or adopted children are raised and living elsewhere
- Foster parents (single or couples) and their children
- A childless brother and sister living together
- A single woman or man who has a child related genetically or through adoption
- A single grandmother raising her grandchildren

The list is not meant to be exhaustive. It mainly illustrates groupings of people that some call families.[4] Note that the situations described only specify genetic or legal ties, or living-together arrangements among "family members." It may be argued, however, that familial relationships need not be marked by any of these empirical features, but must be characterized by a further feature, namely, special caring on the part of family members for one another. If this feature is essential to the meaning of family, some groupings of people who are biologically or legally related to one another do not count as families, and some who are neither biologically nor legally related do count as families. When health professionals consider the therapeutic impact of "family," they are mainly concerned about whether familial caring is available to their patients; whether family members are genetically or legally related to one another is therapeutically irrelevant.

In this chapter, I offer a definition of family that fits all of the above situations so long as the element of caring is added to each. Preliminarily, I discuss the traditional (still influential) notion of family and feminist criticisms of this notion. Several health issues involving families are then examined: teenage pregnancy and parental notification regarding abortion. I conclude with a case that illustrates some of these issues, showing the limitations as well as relevance of family considerations in moral decision making.

The Notion of Family and Feminist Concerns

Philosophers have seldom written about families.[5] By and large they have observed a division between public and private worlds, without considering the terrain that lies between the two. The public world is that of political institutions; the private world is one of individuals. Together these provide the context for all of the major decisions that human beings confront in their lives. If impartiality is credited as an essential criterion for moral judgments, the relationships on which some men, women, and children base their practical decisions are disvalued because of their partiality. Families represent particularity in relationships; they define rights and responsibilities in terms of that partiality. The contrast between an ethic of care and an ethic of justice may thus be recapitulated as a conflict between partiality and impartiality, particularism and universalism.[6] If care and justice are compatible, however, the conflict is resolved. Families define both the need and the possibility for such resolution.[7]

The term *family* comes from the Latin *familia,* which means "household." Originally, the term referred to all of the people who shared the same domicile, including successive generations and servants. Historically, the usual model for the *familia* has been patriarchal, consisting of a male head-of-household, his wife or wives and concubines, all of their children, unmarried daughters, and servants with their children. It has also been patrilineal, which means that family members have been genetically related to the patriarch, and patrilocal, which means that all members reside in the domicile of the patriarch. The fact that servants counted as family members even though they were not genetically related to the patriarch suggests that patrilocality was more important than patrilineality, or that the servants were family members only through their (serving) relationship to those genetically related to the patriarch, or to the patriarch himself.

The patriarch in ancient times held absolute power over members of his family; his decisions were beyond question even if they meant death for family members. Although patriarchal aspects of family life have undoubtedly declined, laws and customs still often assume that husbands and fathers are heads of households. The assumption is hardly supportable in light of statistics regarding single mothers. In the United States, for example, nearly 40 percent of the poor are children and over half of them live in households headed by single female parents.[8] Whether they are heads of households or not, women tend to be more extensively and intimately involved with families than are men. Gender socialization as well as reproductive roles support this tendency.

Not surprisingly, feminist critiques of "the family" focus on the patriarchal structure that attributes absolute power to its single (male) head. Liberal feminists insist on the right to privacy, and therefore avoidance of governmental intervention in women's lives, including their family arrangements.[9] Rejecting the assumption that women are uniquely suited for housework and child care, they argue for social interventions only to the extent that they will maximize the options of individuals. Some propose, for example, that the housework and child care usually performed by wives and mothers should be paid for by their spouses.[10] Men should be equally supported in these roles. The relationship between parents and children is interpreted individualistically, in keeping with an emphasis on private property. Indeed, the notion of ownership is a legitimate context for interpreting relationships expressed by the terms "*my* child," "*my* wife," or "*my* husband." In general, the image of the family is nuclear, consisting of a married heterosexual couple and their children (preferably a son and a daughter, in that order), but alternative lifestyles such as unmarried and homosexual couplings are acceptable also. The emphasis is on choice, with societal interventions recommended only to the extent that they facilitate the choices of individuals.

Radical feminists are stronger than liberal feminists in their critique of the patriarchal family, targeting heterosexual marriage as the principal cause of sexism. Some have encouraged women to forego biological motherhood. The development of means through which men may also bear children is considered necessary in order for women to be truly liberated from the subjugation that marriage and motherhood involve.[11] Other radical feminists distinguish between the institution and experience of biological motherhood, rejecting the former and embracing the latter so long as women enjoy and control the experience.[12] Adrienne Rich suggests that families would be fine for women if men were not involved in running them. Women, she maintains, ought to remain in control not only of childbearing but also of childrearing.[13] Because self-insemination and communities of women are means of accomplishing this goal, feminist separatism is encouraged.[14] Control of reproduction calls for access to technical assistance, but not in a manner that subjects women to the male power establishment.

Socialist feminists take direct aim at the dichotomy between private and public realms that liberalism, including liberal feminism, endorses. This is the terrain that traditional philosophy has ignored in its consideration of ethics and politics as distinct areas of decision making. By insisting that the personal is political, socialist feminism identifies family issues as crucial to both. According to Alison Jaggar, "it is precisely their focus on the importance of the family and 'personal life' in general which distinguishes the social analysis of socialist feminists from that of classical Marxists."[15] The reproductive work of women, for example, is treated as relatively inconsequential in a Marxist framework. For socialist feminists, reproductive work is crucial to social as well as personal development. Abetted by the insights of psychoanalytic and radical feminists, they view the nurturant contributions of women (or men) as "productive work," to be empirically rewarded in a just social system.

While all varieties of feminism critique patriarchal interpretations of family, radical and socialist feminism stand in clearest opposition to the concept and practices of the "nuclear family." As previously described, however, the context in which socialist feminists reject the nuclear family is more broadly egalitarian than the radical feminist critique. Socialist feminists do not advocate separatist communities, or renunciation of marriage or motherhood; they propose something yet more drastic, namely, transformation of the social structure of communities that rely on stereotypic roles for women, men, and children in order to achieve economic "progress." Capitalism is, of course, the most obvious example of a social structure that needs to be transformed, but so are socialist systems in which gender stereotypes are still supported. The latter, for socialist feminists, are not adequately socialist, and their inadequacy signals the necessity of tying genuine socialism to feminist critique.

Each of the feminist critiques of a patriarchal concept of the family is based on a moral argument about injustice toward women. They are reinforced, however, by current data regarding the relative scarcity of such families. The scarcity is mainly due not to the elimination of patriarchal standards in families, but to the virtual absence of men from so many families.[16] Many nonpatriarchal families were originally patriarchal, but became nonpatriarchal because of abandonment or divorce by the patriarch. From an egalitarian standpoint, this is a loss only to the extent that the welfare and autonomy of others is disproportionately compromised. To some women, the psychological gain may be more significant than the economic loss occasioned by the patriarch's absence. For children, too, who are undoubtedly influenced by the gender roles they observe in their parents, loss of a patriarchal parent is not necessarily unhealthy. Women may assume patriarchal roles in the absence of men, or even when men are present in families. Regardless of whether patriarchs are men or women, however, patriarchal practice is not a healthy component of family life. The principal healthy component is the set of caring relationships that family life can and often does provide. By supporting such attachments, a nonpatriarchal concept of family is consistent with various versions of feminism as well as a care-based model of moral reasoning. It is also consistent with the parentalist model described in Chapter 2.

If family is defined as patriarchal, it obviously does not represent a value endorsed by feminists or egalitarians. If family is defined more broadly, eschewing patriarchy while embracing the positive relationships that bind family members together, it is endorsed by them. Consider the following as such a definition: a family is a set of people who are indefinitely committed to care for each other and for those dependent on them. The phrase "dependent on them" refers particularly to children, or those who are not themselves capable of an indefinite commitment to care. However, the definition does not imply that there must be children or other dependents in order for there to be family.

The term *care* suggests emotional attachment as well as supportive actions. How family members express their care for one another is dependent on the abilities and needs of each, and these of course vary over time and circumstances. As with Gabriel Marcel's notion of commitment,[17] the caring it involves is open ended, avoiding predetermination so as to respond optimally to unanticipated events. Caring is an underlying disposition to do whatever promotes the interests of another, with *interests* connoting respect for others' wishes as well as welfare. Because of their caring disposition, family members constitute the most appropriate proxy decision makers for patients. The caring practiced by health caregivers is a more limited expression of care.

Commitment means that a decision has been made, at least by those capable of responsible decision making, and the fact that the commitment is "indefinite" means that no limits are set to the caring that is pledged. Note that indefinite commitment is not equivalent to permanent commitment: indefinite commitments may be initiated or ended, either voluntarily or involuntarily. Marriage and divorce are examples of voluntary changes in commitment; birth and death are examples of involuntary changes. Although friendship may appear to fit my proposed definition of family, many friendships do not involve indefinite commitment in the strong sense I intend; if they do, then the friends *are* a family. In other words, family members are friends but not all friends are family.[18] The law has increasingly recognized that people who live together and care for each other over an extended period of time are a family, or should at least be regarded as one. Health professionals have acted similarly, recognizing that "significant others" are sometimes more appropriate decision makers than those who are related by blood or marriage.

Although biological or genetic linkage is not essential to my proposed definition of family, such linkage constitutes grounds for *assuming* that individuals are members of a family. A commitment to care is typically and naturally supported by the fact that individuals are biologically related to one another. But the assumption can be refuted in situations where commitment to care is transferred or renounced, as occurs in adoption or when parents "disown" their grown children. An adult may leave her family of genetic origin not only physically but psychologically as well. If her family persists in their commitment to care for her, she remains part of their family even though they are no longer part of hers.

Communication, as Aristotle noted, is essential to friendship,[19] and *a fortiori* to family. But communication does not always require living together in the same physical locality. When family members do not live together they usually share a home to which they return from time to time. A resident in a long-term nursing facility, for example, usually has a family and home elsewhere. Or a daughter in college is surely a family member even though she may live most of the time in another locale. If and when the daughter forms another family by indefinitely committing herself to the care of another or others, she may still belong to her family of origin. Individuals can thus belong to several families simultaneously. It is also possible, sadly, that they belong to none.

Such examples illustrate the flexibility of family arrangements. As people mature and their life circumstances are altered by chance or by choice, families are also affected. New reproductive technologies and life-styles have challenged traditional concepts of family by extending the range of possible family arrangements. From a feminist point of view, the challenge is desirable because it provokes a reassessment that could result in more egalitarian practice. But family rearrangements may be no more egalitarian than the preceding arrangements, and may even be less so. In American society, for example, adolescent pregnancy presents the possibility of a family rearrangement that tends to undermine the interests of both women and children, increasing the inequality gap between them and men. In cultures where early pregnancy is expected and supported, the rearrangement is relatively unproblematic. Regardless of the outcome, however, adolescent pregnancy poses a challenge not only to traditional concepts of family, but to families themselves.

Teenage Pregnancy

Adolescent pregnancy is an issue that illustrates the importance of respecting individual differences triggered by nature, nurture, and circumstances along the entirety of the life span. An egalitarian perspective requires respect for all of these differences, whether or not they can be anticipated through generalized knowledge. From a strictly physiological standpoint, pregnancy is particularly hazardous to adolescents younger than 15 years of age. If the pregnancy is not interrupted, the hazards include increased incidence of toxemia, abruptio placentae, and difficulties associated with immaturity of the pelvis.[20] But births for those younger than 15 constitute a very small fraction (0.3 percent) of all live births, whereas births to older teenagers constitute 13.4 percent of all live births.[21] If age alone is considered as the criterion, health hazards for older pregnant adolescents are no greater than those of pregnant women in their twenties, and less than those of pregnant women over 30 years of age.[22] Nonetheless, pregnant adolescents are generally likely to have more children and less education than their nonpregnant peers (male and female) as well as those who are older, with concomitant economic hardships such as welfare dependency.

Reported health hazards for children of adolescent mothers include preterm birth and low birth weight, decreased growth rate, low IQ, suboptimal school achievement, disordered behavior, and interactional difficulties between child and mother.[23] Recent studies indicate, however, that such problems are mainly due to socioeconomic and life-style factors rather than maternal age.[24] Adolescents who abort their pregnancies generally fare better educationally, economically, and emotionally than those who continue their pregnancies to term. According to one study, they even fare better than those who suspect themselves pregnant but find they are not.[25]

Although teenage pregnancy is surely a social problem in the United States, the perception that it is an epidemic is mistaken. Between 1940 and 1983, the number of live births to women between 15 and 19 years of age grew by 61 percent, but the increase was primarily due to the growth of the teenage population. Similarly, between the mid 1970s and the 1980s, while the absolute number of live births born to girls between 10 and 14 years of age increased, the percentage of live births in this age group declined slightly (from 1.3 to 1.1 per 1000).[26]

Not only chronological age, but gynecologic age and social age are relevant factors in evaluating the impact of teenage pregnancy. Gynecologic age defines the biological maturity of adolescents as the number of years that elapse between menarche and pregnancy.[27] For adolescents of the same chronologic age, pregnancy is more medically problematic for those of younger gynecologic age. Social age refers to cultural milestones achieved by the adolescent, such as age at first marriage, age at birth of first child, or age of adopting adult work roles. For those who fail to fulfill these milestones at the expected time, whether that be earlier or later, social costs range from embarrassment to social ostracism, and include educational and employment interruptions as well as economic pressures. Although pregnancy has increasingly been postponed by some women, the delay has not prevalently occurred among those who are socioeconomically disadvantaged. The greater incidence of adolescent motherhood among the poor is in part explained by a difference in social expectations and opportunities.

Among black adolescents, the incidence of pregnancy is higher than for whites, but this does not adequately account for the higher incidence of low birth weight among blacks of all ages. According to Kwang-sun Lee and Maria Corpuz, the incidence of low birth weight among blacks would be reduced by less than 1 percent if births to black teenagers were entirely discounted.[28] Although some believe that the lower infant mortality rate for whites in the United States is due to socioeconomic advantage, the question of whether other factors are also determinative is not yet settled. Non-black women in the same socioeconomic bracket tend to have infants with higher birth weights than their black counterparts.

Psychological differences also influence the outcome of pregnancy in adolescents. Beatrix Hamburg distinguishes three subsets of teenage mothers: those who are "problem prone," those who are depressed, and those who are "competent to cope."[29] The difficulties already encountered by those in the first two groups are exacerbated by pregnancy, whether it leads to motherhood or not. The last group is worth considering because it illustrates the fact that pregnancy and motherhood do not always cut short or limit the development of adolescent potential. In some instances, it may even be a means of maximizing that potential. Hamburg thus describes a normative, alternative life course within the black subculture in the United States:

> An urban, poor, black woman who wishes to have children may find that her best option may be to commence childbearing in her teen years. This is her "time off" from the labor force. Youth unemployment among blacks is high.... With adolescent pregnancy, her children will be old enough to manage substantial amounts of household responsibilities when the mother enters the work force. This may serve to relieve some of the stresses of being a working mother that a poor black mother could not afford to ease through other means.[30]

For such adolescents, full membership in the adult community can begin with childbearing. Early motherhood expands the kin networks on which poor, urban blacks depend for social and economic support. By the time the young mother is ready to enter the work force, she has consolidated her own growth, and network members can provide childcare. Childbirth during adolescence is thus an adaptive mechanism for those who are competent to cope with its demands.

Unfortunately, the cohort of adolescents just described represents a small minority of those who become pregnant. For most, the major difference in outcome is based on whether they choose to terminate or continue the pregnancy, and if they continue the pregnancy, whether they choose adoption or "keeping" the child. As already suggested, those who choose abortion experience fewer problems than those who continue their pregnancies. Attempts by adolescent parents to raise their children clearly limit their own educational and economic prospects. Laurie Zabin and her colleagues studied a large group of urban adolescents who sought pregnancy testing.[31] Those who underwent abortions were far more likely to have graduated from high school or still be in school at the appropriate grade level two years later than those who continued their pregnancies. Following the experience of pregnancy, the abortion group were no more likely than their peers to have psychological problems, but they were more likely to practice contraception.

In the United States, the incidence of adolescent pregnancy is markedly higher than in nations such as Canada, England, France, The Netherlands, and Sweden. In fact, younger teenagers are five times more likely to become pregnant in the United States than in other industrial countries that have a similar rate of sexual activity.[32] The reason commonly cited for this discrepancy is the greater availability of sex education, contraception, and abortion services in other countries.[33] Yet comparable services could be available to all adolescents in the United States if adequate funding were provided.[34] Funding is also needed to strengthen the educational and family supports that pivotally influence the self-concepts of teenagers. Without these supports, adolescents are particularly vulnerable to pregnancy. Rectifying the inequality that places them at a disadvantage vis-à-vis their peers might be accomplished most effectively by those trained and committed to changing such circumstances than by practitioners whose expertise lies in dealing with the medical consequences of social disadvantage.

Because most teenagers live at home when they become pregnant, their decisions regarding abortion or continuation of pregnancy obviously affect other family members as well. Adolescent motherhood introduces another generation into the family mix, but it does not routinely introduce another male adult or teenager. Shotgun marriages are relatively rare for teenagers (as well as for older persons), and those that occur are more likely than others to end in divorce.[35] Although men are as responsible for pregnancies as women, the burden is not equally shared even to the extent that it can be (beyond biology), whether the pregnancy is terminated or continued. In rare instances men have attempted to block abortions of those they have impregnated, basing their position on their relationship to the fetus and their desire to raise the child. While their intent to raise the potential child is admirable, its speculative nature makes it an inadequate reason for coercing someone to maintain a pregnancy and give birth. As mentioned in Chapter 13, among men who pledge at marriage to support their offspring, only a minority do so if and when their partners separate.[36] Regardless of whether impregnating men have rights concerning fetuses, however, family considerations (broadly construed) influence women's decisions about termination of pregnancy. Parental notification concerning adolescents' requests for abortion are defended on this basis.

Abortion and Parental Notification

Arguments regarding the legality and morality of abortion tend to emphasize individual rights such as the right to life or the right to choose. Yet Carol Gilligan's research indicates that it is not individual rights but concerns about relationships and attachments to others that figure most prominently in women's abortion decisions.[37] Individualistic rhetoric surrounding abortion thus misses the broader, more complex context of women's moral reasoning. As a corrective to this narrow focus, Theodora Ooms argues for a family perspective on abortion. Her view extends to endorsement of the legal requirement (in some states) of parental notification concerning abortion requests of teenagers.[38]

While agreeing with Ooms's criticisms of individualism, I think her critique has not gone far enough, in part because her concept of family is inadequate.[39] For Ooms, a family consists of "those persons most closely related by blood, marriage, or adoption to the pregnant woman and the fetus." Such persons include the woman's "spouse or the expectant father, or, if she is an adolescent, her parents."[40] Ooms's use of the disjunctive "or" between references to the "expectant father" and the adolescent's parents may mean that she intends only one or the other side of the disjunction to be recognized as family. In other words, determination of who counts as family members depends on whether the pregnant woman is an adolescent or not. Yet, relationships with individuals on both sides of the disjunction may be morally relevant to adolescents as well as to adults considering termination of pregnancy. For example, responsibilities to an elderly parent may be a significant factor for a pregnant adult; and the responsibility *to* (as well as *of*) the "expectant father" may be relevant to a pregnant teenager. Nonetheless, only about one-fourth of those who obtain legal abortions in the United States are younger than 20 years of age.[41] Among the remaining three-fourths are some adults for whom genetic or legal ties are irrelevant considerations because their primary support systems are not based on blood, marriage, or adoption. There are also adults, both married and unmarried, whose concern for other children is their main reason for choosing abortion. For such individuals, family is defined more broadly than Ooms defines it.

Using Ooms's definition, the spectrum of those who count as a family unit is so narrow that it may comprise only the pregnant woman and her fetus. Consider, for example, a pregnancy that results from a casual encounter between single, independent adults. Unless the fetus is construed as an actual rather than a potential person, "family considerations" are hardly relevant in such a situation. A similar problem arises with regard to possible adoptive parents, because actual persons surely have a stronger claim to moral standing than do merely possible persons. Moreover, even when the pregnant woman is married, her husband may not be responsible for the pregnancy. When the pregnancy is initiated through an extramarital liaison, it is unclear which "family" concerns are paramount.

Because families affected by pregnancies are both multiple and overlapping, we cannot neatly draw concentric circles illustrating a priority of familial responsibilities from the innermost circle outward. True, a pregnant woman and her fetus are at the center, but the possible family configurations that proceed from that center are so numerous that a "family emphasis" is utterly vague beyond the center. Whatever the family configuration, an egalitarian approach calls for consideration of a variety of relationships. So also do the social configurations that arise from other-than-familial reasons: nonfamilial friendships, religious or political commitments, and so on.

Ooms deserves credit for pointing out the antifamily elements in the popular pro-life view of abortion, as well as the inegalitarian or sexist interpretation of family roles that that view assumes. However, her critique does not recognize that an emphasis on a traditional or nuclear family may be a not-too-subtly disguised individualism in its own right. Instead of looking out exclusively for the interests of the pregnant woman or the fetus, as do many on either side of the abortion debate, this extended form of individualism assumes a nuclear family unit, that is, a man, a woman, and their children. Because one family's wealth or welfare may be obtained at the price of another's, all sorts of chauvinistic behaviors may be defended in the name of such family individualism. If the fetus counts as a family member, then abortion is never permissible on the basis of the pregnant woman's wishes alone, or on the basis of social costs (as in cases of severe fetal anomaly). If the fetus does not count as a family member, the abortion of a healthy, viable fetus is permissible even for trivial reasons, such as the convenience of any family member. In either case, family interests override those of others.

An excessive emphasis on the family, narrowly defined, tends to limit the autonomy of individual women deciding for or against abortion. Typically, young women are socialized into thinking that they are primarily, even exclusively, responsible for the fate of the fetus, as well as for any future child. Where their own immaturity, economic instability, or a severe fetal anomaly makes it impossible for them to care for a future child, their range of options is obviously restricted. A family support system *may* expand the options available to the pregnant woman, but reliance on that system places her in a position of dependence on family members. If our social structure and our socialization process were such that pregnancy was generally acknowledged as a contribution and a burden to be shared beyond the family community, her options would then be increased; she would not then be as alone and limited in her decision as individualistic or even family considerations suggest.

From an egalitarian perspective, all moral decisions are social, but they are not always familial, even where family is defined broadly. Especially when the pregnant person is a minor, dependence on family has sometimes exerted considerable pressure to choose a course in conflict with her own moral conviction. I am thinking, for example, of a case in which the parents of a seriously ill young woman threatened to cut off necessary support if she were to continue her pregnancy. Eventually, the woman acquiesced to her family's wishes, but such acquiescence is hardly interpretable as a free decision. As Ooms points out, a pro-family position is not necessarily pro-life. Neither "family" nor "life" is an absolute value.

Ooms's critique of the individualism expressed from either side of the abortion debate may be strengthened by fuller treatment of what such individualism entails, and why it is wrong. The philosopher Josiah Royce once explained a distinction between individualism and individuality that is apropos. "Individuality," he maintained, refers to the unique core of someone's personality, the congruence of characteristics that makes the individual distinctive and unrepeatable. In contrast, "individualism" entails an attempt to extend and enrich that core of uniqueness by pursuing the goal of a "separate, happy self."[42] Unfortunately and ironically, this ideal of individuality is unrealizable because ongoing social interactions are essential to the fulfillment of individuals as such. Their individuality will ultimately be stifled rather than developed through purely individualistic pursuits. Thus, the genuine community that is essential to the full flowering of individuality is at odds with individualism.

Applied to the abortion issue, an exclusive emphasis on the right to life of the fetus or on the pregnant woman's right to choose ignores the social fabric with which both are inextricably interwoven. To treat the pregnant woman as an isolated individual standing alone in her decision strips her of the context that is integral to her identity. It may be better to define her identity in the limited context of her family. The reality to address, however, is that the pregnant woman stands at the center of multiple social relationships that affect her and her decision making. The fulfillment of her individuality is intimately linked with all of these relationships as well as with her fetus. From a care-based perspective, morality is similarly linked with those relationships.

Family advocates on both sides of the abortion issue have supported legislation requiring parental notification of minors' requests for abortions. Communication is obviously essential to families, but it is necessarily a communication that comes from the heart or the spirit rather than from legal notice or mandate. It is also essentially respective of the different language capacities of individuals, that is, their ability to interpret one another.[43] Where such communication occurs, families already know of the pregnancies of their daughters (and preferably, but less probably, of their sons' impregnation of others' daughters). If legislation is necessary to provide such information, the revelation is surely not equivalent to genuine communication. Some maintain that parents who initially react angrily to news of their daughters' pregnancies later rally in their support. But studies suggest that greater divisiveness is likely to occur, especially after the critical period of decision making regarding the abortion has passed.[44]

Those who favor mandatory notification concerning pregnancy and abortion decisions point to a distinction between those dependent on and those independent of others for their support. Perhaps because abortion is still a medical procedure, like others for which parents are responsible concerning their children, parental consent as well as the pregnant daughter's consent ought to be required. But this reasoning is in conflict with the privacy argument of *Roe v. Wade*.[45] Consistency would demand that minors be permitted to make unilateral decisions regarding abortion only to the extent that they are permitted to make such decisions for comparable medical procedures. If parents are responsible for these decisions—morally, legally, or economically—then they have a justifiable claim to be informed about abortion. Their claim, however, is based on fairness rather than on family interest.

A family perspective has also been invoked to support the right of the sperm provider to determine whether the pregnancy will continue. The principal arguments for this position are based on his genetic tie to the fetus and his intention (if present and credible) of raising the child (now a fetus) that the pregnant woman wishes to abort. From an egalitarian perspective, this recommendation ignores the disproportionate effects of childbearing and childraising on women and men. The woman alone, after all, has already contributed part of herself over a considerable period of time in the interests of the fetus, whereas the man merely hopes or plans to spend part of his future in the future child's behalf. To put it crudely, the principle of "equal pay for equal work" may be at stake. Excluding the genetic tie, a similar argument may be made with regard to potential adoptive parents.

However, *asking* is not equivalent to *requiring*. Although it seems wrong to coerce a woman to undergo the risk and pain of childbirth, it may not be wrong to *ask* her to do so for the sake of another (including the fetus). An important proviso to the legitimacy of such asking is that the woman be truly free to decline the request. The relevant distinction here is between legality and morality, or between an ethic of obligation and an ethic of virtue. It is not inconsistent to support the pregnant woman's legal right to choose abortion, while also maintaining that she ought to consider the willingness of others to care for the fetus she is carrying, especially where the pregnancy is relatively risk-free. To the extent that the pregnant woman honors such a wish for the sake of others' interests (for example, those of the potential father or the adoptive parent), she is acting virtuously or supererogatorily. Care for family members as well as others often motivates individuals to act supererogatorily.

Within the parameters considered in Chapter 4, the law should not coercively inform family members about pregnancies or abortion decisions. In the area of morality, however, the individual should take full account of the impact of abortion decisions on her family (defined broadly). Wider social considerations simultaneously provide a vantage point from which family interests themselves must at times be assessed and challenged. In other words, on the basis of an egalitarian framework—the one I have attempted to elaborate—a family perspective on abortion is useful but ultimately inadequate. The following case illustrates both points.

Family Considerations in Moral Decision Making

ADOLESCENT PREGNANCY IN A DYSFUNCTIONAL FAMILY.

Pat Quin was a 15-year-old retarded girl who, along with her 13-year-old brother, was raised by her 70-year-old grandmother from the time she was 2 years of age. Her mother was a prostitute and a drug addict who had not seen her children in ten years. The grandmother had legal custody of both children. Ms. Quin was described as having a mental age of 7 years and was enrolled in a special school.

During a routine visit to the clinic, the 15-year-old underwent a general physical examination. Birth control was discussed, but the grandmother was not interested in obtaining contraception for her granddaughter. Three months later, she was brought to the emergency room with complaints of dizziness and blurred vision. Physical examination suggested sexual activity, and a urine pregnancy test was performed. The result was positive. Told about the pregnancy, the teenager was unable to respond to questions about intercourse. She appeared to have limited comprehension of the situation. The options of termination, adoption, or raising the infant were discussed with the grandmother, who subsequently indicated that her grandson might have had intercourse with his retarded sister. When the state Department of Child Welfare was informed about suspicion of sexual abuse and incest, the department spokesperson said there was insufficient evidence to act on the report.

The risks of birth defect due to incest were explained to the grandmother.[46] On a subsequent visit to the obstetric clinic, ultrasound confirmed the presence of an 11-week fetus. The grandmother stated that she would like her granddaughter to continue the pregnancy and give the baby up for adoption. At this point in her life, she said, she did not feel able to raise another child herself.

The trio of persons mentioned in this case constitute a family, but not a traditional or nuclear family. The grandmother has demonstrated an indefinite commitment to care for her grandchildren, who are dependent on her. Nonetheless, her care has not been adequate to protect one child from sexual abuse, and the other, apparently, from becoming the perpetrator of the abuse. Because the family system has been the source of serious harm to at least one member, it is profoundly dysfunctional. Although the degree of dysfunctionality is rare, this particular case is real and recent.

Health caregivers dealing with this family face a number of complicated issues. The immediate issues concern Pat Quin's pregnancy: whether abortion is in her best interests, and whether adoption is a realistic option if the pregnancy is continued to term. Longer range issues include the possibility of contraception or sterilization, and whether the teenager should remain with her family or be placed in protective care.

If one only considers the demands of pregnancy and childbirth for Pat Quin, abortion is justifiable on grounds of her best interests. At eleven weeks, abortion is a less painful and less risky experience than continuation of pregnancy and childbirth. Ordinarily, it is precisely the child's (or adolescent's) best interests that pediatricians are committed to promote. But children's best interests are inevitably affected by the decisions, abilities, and interests of those who care for them, that is, their family members. The grandmother has indicated her desire that the pregnancy continue. Although her decision is challengeable on grounds of her granddaughter's best interests, the challenge may be defeated on grounds of fetal interests. Moreover, the fact that her grandmother wants her to continue the pregnancy is also relevant to the teenager's best interests, at least so long as she remains in the care of her grandmother. The grandmother is probably better able than others to communicate with her granddaughter, and to assist her in dealing with pregnancy and childbirth. Because the grandmother is unwilling or unable to raise another child, Pat Quin's own "right to have a child" can only practically mean the right to bear a child, and not the right to raise or participate in the child's upbringing. Even if she wishes to raise the child, her disability will interfere with her doing so.

There is an unmet demand for infants to be adopted, but less demand when the infant is black, as Pat Quin is. Further factors that reduce the likelihood of adoption include her retardation and the possibility of incest, which may be confirmed or disconfirmed after birth through DNA testing. If the retardation is due to genetic abnormality (a factor unknown to practitioners in this case), the possibility of bearing a retarded child is increased regardless of whether the pregnancy is due to incest. Although there are few published studies regarding the outcome of incestuous pregnancies, one study reports an incidence of greater than 50 percent birth defects, the majority of which are severe.[47] Few persons are interested in adopting disabled infants. Accordingly, before the abortion option is totally negated, the grandmother should be aware that adoption may not be the outcome of her decision to have her granddaughter continue the pregnancy.

If the grandmother determines that the pregnancy will continue and adoption is ruled out, the interests of the future offspring may be best promoted by being raised in a foster home or institution. A foster home could, in fact, provide a set of family or affective relationships for the 15-year-old; institutional existence could conceivably do that also. Either situation would at least provide means of survival, and this may be better than the nonsurvival brought about by abortion.

Both the grandmother and others have a responsibility to prevent sexual abuse of minors or retarded persons. If the grandson in fact impregnated his sister, he may need help also. The refusal of the Department of Child Welfare to pursue the suspicion of abuse ought to be challenged, or other means found to deal with the issue of incest and questions regarding the grandson. Sterilization should be considered as a means of avoiding future pregnancies, but not as a means of removing the threat of sexual abuse. Arguments for placing her brother elsewhere may be made in behalf of Pat Quin. These may be countered on the basis of his interests, and the supportive aspects of family relationships that might then be lost. From an egalitarian perspective, the needs of each family member, both separately and in the context of their relationships to each other, are relevant to decisions made for individuals. Family (as broadly defined) represents a significant but not absolute value to be weighed against the alternatives. Even a dysfunctional family may be preferable to a situation in which there is no indefinite commitment to care.

This case also illustrates a growing phenomenon in American society: grandmothers parenting their grandchildren.[48] More often than not, this type of family situation provides a nurturant environment for children. Even in this case (which could hardly be called a healthy example), the environment is probably better (less unhealthy) than the only alternatives available. Along the spectrum of possibilities that constitute traditional or nontraditional families, any particular arrangement may be healthy or unhealthy—temporarily, recurrently, or permanently. As used here, the terms *healthy* and *unhealthy* signify a range of well-being that includes mental and physical as well as social and emotional components inseparable from relationships. In healthy families, the relationships are predominantly nurturant.

By insisting on the importance of relationships and attachment to others in moral decision making, a care-based model of reasoning supports a broad interpretation of family, and the crucial role of family considerations in moral decision making. Feminist critiques of patriarchal family structures are compatible with this emphasis. The compatibility stems from the meaning of equality developed in the first chapter. It necessarily includes the notion that differences, including different relationships, should be respected so long as they do not result in harm to others. An egalitarian perspective is not committed to blind impartiality but to *just* caring toward those we affect. More often than not, those we most affect and those who most affect us are family members.

NOTES

1. Eric Kodish, M.D., a pediatric oncologist at Wyler Children's Hospital, University of Chicago.

2. I do not wish to suggest that family medicine, in comparison with other medical specialties, has been widely supported by the medical profession in the United States. Unfortunately (in my view), this has not been the case. Primary care specialists in pediatrics, internal medicine, and obstetrics and gynecology are more likely to receive such support.

3. These types of relationships have been increasingly mentioned in the press, mainly with regard to adoption efforts and custody disputes; for example, Dudley Clendinen, "Homosexual Foster Parents Debated," *New York Times* (May 19, 1985), 12; Mary Beth Lane, "High Court to Decide if Gay Man Can Adopt," *The Cleveland Plain Dealer* (Sept. 14, 1989), 1, 17A: and David Margolick, "Lesbians' Custody Fights Test Family Law Frontier," *New York Times* (July 4, 1990), 1, 10. Regarding the influence of these family situations on children, thirty-five studies conducted in the last fifteen years have shown that children of gay and lesbian parents are just as well-adjusted as other children. See Jane Gross, "New Challenge of Youth: Growing Up in Gay Home," *New York Times* (Feb. 11, 1991), A12.

4. Nontraditional family arrangements have been recognized by courts of law. The New York State Court of Appeals, for example, ruled in July, 1989, that a gay couple who had lived together for a decade could be considered a family under New York City's rent control regulations. Protection from eviction, the court maintained, "should not rest on fictitious legal distinctions or genetic history, but instead should find its foundation in the *reality of family life*" (my italics). Philip S. Gutis, "What Makes a Family? Traditional Limits Are Challenged," *New York Times* (Aug. 31, 1989), 15.

5. Onora O'Neill and William Ruddick, *Having Children* (New York: Oxford University Press, 1979), 3.

6. See also Alan Gewirth, "Educational Universalism and Particularism," *Journal of Philosophy* 85, no. 6 (June 1988): 283–392. According to Gewirth, "ethical universalism can justify certain kinds of ethical particularism ... showing that impartiality toward all persons can justify partiality toward some persons" (p. 283).

7. See Susan Moller Okin, *Justice, Gender, and the Family* (New York: Basic Books, Inc., 1989). Okin stresses the need while theoretically contributing to the possibility of resolution.

8. Ruth Sidel, *Women and Children Last* (New York: Penguin Books, 1986), xvi. The majority of these parents were married when they gave birth. See Okin, 3, 179.

9. Alison M. Jaggar and Paula Struhl, *Feminist Frameworks* (New York: McGraw-Hill, Inc., 1978), 208.

10. Ann Crittendon Scott, "The Value of Housework," in Jaggar and Struhl, 227–31.

11. Shulamith Firestone, *The Dialectic of Sex* (New York: William Morrow and Company, Inc., 1970).

12. Rosemarie Tong, *Feminist Thought* (Boulder, Colorado: Westview Press, Inc., 1989), 89.

13. See Adrienne Rich, *Of Woman Born* (New York: Bantam Books, 1977), 211; and Tong, 88–90.

14. For example, Rita Mae Brown, "Living with Other Women," in Jaggar and Struhl, 245–47.

15. Jaggar and Struhl, 212.

16. Okin, 92.

17. Gabriel Marcel, "Creative Fidelity," in *Creative Fidelity,* trans. Robert Rosthal (New York: Farrar, Straus and Company, 1964).

18. I realize that this insistence that family members be friends opposes the reality of many situations commonly perceived as families, where caring does not prevail. Care is the common denominator between friendship and family, as I define them.

19. Aristotle, *Nichomachean Ethics.*

20. Robert Blum, "Contemporary Threats to Adolescent Health in the United States," *Journal of the American Medical Association* 257, no. 24 (June 26, 1987): 3392.

21. Kwang-sun Lee and Maria Corpuz, "Teenage Pregnancy: Trend and Impact on Rates of Low Birth Weight and Fetal, Maternal, and Neonatal Mortality in the United States," *Clinics in Perinatalology* 15, no. 4 (Dec. 1988): 930–31.

22. Lee and Corpuz, 939.

23. Barry S. Zuckerman, Deborah K. Walker, Deborah A. Frank, Cynthia Chase, and Beatrix Hamburg, "Adolescent Pregnancy: Biobehavioral Determinants of Outcome," *Journal of Pediatrics* 105, no. 6 (Dec. 1984): 857.

24. Lee and Corpuz, 929.

25. Laurie Zabin, Marilyn B. Hirsh, and Mark R. Emerson, "When Urban Adolescents Choose Abortion: Effects on Education, Psychological Status and Subsequent Pregnancy," *Family Planning Perspectives* 21, no. 6 (Nov./Dec. 1989): 248–55.

26. Lee and Corpuz, 930–31.

27. Zuckerman, 859.

28. Lee and Corpuz, 933.

29. Beatrix A. Hamburg, "Subsets of Adolescent Mothers: Developmental, Biomedical, and Psychosocial Issues," in *Schoolage Pregnancy and Parenthood: Biosocial Dimensions,* ed. Jane B. Lancaster and Beatrix A. Hamburg (Hawthorne, New York: Aldine De Gruyer, 1986), 115–45.

30. Hamburg, 121.

31. Zabin, Hirsh, and Emerson, see note 25.

32. Joelle Sander, *Before Their Time* (New York: Harcourt, Brace Jovanovich, 1991), 179.

33. Lee and Corpuz, 940.

34. Lee and Corpuz, 940.

35. Joan Berlin Kelly, "Divorce: The Adult Perspective," in *Family in Transition,* ed. Arlene Skolnick and Jerome Skolnick (Boston: Little, Brown and Company, 1986), 305–06.

36. Okin, 165.

37. Carol Gilligan, *In a Different Voice* (Cambridge, Massachusetts: Harvard University Press, 1982).

38. Theodora Ooms, "A Family Perspective on Abortion," in *Abortion,* ed. by Sidney Callahan and Daniel Callahan (New York: Plenum Press, 1984), 83–107. In 1990, the U.S. Supreme Court ruled that the states may require a teenage girl to notify both parents before obtaining an abortion, as long as the alternative of a judicial hearing is provided for pregnant girls who do not wish to so inform their parents. See Linda Greenhouse, "States May Require Girl to Notify Parents before Having an Abortion," *New York Times* (June 26, 1990), 1.

39. Part of this selection is taken from a commentary I wrote on Ooms' article, published in Callahan and Callahan, 109–15.

40. Ooms, 83.

41. Lisa M. Koonin, Kenneth D. Kochanek, Jack C. Smith, and Merrill Ramick, "Abortion Surveillance, United States, 1988," in Centers for Disease Control, *CDC Surveillance Summaries,* July 1991, MMWR 1991; 40 (No. 55-2), 22.

42. Josiah Royce, *The Religious Aspect of Philosophy* (Boston: Houghton Mifflin, 1885), 201.

43. Royce makes this point with regard to genuine community, basing it on a theory of interpretation among the members. See Josiah Royce in *The Problem of Christianity,* ed. John E. Smith (Chicago: University of Chicago Press, 1968), especially 297–319.

44. Susan Phipps-Yonas, "Teenage Pregnancy and Motherhood: A Review of the Literature," *American Journal of Orthopsychiatry* 50, no. 3 (July 1980): 403–31.

45. See 410 United States Reports 113: "This right of privacy . . . is broad enough to encompass a woman's decision whether or not to terminate her pregnancy."

46. According to Baird and McGillivray, most infants conceived through "close consanquineous matings" are relinquished for adoption despite the high empiric risk of severe abnormalities. See Patricia A. Baird and Barbara McGillivray, "Children of Incest," *Journal of Pediatrics* 101, no. 5 (Nov. 1982): 854.

47. Baird and McGillivray, 854–57.

48. According to the U.S. Census Bureau, 5% of 63 million U.S. children lived in households headed by a grandparent in 1989. In nearly one-third of those households, neither of the child's parents were present. See Ellen Boddie, "Raising Your Children's Children," *New Choices* (March 1991): 18.

15

Just *Caring: Power for Empowerment*

The phrase "*just* caring" recapitulates my effort in this book to integrate an egalitarian critique with an ethic of care. Although health care is by definition an ethic of care, we have seen that health care practice affects men and women, and adults and children, not only differently but unequally. Unequally in the sense that men and adults are advantaged or empowered as compared with women and children, respectively. In other words, despite the cliché "Women and children first," the reality is that women and children tend to be last rather than first in health care. Just caring demands neither first nor last placement for either group as such.

What is it about inequality that is unjust? As I indicated in Chapter 2, it is the ranking that attaches to differences when such ranking is based on irrelevant criteria. In other words, inequality occurs when irrelevant differences are invoked in order to secure power or advantages of one individual or group over others. Although I set out a list of guidelines to be observed for an egalitarian approach to various issues, these are based on criteria ordinarily involved in moral decision making. For example, the possibility of experiencing pain is relevant to decisions made with regard to mature fetuses or animals (guideline 3), and the capacity of human beings to make informed choices about their own health is relevant to decisions affecting them (guideline 4). Respecting differences among individuals calls for avoidance of decisions based on stereotypes (guideline 5) that place obstacles in the way of health as well as moral agency. Gender, race, ethnicity, and social class are not specifically mentioned in the guidelines because these are differences that enrich us (except in the case of disparate socioeconomic classes) *as* differences, without providing a justifiable basis for domination of one individual or group by another.

In what follows, I examine the use of professional power as a means of summarizing various themes, criticisms, and positions already elaborated, including the socialist feminist critique considered in various chapters. The very meaning of *professional power* suggests a criterion through which individuals may assess whether the differences between them and others are consistent with an egalitarian framework. I concentrate on the role of the physician because this epitomizes the problems that professional power raises for an egalitarian perspective regarding health care.[1]

The Meaning and Role of Power in Professional Life

Power is a term that is rarely neutral in its impact. For some, it connotes a positive value to be pursued and upheld, a good in itself. When we talk about Black Power or Woman Power, for example, we affirm a significant good that is promoted by others besides blacks and women. In health care, power involves the ability to heal, to save lives, and to relieve human suffering; in other words, ability to fulfill the basic goals of health care as a profession.

For others, the term power has negative connotations, suggesting a division between the empowered and the powerless or less powerful, providing opportunities for exploitation and inequality. Crimes of violence, after all, are an exercise of power. We attempt to reduce the incidence of such acts by withdrawing power from people through imprisonment or denial of liberty in other ways. In health care, abuse of power may lead to forfeiture of its legitimate use through loss of one's license to practice. The entire system of certification of health professionals, whether by the government or by the professions themselves, is a means of ensuring that their power is exercised positively rather than negatively.

The fact that power can be exercised positively or negatively suggests that power itself is morally neutral. Further illustrating its neutrality is a basic definition of power as ability or capability with regard to someone or something. The ability may be intellectual, physical, or psychological; it may be innate or instrumental. Intelligence, for example, is an innate power; income is an instrumental power. Either of these is a power regardless of whether and how it is exercised. Moreover, the very perception of power implies that someone or something has power over another, whether or not the perception is correctly founded.

Power is a correlative term because it necessarily connotes another over whom, or in behalf of whom, the power may be exercised. The one who has power is superior to the other, at least with respect to the power that may be exerted. Inequality between them thus appears inevitable, and at times, at least, morally problematic. It becomes especially problematic when we consider degrees of power as possibilities for maintaining or increasing inequality, and for obstructing the autonomy of people. At one end of the spectrum the exercise of power involves outright coercion or force; at the other end it involves minimal and often subtle influence, such as education or dissemination of information. Between these are various degrees of power, such as the power to persuade or to manipulate others. The lives of health care workers, especially physicians, are filled with opportunities for expressing the full range of power. Since most lower level health care workers are women, and most patients are women, this makes women in the health care system particularly subject to domination by physicians.

The sources of power are in some ways mysterious. They include luck, divine gratuity, effort, and merit. Physicians are naturally and circumstantially gifted with ability and opportunity to learn the art and science of medicine; most of their patients are not similarly, or as extensively, gifted. But most physicians invest a great deal of time and effort in developing and maintaining their professional expertise; the power derived is thereby earned.[2] So power is at times a function of chance or inborn abilities, and at other times it is acquired over time through specific endeavors. Rarely, if ever, can we know for sure whether our own or others' power is deserved or gratuitous. In many cases it is probably both. Nonetheless, power is often interpreted as a sign of success, and thus associated with human achievement. There seems to be a human tendency to consider our own successes as earned and our failures as inevitable, while considering others' successes lucky and their failures deserved. We thus assume the good fortune of others while remaining unselfconscious about our own.

To understand power as it applies to the physician, we should first examine, at least briefly, the meaning of a profession. From different accounts, three traits are usually mentioned: an intellectually-based expertise, independence in exercising that expertise, and a humanitarian or human service orientation.[3] The first two traits are forms of power in themselves; the third mark of a profession, its service feature, distinguishes it from a business and coincides with the physician's commitment to health *care*. The purpose of business, as defined by our capitalistic system, is profit to oneself or to one's shareholders;[4] the purpose of a profession is (in some respect at least) service or care for others.[5] Eliot Freidson remarks that medicine "is an occupation first and only on occasion a profession."[6] This may also be true for other health care professions. Obviously, self-interest and a desire to promote others' interests are often present and morally appropriate in the same individual, regardless of her occupation. It needs to be recognized, however, that these purposes are at times at odds. In fact, professionals sometimes exercise their power to increase the disparity between themselves and others. Doing so constitutes an impediment to achievement of an egalitarian ideal.

Moral Assessment of the Exercise of Power

Because moral mischief may be associated with the exercise of professional power, a criterion is needed to assess the nuances of particular cases in this regard. Accordingly, I wish to propose a criterion by which to determine whether a professional activity or behavior truly fulfills the third mark of a profession, its caring or human service orientation, despite the fact that the person thereby accrues profit or prestige. The criterion looks to a crucial link between the moral neutrality of power and its moral or immoral uses. This link is the possibility of domination. Relating that possibility to the requirement that a profession involve a humanitarian orientation, I propose that professional power is morally exercised when it reduces domination by empowering the individual or group it serves. To the extent that its use dismantles their power or increases their domination, it is immoral; to the extent that its use fails to improve the status of the client or patient, it is amoral. Professional power is thus morally exercised as *power for empowerment;* it fulfills its essential purpose only to the extent that its exercise enhances the lives of those on behalf of whom the profession is practiced. If individual professionals enhance *only* their own lives through professional practice, an essential feature of the profession, its human service orientation, is betrayed.

While using this criterion to assess the uses of professional power, we need to take account of the specific tasks of each profession as ways in which its care or service function is fulfilled. The power of the physician for accomplishing the tasks of healing, saving lives, and alleviating human suffering is different for different specialties within medicine. For example, the obstetrician has the power to facilitate healthy births where this might not otherwise occur through advances in treatment of high-risk pregnancies. Gynecologists have power to prevent conception and birth, and to facilitate both through provision of infertility drugs, artificial insemination, in vitro fertilization, and embryo transfer. Neonatologists have power to maintain the lives of very early or extremely disabled newborns, or to allow their deaths by declining to treat them, with or without parental consent. Pediatricians have power to challenge and even override the parental prerogative of obtaining or refusing treatment for their child, basing this on their professional commitment to provide optimal care for children. Today's transplant surgeons have power to remove organs from brain-dead patients, and use these to save, prolong, or improve the lives of others. Intensive care and emergency specialists have power to prolong life or allow patients to die through use or refusal of resuscitation procedures.

The ordinary practice of medicine represents legitimization of power over people to an immense degree. Consider, for example, the violation of bodily integrity that surgery in all its forms requires, the assault upon natural body functions that drug management involves, and the ways in which psychopharmacology or psychosurgery can transform the human psyche. Consider also the intrusions on time and privacy that patients experience through routine medical examination and hospitalization. In the name of therapeutic privilege, physicians are even empowered at times to withhold information from the very source of that information, and to undertake treatment without the patient's consent.[7]

Beyond these powers of medicine and its diverse specialties are powers that accrue to most physicians as common accoutrements of their profession. Personal income is generally well above average, even after high malpractice premiums are paid. A certain level of prestige is another common feature, as attested by the fact that physicians, more than any other group whose members earn a doctoral degree, generally use and expect to be referred to by their title.[8] The title itself invokes respect, the rendering of which is acknowledgment of power.

As physicians recognize their power to influence the public at large, some are attracted to political roles as means of effectively fulfilling their therapeutic commitment. Some are elected, some are appointed to public office, and some assist others concerned about public health to be elected. I think, for example, of Carla O'Day, a physician who directed a busy emergency department (ED) in an urban setting. After a full day of lobbying at the state capital for increased funding for emergency medical technicians (EMTs), Dr. O'Day told a gathering of medical students:

> Medicine is political. I can save more lives by spending my day getting the legislature to increase the budget for EMTs than by spending the day resuscitating patients in the ED. If we have better and more EMTs in the field, there will be fewer patients to resuscitate, and more that I can resuscitate successfully.[9]

Dr. O'Day also served as campaign manager for a candidate for judgeship. Her rationale for assuming this role was consistent with the notion that medicine is political. She thus used her power to promote the candidate's priorities concerning health care, which she (obviously) supported.

A well-known example of physicians using their power to promote health-related goals is former Surgeon General C. Everett Koop, whose positions and statements on smoking and AIDS have had considerable impact on the promotion of public health.[10] As the history of the American Medical Association (AMA) amply illustrates, medical organizations are also influential. Physicians for Social Responsibility (PSR), for instance, has utilized the power of the profession to persuade the public of the importance of addressing what has been called the most important therapeutic goal of our time, the prevention of nuclear holocaust.[11] Worldwide membership in this organization of physicians has risen with the recognition that physicians have greater responsibility than others for issues to which their professional expertise and commitment are relevant. Increasingly, physicians themselves call for involvement of their profession in such efforts.[12]

The Social Power of American Medicine

The social power of American physicians has been chronicled by Paul Starr as a complex relationship between public image and economic status.[13] According to Starr, in the early nineteenth century America was mostly rural and sparsely populated, and nonscientific views of illness prevailed. Health problems were mainly dealt with in the home by family members. During the latter part of the century, as transportation and communication improved and medicine became more scientific, both the prestige and the income of physicians escalated. As education became more standardized, physicians developed "mechanisms of dependency" such as control over access to drugs, hospitalization, entry into medical schools, and their own fees. In effect, they assumed the dominant role in the health care industry. The subordinate role of nurses and other health care professionals was evidenced by the fact that they remained salaried employees, while the AMA insisted on physicians' maintaining power or control over their own income. Then, as now, the vast majority of subordinate positions within the health care system were occupied by women. Starr attributes this to a perception on the part of physicians that they might thus maintain control of the industry.[14]

During the middle part of the twentieth century,[15] the ascendancy of power of American medicine was accompanied (and probably incremented) by its success in avoiding government intervention. The profession nonetheless relied on government funding for its institutions and research programs. Although private health insurance expanded, physicians' fees continued to be set by the profession itself. Despite a shortage of physicians, there was increased recognition of unmet health needs of the poor, and increased suspicion of those who held positions of authority and high income. Personal, professional, and political power of physicians continued, but the power generated by their prestige diminished. Concomitantly, the power of the AMA declined through reduced membership, infighting, and factional argument within the organization. A greater number of physicians were politically radicalized to speak out against "the system," and the women's movement articulated criticisms of the ways in which female patients were treated.[16] Ivan Illich wrote his iconoclastic critique of medicine as causing more medical problems than it solves.[17] The rise of the holistic health movement de-emphasized the physician's role by stressing the role played by other health care practitioners and by patients themselves.[18]

The 1980s saw a continuation of the complex relationship between the public image and economic status of physicians. Starr points to two factors that further limited the power of physicians: a rapidly increasing supply of physicians, and ongoing efforts by government and business to control medical costs.[19] Abortion laws, dialysis funding, and Baby Doe regulations are but a few examples of governmental interventions that have significantly affected medical practice. Various methods of government reimbursement and prospective payment schemes, designed mainly to reduce health care costs, have notably influenced the way physicians manage patients' care.[20] No doubt the desire to avoid government interventions and malpractice litigation have reduced the level of independence of physicians. Escalating malpractice claims have encouraged physicians to choose modes of treatment that will reduce the likelihood of suit. Generally, this means prolonging treatment, even in cases where the physician believes it is no longer in the patient's own interests. As one neonatologist observed, "I never write a DNR order (an order not to resuscitate a patient) even when I know that a patient is dying because I'm more likely to be sued for that than for continuing treatment."[21] The climate of litigiousness has resulted in the practice of defensive medicine, a situation which is costly and unfortunate for everyone involved, except perhaps for some attorneys. It has also prompted the departure of highly qualified individuals needed in specialties such as obstetrics and intensive care to specialties less likely to provoke litigation.

It seems clear, then, that in certain respects the power of American medicine has declined over the last few decades, even while medical expertise has developed to an incredible degree. The diminishment may be inevitable because increased specializations have brought about dependencies within the profession itself and on the service of nonphysician specialists. An important benefit of this type of "decentralization" of medical power is that the increased range of specialization promotes a higher quality of care, at least in situations where the experts collaborate to effect optimal integrated decisions. Another reason for decreased power is that decisions about patients have extended beyond the medical arena, from questions about how to treat patients to whether treatment ought to be provided. Fortunately or unfortunately, each new advance in medical science affirmatively answers the question of whether something can be done, but simultaneously raises the question of whether it should be done. Since the latter is a moral rather than medical question, medical expertise alone is inadequate to answer it. An adequate response requires moral considerations available to patient and physician alike, as well as to others involved in the moral dilemma. Addressing complicated and momentous questions effectively deserves collaborative input.

Collaboration and Parentalism

In Chapter 2 I elaborated a model of the health professional's role by examining paternalist (beneficence-based) and maternalist (patient-autonomy-based) models, finding both of them wanting, and taking valid elements from each to describe a model of caregiving that emulates the relationship between parents and children of all ages. That relationship, I claim, is one that involves mutual caring or nurturance as well as protection from harms, based on changing needs and abilities throughout the life cycle.

It may be objected that the parentalist model is inappropriate to the physician–patient relationship because it assumes a meaning of care expected of those related by familial ties but not by others. Consider, however, the definition of family proposed in Chapter 14: a set of people who are indefinitely committed to care for each other and for those dependent on them. If care or nurturance includes the notion of facilitating independence (maternalism), the aim of health care is to render the caregiver unnecessary, even as the aim of motherhood (epitomized in the push of childbirth) is to help children be "on their own" when they are ready for such independence. Just as parental restrictions should be introduced only insofar as they are necessary in order to protect children from harming themselves because of their lack of maturation, so the restrictions of medical treatment should be introduced only to the extent that they represent a corresponding situation for patients. The corresponding situation is one where patients are unable to provide their own free and informed consent to medical treatment. In other words, the parentalist model involves care for a patient's capacity for autonomy as well as health needs.

Note, however, that the parentalist model is meant to work both ways, signaling the mutuality of responsibilities throughout the life span. To the extent that this is possible, patients are also obliged to facilitate the independence of caregivers and avoid harming them. Facilitating the independence of health professionals may be demonstrated by providing them with honest and full disclosure of conditions relevant to their professional decision making. Avoiding harms may be demonstrated by refraining from unjustified malpractice suits or by voluntary disclosure of exposure risks such as HIV status.

Although the interchangeability of roles in the parent–child relationship (e.g., when a child parents a parent) is not really replicated in the physician–patient relationship, both situations demand mutual recognition and response to changing features of the relationship. In this respect, both relationships are collaborative or interactive. From an egalitarian point of view, the responsibility for collaborative interaction extends beyond the physician–patient encounter to all of those involved, and is proportionate to the degree of power possessed by those involved in relevant relationships: family members, other patients, nurses, therapists, social workers, administrators, and so on. Among physicians themselves, collaboration has long been effective in the practice of obtaining consults or making referrals. The newer concept of the health care *team*, involving caregivers from various disciplines, participating as appropriate in case management conferences on particular patients, is further evidence of collaboration.[22] The same model is applicable and, in fact, preferable for ethical decisions. During the 1980s, hospitals were strongly encouraged by the government to extend their decision base further through establishment of hospital ethics review committees.[23] These committees include non-health care professionals in their membership. The reason for extending the decision base is twofold: to acknowledge the fact that others besides physicians and patients are responsible for ethical decisions in health care, and to improve the quality of those decisions. Obviously, shared decisions mean shared power.

Collaboration in ethical decision making involves acknowledgment that every agent has a distinct responsibility stemming from a distinct set of relationships as well as professional expertise or power. These responsibilities yield different answers to the question "What should I do?" In other words, in each case, the moral responsibility of the physician is different from that of the nurse, different from that of the patient, and different from that of family members. Paraphrasing the rule of thumb referred to in Chapter 3, in cases of conflict the power of the one most affected should count most. Ordinarily, this means that the autonomy of the competent patient is paramount. The physician empowers the patient by respecting the patient's choice. Towards the nonautonomous patient, however, the exercise of beneficence is paramount. The physician empowers that patient by attempting to optimize the patient's interests. Occasionally, this may mean allowing (and allowing is a kind of empowering) the patient to die.

Although the power of the autonomous patient and the physician's obligation to empower the patient are primary considerations in moral decision making, neither consideration implies that the power of patient or physician is absolute. Ideally, just as parental care aims to facilitate the child's eventually becoming capable of parenting or caring for others, so the practitioner's power to care aims at empowering the patient to care as well. There is thus a complex set of interactions and relationships rather than simply a practitioner–patient relationship, with caring expressed in different ways by those involved. For practitioners, collaborative interaction occurs among colleagues as well as patients. In either case, the criterion for its assessment remains the same: power for the sake of empowerment. From an egalitarian perspective, the professional's commitment to those who are disadvantaged is to empower them to the point where they are equally advantaged, which means empowering them to empower others, so that the cycle continues.

Conclusion: Toward a More Egalitarian (Less Inegalitarian) Future

At the outset I indicated that a care-based model of moral reasoning seems at odds with the feminist view that equality should prevail between women and men. The critical lens through which I have examined a variety of topics has stressed equality rather than care, in part because I believe an egalitarian perspective has been neglected in ethical considerations of health care. An important exception to this is the range of issues associated with the scarcity of health care resources, that is, questions of microallocation and macroallocation. Consideration of both types of questions involve distributive justice or equality in the relationship between the health care system and the wider society. In other health care issues, the emphasis is on beneficence and respect for autonomy as part of care, and the ethical questions are primarily addressed through consideration of the practitioner–patient relationship. Even then, conflicts occur that require an egalitarian perspective in order to maintain a caring balance between respect for autonomy and beneficence on both sides of the relationship.

Critics might bemoan the fact that I have not presented a systematic method of resolving ethical disputes. I have, however, presented guidelines that I consider more useful than rules and theories in facilitating decisions. Even prima facie rules often prove inadequate to complicated cases and issues. Guidelines are more useful than rules and theories because they represent more flexibility in the face of differences and are thus less likely to obstruct progress toward a clearly defined ideal such as equality. Just as William James defined pragmatisim as a method of reaching truth by straddling the diverse orientations of traditional philosophies,[24] so the use of guidelines rather than rules permits us to straddle the different and sometimes conflicting values at stake in modern health care.

Another complaint that critics may raise is that my views are too "idealistic." I do not dispute the descriptive legitimacy of the term, but I do dispute its being used pejoratively on grounds that my so-called idealism constitutes an impractical or useless approach to the issues. Like Charles Sanders Peirce, who introduced the term *pragmatism* into philosophical circles, I believe strongly in the practicality of generality.[25] In fact, twenty years after publishing a work entitled *An Idealistic Pragmatism,*[26] I still subscribe to the belief that underlies its title, namely, that pragmatism and idealism are compatible, as compatible as theory and action.

Some people, perhaps even most people, deny the compatibility, often critiquing others (especially those in positions of leadership) for not putting their ideals or theories into practice. My response to this criticism is based on H. S. Thayer's *Meaning and Action: A Critical History of Pragmatism.*[27] After analyzing the pragmatisms of Peirce, James, John Dewey, George Herbert Mead, and Clarence I. Lewis, Thayer concludes with a definition of pragmatism as a philosophical doctrine that "emphasizes the practical character of thought and reality."[28] One of the essential ideas of pragmatism, he claims, is the inseparability of theory and action. My approach in this book illustrates that inseparability. Its egalitarian perspective represents in many respects an idealistic theory (in a teleological rather than metaphysical sense) applied to practical decisions or actions. The theoretical and practical components are intermingled with cases and issues illustrating the theory, and theory explaining the cases and issues.

Peirce maintained that the way through which we clarify ideas is by examining their empirical consequences and formulating a plan of action on that basis. The veracity of our beliefs is tested by experience, and truth as an absolute ideal is something we best approximate collaboratively. Note the term *approximate,* as contrasted with *achieve.* Ideals as such remain elusive, but this does not imply their impracticality. All of our scientific knowledge is partial, yet few dispute its usefulness. We simply do not and cannot know everything about anything, and this is not only because of our intellectual limitation, individually and collectively, but also because the world is not finished. It is in flux. Our understanding of abstract ideas, if their meanings accord with reality at all, is therefore partial. Abstract knowledge is useful but inadequate precisely because it removes the object studied from its relational context. This is the realization that supports the clinical advice: "When in doubt, look at the patient." It also supports the emphasis on context in a care-based model of moral reasoning.

As Gabriel Marcel put it, human beings are mysteries rather than problems.[29] A problem can be completely known by analysis of its unchanging parts. Consider the cadaver that the medical student studies as an example of a problem. A mystery, on the other hand, is inexhaustibly knowable because it is always changing. Consider the actual patient that the health caregiver treats as an example of a mystery. Although the medical student can learn all there is to know about a cadaver by measuring and studying its various parts, the same cannot be said for persons with whom the students deal in personal or professional life. The relational and dynamic character of human beings is inevitably missed in the process of analysis based on abstraction. Nonetheless, we must treat one another as problems from time to time in order to find answers to the questions that are crucial to practical caring.

Reverence or respect is the proper attitude to observe toward persons as mysteries.[30] The possibility of moving beyond analytic knowledge in our understanding of persons is through relationships with them. Some relationships are based on the correspondence between need and expertise, as in the professional–client relationship; some are based on affection, as in the commitment between lovers; and others are based on choice or chance, as in marriage or citizenship. In some situations, attachment to others is based on more than one factor, as in a loving marriage, but all relationships involve responsibilities for those capable of exercising them. Responsibilities are responses to lived relationships.

The lived relationships of those involved in health care involve possibilities for inequality or domination as well as empowerment. As the preceding chapters have shown, women and children are more likely than men to experience the "downside" of inequality. In recent years, the apparent demise of Marxism suggests that a thoroughly egalitarian society is an impossible goal. But if equality is a good, then more equality is better than less, or less inequality is better than more. In that light, consider the impracticality of not having an ideal such as equality, or of not being idealistic in the sense that one postulates ideals and pursues them even while they may not be achievable. Since we inevitably make choices, not to direct our choices toward a desirable overall end is inefficient. If equality is desirable, then an account that points to ways of reducing the inevitable inequalities that are present in society is useful despite, and in part because of, its "idealism."

So what do I propose as concrete means to reduce inequality or promote equality? The main thing I propose is attention to differences. In contrast to this proposal, consider a rather widespread trend among "liberal"-minded individuals to speak and write in a gender neutral manner. Corresponding trends exist with regard to race, ethnicity, and religion. In fact, though, differences in sex, race, culture, and religion have important positive and negative consequences in people's lives. An ethic of care and socialist feminism converge in their concern about these differences. Some differences are permanent, and some are changeable. Some are appropriately ignored because they are irrelevant to judgments that incorporate them. Ignoring relevant differences increases inequality. For example, if we pay no attention to the fact that some persons are disabled, we will do nothing to help them experience opportunities that those who are advantaged commonly enjoy. If we pay no attention to the fact that minorities and women have been discriminated against historically, we will not notice that many still live with the effects of past discrimination. If we disregard the fact that children have no voice in major decisions affecting them, we will go on listening only to adults who may place their own interests first.

One practical means of attending to relevant differences in an egalitarian fashion is to ensure that those who are differentially affected are proportionately represented among the decision or policy makers. An obvious example of the need for this representation is the area of reproductive health. Because women's bodies are more affected by reproductive decisions than men's bodies, more women than men ought to be included in such decision making. Moreover, proportionate representation should be determined not only by the number of those affected, but also by the differential impact of decisions to be made. Majority-based decision making may be labeled democratic even though it permits injustice. By that simplistic measure, minority interests may always be overridden by those of the majority.

Similar points may be made for children and adolescents. Although those who are intellectually or volitionally incapable of competent decisions require proxies (usually their parents), many are sufficiently mature to participate in decision making that affects them. Adults whose competence may have declined with age or because of illness or disability are also often capable of participation in decision making. Proportionate representation of any of these groups requires not just surrogate decision makers for those who are incompetent to decide for themselves, but representation based on the extent to which such individuals will be affected by the decisions. Such participation in decision and policy making is a means of increasing attention to differences.

Martha Minow has argued for attention to differences from a legal perspective.[31] Obviously, the law attempts to prevent discrimination based on failure to recognize relevant differences among groups. Legislation involving affirmative action, school integration, and equal employment opportunities are examples in this regard. But attention to differences is an egalitarian strategy to be observed not only with groups distinguishable by gender, age, race, ethnicity, or religion. It is also applicable to the uniqueness of individuals in any group. While the generalizable differences signified by categorizations of people are important cues to what individuals wish, individuals do not always embrace the priorities of the groups with whom they are identified. For example, consider the abortion question. Although the Roman Catholic Church has a clear position opposing legalization of abortion, many church members believe that it should be legal.[32] Moreover, many Catholic women obtain abortions,[33] and many Catholic couples practice birth control despite knowledge of their Church's prohibitions concerning these practices.[34] To assume that a particular woman would not want an abortion or birth control because she is Catholic, without inquiring about her actual preferences, involves inattention to differences that inevitably prevail among individuals.

An egalitarian perspective involves not only attention to the differences themselves, but also to the fact that differences sometimes become the basis for ranking individuals or groups so that one is more advantaged than the other. It requires, as I have attempted to illustrate throughout this book, a persistent critique of ways in which relevant differences are ignored and irrelevant differences are invoked in defense of inequality. Determination of relevance is a complicated task involving the assessment of burdens, benefits, and preferences in relation to an ideal of social equality. It thus demands a critical focus on power discrepancies.

Because sickness or disability render individuals vulnerable and dependent on others, the relevance of power to health and health care is obvious. The feminization of poverty, for example, is undoubtedly relevant to the power disparity that results from the compromised health status of women and children, whereas the provision of synthetic growth hormone therapy to financially advantaged adolescents is irrelevant to their health and possibly detrimental to an egalitarian ideal. Even then, other factors such as cost and feasibility of solutions are appropriate considerations in critiquing discrepancies in power, and for determining long-term means of reducing inequality. When doctors and hospitals charge the wealthy more for elective treatments, using the overcharge to cover part of the cost of necessary care for the poor, they appear to be moving in this direction.

Just what constitutes necessary care is a complex topic in its own right. Although most people agree that human health is a basic value, there is widespread disagreement about what this concept involves. No matter how health is defined, however, it cannot be equally maintained or promoted apart from consideration of the other values with which it sometimes competes or overlaps, such as faithfulness to one's commitments, autonomy, happiness, and life itself. An egalitarian perspective calls for acknowledgment of these values, while providing a strategy for prioritizing them when they are not simultaneously supportable. Diverse values often give rise to different positions, and not all of the positions are compatible with an ideal of equality.

In applying an egalitarian perspective to various issues, I have described a range of positions that fulfill requirements of gender justice and age justice with regard to health care. My concentration on women and children has brought to the fore two groups of persons whose differences have been ignored where relevant, or who have been treated differently on the basis of irrelevant criteria. Changes in thinking and practice are necessary in order to provide the care to which they are entitled on the basis of their equality with men.

The changes in thinking involve transformation of gender stereotypes applicable across the life span, in personal as well as public spheres of life. Although such transformations are more difficult to accomplish than behavioral changes, the latter can facilitate the former. In other words, acting as if equality is possible helps to reduce attitudinal inequalities among people. As James put it, faith in fact can help create the fact.[35] In my more pessimistic moments, I doubt that sexism or racism or classism or ageism will ever be eradicated because prejudice is an inevitable element of human experience. Every once in a while, however, I meet someone or observe something that helps me believe that the ideal of equality is achievable in at least some aspects of our lives. For example, as I observe the different expectations of our children with regard to gender roles, I believe (rightly or wrongly) that there is less gender inequality now than there was when I was their age. And as I listen to clinicians struggling with ethical issues, I believe (rightly or wrongly) that there is less paternalism than in times past in clinical attitudes toward patients. Of course, it is possible that having worked within the health care system for ten years, I have lost some of the critical distance that I might have had if I had remained outside the institution. While proximity has vastly increased my sense of the empirical and moral complexity of issues in health care, that very involvement signals the need for sustained self-consciousness about the possibility of cooptation, assimilating the same priorities as the institution.

Obviously, my own thinking along lines of equality not only informed but motivated the critical perspective developed in this book. The concepts of self-consciousness, parentalism, and feminism considered in various chapters can be best understood within that context, and my view of professional relationships and interpersonal attachments in an ethic of care is also inseparable from that context. If the equality of women, men, and children is a worthwhile social goal, it relies rudimentarily on the nurturance or caring that is essential to individual as well as social health.

Ironically, the term *care* has lost much of its original meaning in the modern health care system. The dictionary defines it as "watchful regard or attention."[36] As I remarked in Chapter 2, because (most) nurses are consistently attentive while (most) physicians are only episodically involved with patients, the two professions have sometimes been distinguished on the basis of a commitment to caring or curing, respectively. Nonetheless, in the modern hospital complex, what is typically meant by "health care" is medical treatment or technology, and attentive regard is frequently provided through impersonal (but effective) monitors rather than through direct contact with clinicians of any sort. That "care" is often equated with "treatment" was vividly illustrated to me several years ago when I was involved in the development of a hospital policy for discontinuing or foregoing treatment of patients near death. Until someone pointed out the inappropriateness of the phrase, the policy was entitled "Limitation of Care." The title was then changed to "Limitation of Treatment," and the policy was reworded to express the hospital's ongoing commitment to provide optimal *care* for all patients, with the acknowledgment that this occasionally means that *treatment* should be withdrawn or withheld.

Optimal care of all patients is one way of defining equal care or equal treatment, but it does not imply that every patient should be given the same treatment. Optimal care requires optimal attention not only to the differences between individuals, but also to the possibilities for injustice or domination in our relationships with one another. Just health care calls for both kinds of attention.

NOTES

1. Much of the ensuing discussion of power and its moral exercise by physicians is taken from my article "The Physician," in *The Power of the Professional Person,* ed. Robert Lawry and Robert Clarke (New York: University Press of America, 1988), 119–31.

2. Another term that might be used here is *authority,* which is synonymous with at least some kinds of power.

3. Michael Bayles, *Professional Ethics* (Belmont, California: Wadsworth Publishing Co., 1981), 7. See also Abraham Flexner, "Is Social Work a Profession?" *National Conference of Social Work* 42 (May 1915): 576–90; Alan H. Goldman, *The Moral Foundations of Professional Ethics* (Totowa, New Jersey: Rowman and Littlefield, 1980), 18; and Walter Metzger, "What Is a Profession?" *Seminar Reports* 3 (Program of General and Continuing Education, Columbia University, 1975): 2. Metzger cites Flexner's famous list of criteria for a profession: "a qualifying occupation was one that applied its theoretical and complex knowledge to the practical solution of human and social problems."

4. As Milton Friedman put it "there is one and only one social responsibility of business—to use its resources and engage in activities designed to increase its profits as long as it stays within the rules of the game...," *Capitalism and Freedom* (Chicago: University of Chicago Press, 1982), 133.

5. It is the ideal of service or of a "calling," as D. H. Hardcastle points out, "with the practitioner standing above the sordid considerations of the marketplace, that separates the professional from occupations." Quoted in Paul Wilding, *Professional Power and Social Welfare* (Boston: Routledge and Kegan Paul, 1982), 4.

6. Eliot Freidson, *Profession of Medicine* (Chicago: University of Chicago Press, 1988), 5. Freidson would consider practitioners who are not physicians not to be professionals because they lack the autonomy required of professionals as such. See, for example, p. 71: "a profession is distinct from other occupations in that it has been given the right to control its own work."

7. See Ruth R. Faden and Tom L. Beauchamp, *A History and Theory of Informed Consent* (New York: Oxford University Press, 1986), 36–38.

8. Samuel Gorovitz, *Doctor's Dilemmas* (New York: Dodd, Mead, 1970), 7–8.

9. The gathering was scheduled in conjunction with a medical school elective on Women and Medicine, taught by Miriam Rosenthal and me at Case Western Reserve School of Medicine, Cleveland, Ohio. Dr. O'Day had been invited to speak to the students.

10. See C. Everett Koop, "Surgeon General's Report on Acquired Immune Deficiency Syndrome," *Journal of the American Medical Association* 256, no. 20 (Nov. 28, 1986): 2784–88, and "A Parting Shot at Tobacco," *Journal of the American Medical Association* 262, no. 20 (Nov. 24, 1989): 2894–95.

11. For example, Christine Cassel and Andrew Jameton, "Medical Responsibility and Thermonuclear War," *Annals of Internal Medicine* 97 (1982): 426–32.

12. For example, Michael McCally and Christine K. Cassel, "Medical Responsibility and Global Environmental Change," *Annals of Internal Medicine* 113 (1990): 467–73.

13. Paul Starr, *The Social Transformation of American Medicine* (New York: Basic Books, Inc., 1982), 60–78.

14. Starr, 221.

15. Starr, 9–144.

16. Suzanne Arms, *Immaculate Deception* (New York: Bantam Books, 1975); and Diane Scully, *Men Who Control Women's Health* (Boston: Houghton Mifflin, 1980).

17. Ivan Illich, *Medical Nemesis* (New York: Pantheon Books, 1976).

18. Sally Guttmacher, "Whole in Body, Mind & Spirit: Holistic Health and the Limits of Medicine," *Hastings Center Report* 9, no. 2 (April 1979): 15–21.

19. Starr, 421.

20. For example, diagnostic-related groups or DRGs. See E. Haavi Morreim, "The MD and the DRG," *Hastings Center Report* 15, no. 3 (June 1985): 30–38.

21. Despite this comment, most lawyers advise that DNR orders should be written if resuscitation is not in the patient's interest, or if a competent, informed patient has refused resuscitation.

22. See Edmund D. Pellegrino and David C. Thomasma, *A Philosophical Basis of Medical Practice* (New York: Oxford University Press, 1980), 256–59.

23. Mary B. Mahowald, "Hospital Ethics Committees: Diverse and Problematic," *Newsletter on Medicine and Philosophy,* American Philosophical Association 88, no. 2 (March 1989): 88–94, reprinted in *HEC Forum* 1 (1989): 237–46 and in *Bioethics News* 2 (1990): 4–13.

24. For example, see William James, *Pragmatism* (Cleveland: World Publishing Co., Meridian Books, 1965), 20–26.

25. See James, *Pragmatism,* 43. Peirce characterized the belief through which we settle our doubts after a process of inquiry as a habit or plan of action. See Charles Sanders Peirce, *The Essential Writings,* ed. Edward C. Moore (New York: Harper and Row Publishers, 1972), 125–26.

26. Mary Briody Mahowald, *An Idealistic Pragmatism* (The Hague: Nijhoff, 1972).

27. H. Standish Thayer, *Meaning and Action: A Critical History of Pragmatism* (Indianapolis: Hackett Publishing Company, Inc., 1981).

28. Thayer, 425.

29. Gabriel Marcel, *Being and Having,* trans. Katharine Farrer (Westminster: Dacre Press, 1949), 100–101.

30. Gabriel Marcel, *The Mystery of Being,* I. *Reflection and Mystery* (Chicago: Henry Regency Company, Gateway ed., 1950), 217. Marcel relates "mystery" to "presence," and refers to it as "sacred" (216–17).

31. Martha Minow, *Making All the Difference* (Ithaca, New York: Cornell University Press, 1990).

32. According to a Gallop poll taken October 5–8, 1989, 56% of American Catholics would *not* like to see the U.S. Supreme Court completely overturn its *Roe v. Wade* decision. See Diane Colasanto and Linda De Stefano, "'Pro-Choice Position Stirs Increased Activism in Abortion Battle,'" *The Gallop Report,* no. 289 (October 1989): 17.

33. According to the *Women's Action Almanac,* "evidence from several states indicates that Catholics are obtaining abortions in excess of their proportional representation in those states." The states from which data have been obtained include Maryland, Colorado, New York, Connecticut, Hawaii, and California. See Jane Williamson, Diane Winston, and Wanda Wooten, eds. (New York: William Morrow and Co., Inc., 1979), 21–22.

34. According to Richard A. McCormick, 76% of American Catholic women use some form of birth control, and of those, 94% use methods condemned in official formulations. See his *Health and Medicine in the Catholic Tradition* (New York: Crossroad Publishing Co., 1984), 97.

35. William James, "The Will to Believe," in *Essays on Faith and Morals* (Cleveland: Meridian Books, The World Publishing Co., 1962), 56.

36. *Webster's New World Dictionary,* 2d college ed. (New York: Simon and Schuster, Inc., 1984), 214.

Name Index

Subject Index

276